Food and Development

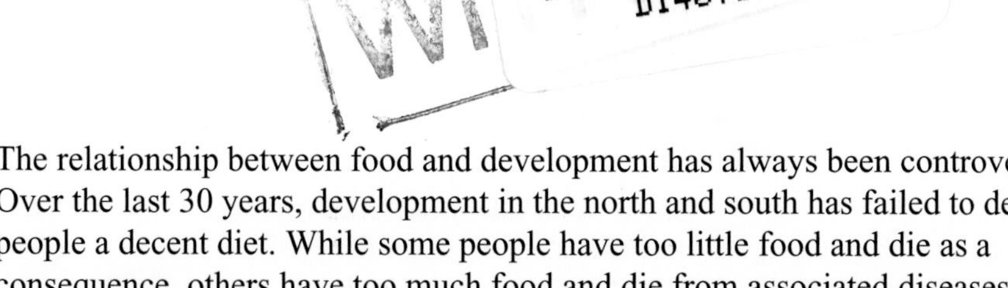

The relationship between food and development has always been controversial. Over the last 30 years, development in the north and south has failed to deliver people a decent diet. While some people have too little food and die as a consequence, others have too much food and die from associated diseases. Furthermore, some methods of food production create social dislocation and deadly environments where biodiversity is eroded and pollution becomes rampant. While guaranteeing enough food for the world's inhabitants continues to be a serious challenge, new issues relating to food have emerged.

Food and Development is a lively and lucidly written book which provides a clear and accessible introduction to these complex and diverse food-related problems. It explores the continued prevalence of mass undernutrition in the developing world, acute food crises in some places associated with conflict, the emergence of overnutrition in the developing world and the vulnerability of the contemporary global food production system. The text identifies the major problems and analyses factors at international, national and local levels to understand their continued prevalence. The book concludes by evaluating the potential of some oppositional forces to challenge the hegemony of the contemporary food system.

This timely and original book will be invaluable to undergraduates interested in the challenges surrounding food and development. The text is richly filled with case studies from the global north and south, illustrating the extent of these urgent issues and their interrelated nature. Each chapter contains a range of features to assist undergraduate learning, including learning objectives, key concepts, summaries, discussion questions, further reading, useful websites and follow-up activities.

E. M. Young is a senior lecturer in Geography at Staffordshire University, UK. Her core research interest is globalization and its implications for sustainability, food security and f

Routledge Perspectives on Development

Series Editor: Professor Tony Binns, *University of Otago*

The *Perspectives on Development* series will provide an invaluable, up-to-date and refreshing approach to key development issues for academics and students working in the field of development, in disciplines such as anthropology, economics, geography, international relations, politics and sociology. The series will also be of particular interest to those working in interdisciplinary fields, such as area studies (African, Asian and Latin American studies), development studies, rural and urban studies, and travel and tourism.

If you would like to submit a book proposal for the series, please contact Tony Binns on j.a.binns@geography.otago.ac.nz

Published:

Third World Cities, Second Edition
David W. Drakakis-Smith

Rural–Urban Interactions in the Developing World
Kenneth Lynch

Environmental Management and Development
Chris Barrow

Tourism and Development
Richard Sharpley and David J. Telfer

Southeast Asian Development
Andrew McGregor

Population and Development
W.T.S. Gould

Postcolonialism and Development
Cheryl McEwan

Conflict and Development
Andrew Williams and Roger MacGinty

Disaster and Development
Andrew Collins

Non-Governmental Organisations and Development
David Lewis and Nazneen Kanji

Cities and Development
Jo Beall

Gender and Development, Second Edition
Janet Henshall Momsen

Economics and Development Studies
Michael Tribe, Frederick Nixson and Andrew Sumner

Water Resources and Development
Clive Agnew and Philip Woodhouse

Theories and Practices of Development, Second Edition
Katie Willis

Food and Development
E. M. Young

An Introduction to Sustainable Development, Fourth Edition
Jennifer Elliott

Forthcoming:

Global Finance and Development
David Hudson

Natural Resource Extraction and Development
Roy Maconachie and Gavin M. Hilson

An Introduction to Sustainable Development, 4th Edition
Jennifer Elliott

Politics and Development
Heather Marquette and Tom Hewitt

Children, Youth and Development, 2nd Edition
Nicola Ansell

Climate Change and Development
Thomas Tanner and Leo Horn-Phathanothai

Religion and Development
Emma Tomalin

Development Organizations
Rebecca Shaaf

Food and Development

E. M. Young

LONDON AND NEW YORK

First published 2012
by Routledge
2 Park Square, Milton Park, Abingdon, Oxon OX14 4RN

Simultaneously published in the USA and Canada
by Routledge
711 Third Avenue, New York, NY 10017

Routledge is an imprint of the Taylor & Francis Group, an informa business

British Library Cataloguing in Publication Data
A catalogue record for this book is available from the British Library

Library of Congress Cataloging in Publication Data
Young, E. M. (Elizabeth M.)
Food and development / E.M. Young.
p. cm. — (Routledge perspectives on development)
Includes bibliographical references and index.
1. Food supply. 2. Food. 3. Economic development. I. Title.
HD9000.5.Y68 2012
338.1'9—dc23
2011035370

ISBN: 978-0-415-49799-2 (hbk)
ISBN: 978-0-415-49800-5 (pbk)
ISBN: 978-0-203-87748-7 (ebk)

Typeset in Times New Roman
by Cenveo Publisher Services

Printed and bound in Great Britain by
CPI Antony Rowe, Chippenham, Wiltshire

John Wilson Young
13 March, 1946–23 January, 2005

Contents

Plates

Figures

Tables

Boxes

Acknowledgements

I would like to thank my colleagues at Staffordshire University for covering some of my workload while I devoted myself to writing this book. I particularly want to thank our cartographer Rosie Duncan, who has been a real star, always supportive and unfailingly competent, and also for some good photographs! Several of my colleagues contributed photographs, my thanks to Louise and Hamish. Mambu Sanduku was also generous in sending me photographs from her work in Kinshasa, thank you. Some students in levels 2 and 3 read this material in a very provisional form; several of them made very helpful suggestions, and the material is better for their contributions. I must thank the referees for their very insightful comments on the first draft of the text; Rebecca Elmhirst was particularly conscientious, and the book is better for her input. I was not able to address all their concerns, but I hope I managed to deal with some of the most important weaknesses. I would also like to thank Faye Leerink at Routledge for her help, especially in the crucial last few weeks. Jim deserves my thanks too, for his support and encouragement, especially after some long days in the latter stages of this project. Every effort has been made to contact copyright holders for their permission to reprint selections in this book. The publishers would be grateful to hear from any copyright holder who is not here acknowledged and will undertake to rectify any errors or omissions in future editions of this book.

Acronyms

AFN	Alternative Food Network
AIDS	Acquired Immune Deficiency Syndrome
BMI	Body mass index
BRICs	Brazil, Russia, India and China
BWI	Bretton Woods Institutions
CAP	Common Agricultural Policy
DFID	UK Department for International Development
ETDZ	Economic Technical Development Zones
EU	European Union
FAC	Food Aid Convention
FAO	Food and Agricultural Organization
FDI	Foreign Direct Investment
GDI	Gender development index
GMO	Genetically modified organism
HDI	Human Development Index
IDRW	International Day for Rural Women
IEA	International Energy Council
IFAD	International Fund for Agricultural Development
IFIs	International Financial Institutions
ILO	International Labour Office
IMF	International Monetary Fund
INGO	International Non-governmental Organization
ISI	Import Substitution Industrialization

LIFDC	Low-income Food Deficit Countries
MDG	Millennium Development Goals
MERET	Managing Environmental Resources to Enable Transitions to More Sustainable Livelihoods
MNC	Multinational Corporations
NGO	Non-governmental organization
NIC	Newly Industrializing Country
OECD	Organisation for Economic Cooperation and Development
SEWA	Self-employed Women's Association
SEZ	Special Economic Zone
SHG	Self-help Groups
SOFI	State of Food Insecurity
SSA	Sub-Saharan Africa
TB	Tuberculosis
TNC	Transnational Corporation
UNCTAD	United Nations Conference for Trade and Development
UNDR	United Nations Development Report
UNDP	United Nations Development Programme
UNICEF	United Nations Children's Fund
UNHCR	United Nations High Commissioner for Refugees
VAD	Vitamin A deficiency
WB	World Bank
WDM	World Development Movement
WFP	World Food Programme
WHO	World Health Organization
WIEGO	Women in Informal Employment: Globalizing and Organizing
WTO	World Trade Organization

1 Introduction: food, politics and power

Learning outcomes

At the end of this chapter, the reader should be able to evaluate the following:

- **The global food crisis: symptoms**
- **Proximate causes of the global food crisis**
- **Structural problems of the global food crisis**
- **Contrasting solutions to the food crisis**
- **Core concepts: food security, entitlements and food sovereignty**
- **Politics and food in the past and present**
- **Scales of analysis and the food crisis**

Key concepts

Global food crisis; proximate and structural causes; food security; entitlements; food sovereignty; the politics of food

Introduction

This introductory chapter establishes the subject of this book, namely the nature of the global food system and its political and economic characteristics. The current food system has evolved in response to specific historical, political and economic circumstances; it is not a natural system but a socially constructed one which reflects patterns of power and privilege. Analyses of what we eat and how our food is produced expose our 'dependence upon a whole world of social labour conducted in many different places under many very different social relations and conditions of production' (Harvey, 1990: 422). The objective of this text is to expose some of the most vital of these connections. It cannot do justice to the complexity of economic, political and cultural components of the global food system, but it investigates some of its most potent ligaments. This analysis maintains that the food system has become the most important globally embedded network of production and consumption; its integral connections with the petroleum industry in recent decades only serve to confirm its centrality and significance.

This chapter opens with a review of the symptoms of the contemporary global food crisis and suggests some of its proximate and structural causes. Next, some core concepts are introduced; namely, food security, entitlements and food sovereignty. Then, the chapter considers the politics of food in the past and present. Finally, I explain why I have employed different scales to investigate the food system.

A global food crisis: symptoms

> We're in an era where the world and nations ignore the food issue at their peril.
>
> (Josette Sheeran, Head of the World Food Programme, 24 January 2011)

World food prices are high and are predicted to go higher. Many experts claim that food riots will return in the spring and summer of 2012 and warn that they will be similar or more serious than those experienced across the globe in 2008.

News reports about the state of global food security are peppered with apocalyptical warnings about the looming food crisis. Robert Zoellick, president of the World Bank, asserts that 'rising food prices are a threat to global growth and stability' and warns that 'we are only one poor harvest away from chaos' (2011). Even before the current crisis (2011), there were more malnourished people around the world than ever before, and the numbers had increased due to the global economic recession ushered in by the banking crisis of 2008. Most of the desperately undernourished live in Asia and sub-Saharan Africa, and although most of them still live in rural places, urban populations now form a significant proportion of the world's hungry people.

That such large numbers of people still suffer from malnutrition is a global tragedy that should not exist in the twenty-first century. The worsening situation is shown in Box 1.1, which also establishes that the problem is not restricted to the developing world. This also details some of the various International Declarations made in recent decades where political leaders confirm their dedication to ending poverty

Box 1.1

Hunger's timeline

- **1974**: *500 million hungry people in the world.* The World Food Programme Conference pledges to eradicate child hunger in 10 years.
- **1996**: *830 million hungry people.* The World Food Summit pledges to reduce the number of hungry people by half by 2015.
    - 12 per cent of the US population is hungry. The US Farm Bill increases food nutrition programs (food stamps, women and children in need), and food banks augment donations of government surplus with local and industry-donated food.
- **2000 Millennium Summit**: World Leaders pledge to reduce extreme poverty and hunger by half by 2015.
- **2002:** *850 million hungry people.* The World Food Summit+5 admits to poor progress on the Millennium Development Goals.
- **2008**: *862 million hungry people.* The FAO High-Level Conference of World Food Security announces that instead of reducing the ranks of the hungry to 400 million, hunger has increased. The World Bank re-calculates its projections for extreme poverty upwards from 1 billion to 1.4 billion. Over 3 billion people live on less than US$2 a day.
    - 12 per cent of the US population is still hungry. Despite US$60 billion spent yearly in government food nutrition programs and the explosion of over 70,000 food banks and emergency food programs across the nation, one in six children in the United States go hungry each month, and 35 million people cannot ensure minimum caloric requirements.
- **2009**: *1.023 billion people hungry.* The sudden spike in numbers of hungry is connected to the global economic crisis, when millions were plunged into poverty.
- **2010:** *925 million people hungry.* Slight reduction in numbers of people hungry, but still higher than the figure prior to the global economic crisis of 2008–2009, when it stood at 862 million.

Source: Holt-Giménez, 2008: 2; data for 2009 and 2010 from FAO (2010).

and hunger; unfortunately, such declarations have proved empty of political will or commitment.

A recent diagram published by the World Food Program shows how volatile food prices have been in recent years, and details some associated problems and consequences. It is important to realize that even small price fluctuations can have serious consequences for food security for poor families, some of whom might spend up to 70 per cent of total family income on food, leaving very little for other necessities. Increases in the number of hungry people are

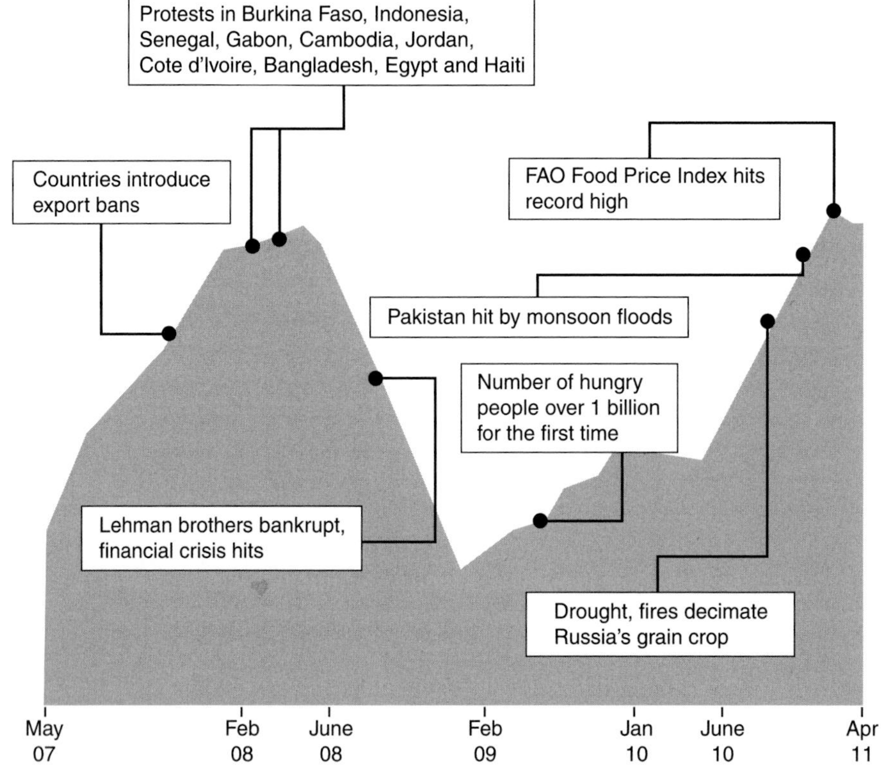

Figure 1.1 Food price rollercoaster (adapted from: World Food Program at: www.wfp.org/stories/rising-food-prices-infographic?gclid=CLy6zciaxKkCFYob4Qodlyv4EA (accessed on 20 June 2011).

associated with economic recessions when vulnerable populations find that their income collapses.

The information in Table 1.1 details some statistics for basic food crops between 2007 and 2008. It is not hard to appreciate the impact that such price changes would have on poor people with limited resources and why food riots occurred.

Why do millions of people still die from hunger and hunger-related diseases at the beginning of the twenty-first century? In a world where so many millions enjoy a more varied diet than ever before and where affluent consumers waste nearly as much as they eat, why does food scarcity still haunt so many millions? This book attempts to answer these questions. The analysis within is contentious and may be challenged, but that too is an objective of this text; that is, to encourage students of social problems to recognize their complexity

Table 1.1 ***Increase in price of staple foods in selected countries, 2007–2008***

Country	*Staple food*	*Price increase (%)*
Bangladesh	Rice	66
Cambodia	Rice	100
Cote d'Ivoire	Rice	>100
Haiti	Basic foods	50–100
Indonesia	Palm oil	100
Mexico	Tortilla	66
Pakistan	Wheat flour	100
Philippines	Rice	50
Sri Lanka	Rice	100
Tajikistan	Bread	100

Source: Action Aid (2009) Food Foremost: A Call for Action at the ASEAN Summit, available at: www.actionaid.org/assets/pdf/ASEAN%20FOOD%20REPORT.pdf (accessed 13 January 2011). Now at http://actionaidusa.org/news/related/food_rights/food_prices_endanger_millions_in_southeast_asia

and the political character of both explanations and proposed solutions.

Traditionally, analysts explain the existence of hunger with reference to simple differences between demand and supply, asserting that hunger existed because there was a shortage of food, locally, nationally or internationally. This simplistic assumption has been successfully challenged since the publication of Sen's work (1981) and is rejected as playing any major role in my analysis because, among other reasons, food production has been higher than population growth in recent decades. Other political and economic factors are more relevant to explanations of the contemporary food crisis and are considered in detail in the following chapters.

In addition to the continuing crisis of undernutrition, the world is facing another food-related health crisis, a global obesity pandemic. An apparently bizarre situation emerged at the end of the twentieth century; there is approximately the same number of people suffering from undernutrition as from overnutrition. The alarming rise in obesity levels is not confined to affluent Western populations anymore; it is now recognized as a burden that threatens to overwhelm health systems everywhere, in developed and less developed countries alike. Deaths from so-called 'diseases of affluence' – heart attacks, cardiovascular failure, type 2 diabetes and various forms of cancer – are increasingly implicated in the deaths of many of the world's poor populations.

Diets rich in sugar, salt and fats, once considered a phenomenon of Western countries, have become global in recent decades.

Plate 1.1 Children eat more processed foods. (Source: Alex Moore at: www.alexmoorephotography.com/F_00d.html)

Children across the globe are beginning to eat highly processed foods that are high in fats and sugars. In some countries that were known less than 30 years ago for high levels of undernutrition, public health workers are now more concerned about childhood obesity. Places as diverse as Mexico and China are facing a problem of crisis proportions which is threatening to overwhelm their public health systems.

Hence, the global food crisis has two contradictory manifestations. Undernutrition in many places remains serious and has worsened as food prices have increased in the last few years. In addition, in some communities, food has never been cheaper and is available in great abundance and in myriad processed confections, which helps explain the emergence of the obesity pandemic. The world food crisis is a complex phenomenon manifested in the continuation of the very old problem of food security and a very novel problem associated with the spread of Western diets and lifestyles.

Although superficially very different problems, the analysis presented in this text argues that undernutrition and overnutrition are both manifestations of a malaise at the heart of the global food system

(Young, 2004). Both forms of malnutrition are symptoms of a broken food system, one which fails to deliver health and well-being to approximately two million people; these people live in countries as varied as Britain, Brazil, India, China and Columbia. Pockets of undernutrition and overnutrition are found in all countries across the globe, in so-called developed countries, in newly industrializing countries such as the BRIC (Brazil, Russia, India and China), as well as in the so-called developing countries across the global south. In the following text, we begin to understand the crisis by considering some important, proximate factors that help explain recent food price rises and increases in obesity.

Proximate and structural factors

Proximate factors are those which are obviously implicated in the food crisis. As becomes clear later in the book, however, we must understand the history and evolution of the global food system and its political and economic contexts to deliver a comprehensive analysis. More profound changes in the structure and nature of global political economy are referred to as *structural factors*; these are not necessarily immediately obvious, but require us to examine the evolution of the world economy to appreciate how changes have differential impacts in different places and communities.

To illustrate the difference between proximate and structural explanations, consider the following examples. A baby in a slum dies and the doctor cites diarrhoea as the cause of death on the death certificate. This is correct, but it is only a partial description, a proximate explanation of the baby's death. A fuller, more satisfactory explanation would examine why that baby, in that community, in that country, in that region of the world, was living in poverty without access to fresh water. To explain why the baby is poor requires a much deeper analysis, because poverty has spatial and temporary patterns that transcend individual circumstances. However, to open our debate, we may identify some proximate factors in recent years that have plunged an already problematic food system into crisis.

The world food crisis: proximate causes

Among the most important proximate causes for the world food crisis is the recent food price inflation. In March 2008, 'average world wheat prices were 130% above their price a year earlier, soy prices were 87% higher, rice has climbed 74% and maize was up 31%' (Holt-Gimenez, 2008: 1). However, why have food prices been rising so dramatically and so suddenly? We can identify several factors that have driven food price increases. Most experts, from diverse bodies such as the International Food Policy Research Institute and the World Bank, suggest that price hikes are caused by large volumes of cereals being used as agrofuels. Apparently, our eagerness to 'go green' has had serious impacts on some of the world's poorest populations and their access to affordable food.

'Carnivorous cravings' (Holmes, 2001) are also part of the problem. Affluent populations everywhere, in the global north and global south, are demanding more meat, and meat production requires large volumes of grain, which means that cereal prices increase. Meat consumption is correlated to growing affluence, and there have been dramatic increases in meat consumption in many of the countries with growing middle-class populations. The environmental as well as social implications of the continued expansion of meat consumption are recognized as important elements of the contemporary food provisioning crisis and are considered in detail in the forthcoming chapters.

In the past, food prices fluctuated as yields varied because of weather; today, food price volatility is correlated to other equally unpredictable variables. With the emergence of industrial food production after the Second World War, food production has become heavily dependent on petroleum and petroleum-based inputs. In almost every part of the food chain, oil is essential, from agrochemicals to cool chains and transport. Oil prices have been increasing, and this means food price rises. Such is the related nature of these problems that '[W]estern countries have upgraded the food and fuel crisis into a national security concern as they fear record high energy prices and agricultural commodity costs are destabilising key developing regions of the world' (Hoyos and Blas, 2008: 1). As oil exploration expands into more inhospitable environments, and demand increases associated with rapid industrialization in emerging world markets become apparent, we can reasonably assume that oil prices will continue to rise with implications for the price of food. Although oil prices are volatile, we may confidently predict that the general price trend will be upwards as accessible deposits are exhausted and global consumption increases.

Another factor considered to be an important explanatory variable for food price fluctuations is financial speculation. A recent report argues that 'speculation, hoarding, and hysteria' (International Food Policy Research Institute, 2009) also played a part in the sudden food price rises after 2008. Although this report is a measured consideration of the role of financial speculation, others are convinced that speculation has seriously exacerbated the food crisis in recent years. The World Development Movement (WDM), a British-based development NGO, declares:

> During the 1990s and early 2000s, aggressive lobbying by investment banks and hedge funds led to weaker regulations on food speculation. Banks like Barclays Capital created special investment products to help financial companies make money from food prices, just like they do from share prices. As a result, hundreds of millions of pounds have been poured into food commodity markets, pushing up prices and resulting in increased hunger and poverty.
>
> (Source: WDM at: www.wdm.org.uk/stop-bankers-betting-food/regulate-food-speculation; accessed on 20 June 2011)

The WDM has launched a campaign to increase financial regulation to reduce speculation of food commodities; people are encouraged to send the following letter (Box 1.2) to the British Treasury. There is, quite understandably, outrage that financial deregulation initiated in the 1980s has allowed food commodities to become just one more financial opportunity in capitalism's 'global casino' (Strange, 1997). Traditionally, governments have intervened to stabilize domestic food prices to avoid the drastic gluts and shortages that have plagued so much of human history. Even when free trade ideologies were largely unchallenged, governments accepted that one of their principal functions was to guarantee relatively low food prices that protected their populations from famines and food scarcity.

The latest food crisis was undoubtedly exacerbated by droughts and floods in some major wheat-producing countries that drove global prices higher and precipitated some policies designed to protect domestic consumers (Russia, for example). This draws attention to another variable, the prospect of climate change in the next 50 years and how this might increase food price volatility. Box 1.3 considers some of the issues raised by climate change.

That the global food system fails to feed some adequately and overfeeds others is only one of its multiple failures. It is also

Box 1.2

Stop banks betting on bread

Dear Mark Hoban, Financial Secretary to the Treasury

Excessive speculation on food by financial institutions has pushed up food prices and left millions across the world facing hunger and malnutrition.

In 2007 and 2008, the IMF food price index rose by over 80%, pushing the total number of people living in hunger to over 1 billion and increasing the number living in poverty by over 100 million. The dangers of high food prices have not gone away – food prices have been rising steadily over the last year and remain higher than seen during the food crisis of 2008.

Commodity futures markets need better regulation to ensure stable and affordable food prices for the benefit of consumers, producers and businesses in the UK and globally. This is especially important for the world's poorest countries, where high food prices can lead to increased hunger, deeper poverty and social and political unrest. Calls for better regulation of commodity futures markets have been supported by a range of experts, including the Head of the UN Food and Agriculture Organisation, the UN Special Rapporteur on the Right to Food, the CEO of Starbucks and the Chief Executive of Unilever.

I would like the Treasury to support European proposals that:

- Ensure transparency on commodity futures markets, by requiring that all deals on food derivatives take place on regulated public exchanges.
- Prevent excessive speculation from distorting food prices by introducing position limits to restrict the number of food futures contracts that can be held by financial institutions at any given time.

Yours sincerely,

____

Source: World Development Movement at: www.wdm.org.uk/stop-bankers-betting-food/regulate-food-speculation (accessed 27 June 2011). Read more about food and speculation at: www.wdm.org.uk/food-speculation (accessed 27 June 2011)

implicated in many of our most serious global environmental problems: industrial food production is a major source for global warming gases; agricultural intensification is associated with severe water shortages and chemical pollution across the world; modern agriculture methods are largely responsible for the catastrophic loss of biodiversity experienced in recent decades; industrial food production

Box 1.3

Food prices driven up by global warming, study shows

Clearly, the specific impacts of climatic change are very contentious and debateable as some regions may increase their food production capacity, but if indeed global weather patterns change dramatically, the implications for feeding the world population in the future is even more challenging than the situation we face at present. Although not discounting the potential implications of climate change for the global food crisis, I argue that changes in the political and economic realms will remain more relevant to policy interventions for the foreseeable future. In other words, climate change may be a proximate factor, with significant regional relevance, but structural factors are considered more relevant in my analysis. Perhaps analyses that prioritize climate change and its significance for the global food crisis detract attention from more politicized variables which the establishment finds more uncomfortable to address? Of course, climate change is politically contentious too, but perhaps less so than the issue of global governance in the post-2008 era? Climatic change has certainly been usurped by the global economic crisis on international political agendas since 2008.

What proximate variables help explain the emergence of the obesity pandemic? People across the globe are embracing Western diets and lifestyles. Adults and children have ready access to fast foods and processed foods, which when eaten too much cause weight gain. Allied changes in lifestyle, more sedentary jobs, car travel and urban lifestyle also help explain the pandemic. In some emerging markets, China and India, for example, a large relatively affluent middle class has emerged since the 1980s. These populations are eager to assume Western habits which are marketed as 'modern'. Increased alcohol consumption and smoking are also increasing in these countries, and are another example of the adoption of Western habits with serious negative health implications.

systems have transformed the global fishing industry and are threatening the collapse of the global marine ecosystem; soil erosion and degradation has increased alarmingly with intensive farming; and public health issues have emerged associated with the increased use of antibiotics in intensive animal husbandry and fears about cross-species diseases. This is a partial catalogue of its failings; if we included animal welfare issues, it would become much longer.

Critics also describe how industrial food production has replaced more benign sustainable systems and, in doing so, has marginalized small producers and concentrated ownership and control in the hands of a few global corporations. Many argue that such concentration makes the system vulnerable to manipulation and collapse. It is imperative that we understand the relatively recent emergence of this way of feeding ourselves and the forces that benefit from its operations. The food system produces more food than ever before in human history but at a very high cost both to some communities and global and local environments. There is a strong case to accuse our modern food production system of creating diets that are deadly in several respects (Young, 2010).

The world food crisis: structural causes

All the proximate factors considered in the preceding text help explain recent food price rises and the emergence of a global food crisis and new fears surrounding the global obesity pandemic. However, the analysis offered in this text considers the issues detailed earlier as the symptoms of a more profound malaise. The history and contours of this malaise is the subject of the book. In summary, this analysis holds that the nature and structure of the global food system that has emerged in the late twentieth century is the core problem. This food system is dominated by a few major corporate players who enjoy immense power over producers, consumers, national policy-makers and international institutions of global governance. Their power is insidious and growing and, as this book reveals, involved at every point in the food system, from field to plate. The power of these multinational or transnational corporations (MNCs or TNCs) has intensified as globalization, generated by neoliberal ideologies, has extended and deepened since the 1980s. Holt-Gimenez (2008) cites some of the factors that explain the rise of the industrial agri-foods complex:

> Built up over the last half century – largely with public funds for grain subsidies, foreign aid, and international agricultural development – the industrial agri-foods complex is made up of multinational grain traders, giant seed, chemical and fertiliser corporations, processors and global supermarket chains.
>
> (Holt-Giménez, 2008: 4)

Just exactly how a few corporate giants triumphed nationally and globally is examined in Chapters 3, 4 and 5, but here we will simply mention a few of the means through which they secured such dominance. Domestic agricultural support policies (massive tax-payer-funded subsidies) in the north, especially the United States and European Union, allied to favourable international financial and trading arrangements, have helped northern transnational corporations secure access to southern markets, and this helps explain their global pre-eminence. Asymmetrical power relations in the food system, already present in the 1970s, have intensified in the last 40 years as corporate concentration has increased through mergers and acquisitions.

The food sector is now highly integrated, both vertically and horizontally, and includes some of the behemoths of the petrochemical and natural science sectors, such as the Bayer Corporation. Now a major global player, Bayer invested €722 million in 2010, or roughly 24 per cent of the Group's research and development budget, on Bayer CropScience. The company boasts of its success with a global presence in almost every country (Source: www.bayer.com/en/Bayer-Worldwide.aspx; accessed 27 June 2011).

Major agri-business corporations (with names that are rarely mentioned in debates that surround the world food crisis or the obesity pandemic) exercise enormous power, which explains their influence at the highest levels of government across the world and in important institutions of global governance such as the World Bank (WB), the International Monetary Fund (IMF) and the World Trade Organization (WTO). Their influence in these global institutions also helps explain the disenchantment of many southern countries in these institutions and suggests why the recent agricultural trade talks, known as the Doha Round of WTO negotiations, have floundered. It is relevant to review a few factors that help explain the dominance of corporate agriculture in the last 40 years.

One of the most relevant factors in understanding the relationship between food and development is to appreciate a central element of recent development theories. Development policy since the 1980s has encouraged southern countries to invest in export crops and to neglect domestic food production. In theory, these countries would import 'cheap food' from northern producers and specialize in high-value export crops, and selling these would generate the income to pay for imports. So, following guidelines dictated by WB and IMF, many southern countries, once almost self sufficient in food, became food-deficit countries. As international food prices rise, due to shortages, real or created, people in these food-deficit countries suddenly find themselves with very inflated food prices. Their vulnerability to international food price rises becomes correlated to increases in malnutrition among their poorest communities. These countries are classified as low-income, food-deficit countries (LIFDCs) and are the most at risk from serious food insecurity. For a list of these countries, visit the FAO page at: www.fao.org/countryprofiles/lifdc (accessed 27 June 2011). These development policies have become a 'recipe for disaster' for millions of poor people:

> So the facile ideas underpinning the agreement on agriculture – that farmers could simply shift to cash crop cultivation to increase their incomes and that developing countries could simply choose to export cash crops and import food whenever needed, without implications for food security – have been a recipe for disaster in many countries.
>
> (Ghosh, 2011)

Not a disaster for all involved, however, because profits of major food corporations have proved very resilient throughout the economic crisis of 2008 and the ensuing global economic downturn; indeed, in many cases, they have soared.

A less obvious mechanism has helped establish the power of the corporate sector in the developing world. Food aid has been a conduit through which power and control is exerted. As recognized by Millstone and Lang (2003):

> Well-managed food aid clearly benefits people suffering from a shortage of food. Conversely, poorly managed food aid (which may be culturally or nutritionally inappropriate) contributes to long-term food insecurity by discouraging local food production, thereby depressing prices and disrupting local markets.
>
> (Millstone and Lang, 2003: 28)

Most food aid is from the northern countries and is destined for southern countries. Exactly who receives aid, how they receive it and why they are selected as recipients, is interesting and worthy of fuller discussion elsewhere (see Chapter 8), but it is important to understand that its provision is not always a response to simple philanthropic instincts.

Massive injections of food aid can create food dependency and consequently a market for exports. Frequently, food aid is subject to certain conditions, so it may be used for political purposes as well as humanitarian relief. Holt-Giménez (2008: 6) describes one of the several perversities of the international food aid regime; when international cereal prices are low, food aid increases, but when food prices are high and food-deficit countries are in most need, food aid declines. Giménez concludes that the system is more sensitive to international food prices than the needs of poor countries. The system is certainly responsible for creating new markets for northern grain producers and increasing food dependency in the some southern countries.

Now established in the global south, the influence of transnational corporate players is expanding and obvious everywhere. Their presence helps explain the diffusion of the obesity problem. Their marketing strategies have proved very effective, and national governments are loath to tackle their influence; more often, they are embraced as innovative, dynamic investors in the food sector. Diets across the south have changed dramatically in a few decades, and now, certainly among urban populations, they too often replicate the worst examples of the Western diet.

Solutions: more of the same or a radical change of direction?

The discussion here alerts readers to two very different perspectives on how to change the world food system. It introduces material which is dealt with in more detail in Chapter 9. Everyone accepts that things are seriously wrong with the global food system, but there is no consensus about how or why it is wrong, or how it might be 'fixed'. Conceptualization of the problem and their associated policy recommendations are, in fact, diametrically opposed.

At one extreme of the spectrum are those who advocate more industrial food production and urge us to embrace new technologies emerging from corporate research laboratories. This vision triumphs the technological revolution that has transformed agricultural production since the end of the Second World War and argues that the revolution should be further diffused, extended and intensified. At the other extreme are those who argue that the nature of contemporary food production is the *cause* of the current crisis, and that its promotion will simply exacerbate the problems we face now and in the future. This vision calls for a fundamental rethinking of global food provisioning, and argues for changes at every stage of the food commodity chain.

These contrasting visions have been promulgated by two leading British scientists in 2011. One urges us to embrace 'genetically modified crops and cutting-edge developments such as nanotechnology to avoid catastrophic food shortages and future climate change' (Beddington, 2010). This perspective is allied to Malthusian perspectives (reviewed in Chapter 3), which emphasizes the importance of increased food production. This argument tends to ignore aspects of food distribution and consumption which others consider vital to explanations of poverty and associated hunger. Simplistic correlations between hunger and food availability were effectively challenged by Sen in his seminal work on famines (1981).

However, naive correlations between food production and hunger remain effectively embedded in popular hunger discourses, political debates, in the media and, most importantly, in corporate food headquarters (see Chapter 3). Conceptualizing 'the food problem' as a simple 'production' problem where 'we need more food' is essentially an argument that ignores the multiple injustices in the food system as presently organized. Such an interpretation justifies an increased role for corporate agriculture as it is viewed as the means through which we can most effectively produce more food. This book challenges this interpretation of the problem and suggests that the problem is more complex than this perspective allows.

The perspective employed in this book shares more in common with another British scientist, namely Watson (2011), who asserts that:

> governments and industry focused too narrowly on increasing food production, with little regard for natural resources or food security. "Continuing with current trends would mean the earth's haves and have-nots splitting further apart", he said. "It would leave us facing a world nobody would want to inhabit. We have to make food more affordable and nutritious without degrading the land".

Watson and others (Shiva, 2000; Madeley, 2002; Lang and Heasman, 2004; Lawrence, 2008; Holt-Giménez and Patel, 2009) understand 'food issues' to be about corporate control and the manipulation of trade and subsidy regimes. At present, trade and public subsidies help generate more corporate profit rather than assist in creating ethical and environmentally sound food systems. An allied vision is available in the most recent World Watch Report (2011). Its authors suggest that, instead of producing more food to meet the world's growing population, a more effective way to address food security issues and climate change would be to encourage self-sufficiency and waste reduction, in wealthier and poorer nations alike. It rejects the narratives emanating from the agri-business interests and its allies in governments across the globe.

The analysis in this book examines the global food system and its beneficiaries, predominantly major agricultural corporations, and finds them guilty of creating myths and disseminating misinformation. They maintain that corporate interests are synonymous with our interests as consumers, and that their vision will deliver us from the diverse malign symptoms of the contemporary food system. My analysis asserts quite the contrary, that these problems will be *exacerbated* by more corporate control over the food we eat. Instead, I argue that food sovereignty must be reasserted at the international, national and local levels to deliver a healthier food system. The concept of food sovereignty is at the heart of this analysis and has largely replaced the more restricted concept of food security in critical analyses of the food crisis. The next task is to review some core concepts.

Food security

The concept of food security has been at the centre of debates about malnutrition for decades and is still widely used in the literature. The most useful definition is:

> Food security, at the individual, household, national, regional and global levels is achieved when all people, at all times, have physical and economic access to *sufficient*, *safe* and *nutritious* food to meet their dietary needs and food preferences for an active and healthy life.
>
> (World Food Summit, 1996)

This definition has the merit of introducing several complexities associated with the concept. First is the problem of unit of analysis. There are two extremes suggested here; we may be interested in food security at the individual or global level or any scale in between. Most accessible data continues to measure kilocalories per capita per day and are mapped at the national level. This material and an evaluation of trends are published annually by the Food and Agricultural Organization (FAO) in their State of Food Insecurity Reports (see FAO, 2010).

Clearly, we have failed miserably to deliver sufficient and safe food to millions of people worldwide. Conventionally, the main objective was to reduce undernutrition; this has been the main focus of food security efforts since the 1950s and before. A more recent phenomenon, the global obesity pandemic, has shifted the nature of the debates, and now it is vital to assess how to ensure a 'nutritious' diet too. So, the concept of food security has been modified in recent years to include a 'nutritious' diet as well as the traditional expectation, that all should access a 'safe and sufficient diet'. We can begin to appreciate that simply producing more food is not the answer.

Food security approaches have some limitations. Transformations of the global system over the last 40 years means that the food security concept has limited purchase to tackle the myriad problems associated with the food system today; namely, international trade and aid policy, global governance and development policy, equity, access and ownership, public health and environmental issues, for example.

Maxwell and Slater (2004) review old and new issues on the food policy agenda and discuss how the agenda has shifted from a narrowly conceived food security perspective; their summary is shown in Table. 1.2. Many of the themes identified are considered in this book. Their diverse nature and breadth indicates the complexity and interdisciplinary nature of food analyses today. It is unsurprising that Maxwell and Slater conclude that '[A] preoccupation with food security is no longer sufficient. It is necessary to rediscover food policy' (2004: 3). A further limitation of food security perspectives is exposed when we examine the concept of entitlements.

The entitlement concept

Traditionally, the emphasis has been on ensuring that there were sufficient amounts of food available. This theory is perpetuated by productionist approaches to the food problem; such analyses continue to argue that, by producing more food, we will 'solve' the problem of hunger. However, since Sen's intervention (1981), such simplistic assumptions have become untenable. Sen (1981) introduced the concept of entitlements to theories of food security and has

Table 1.2 ***Food policy: old and new***

Theme	*Food policy old*	*Food policy new*
Population	Mostly rural	Mostly urban
Rural jobs	Mostly agriculture	Mostly non-agricultural
Employment in food sector	Mostly food production and primary marketing	Mostly on processing and retail
Actors in food marketing	Grain traders	Food companies, many are global in scope
Supply chains	Short–small number of food miles	Long–large number of food miles
Typical food preparation	Mostly cooked at home	High proportion of pre-prepared meals, food eaten out
Typical food	Basic staples, unbranded	Processed foods, branded products, many global
Packaging	Low	High
Purchased food	Local stalls or shops, open markets	Supermarkets, increasingly global in scope
Food safety issues	Pesticide poisoning of field workers; toxins associated with poor storage	Pesticide residues in food, adulteration, bio-safety issues in processed foods (salmonella, listeriosis)
Nutrition problems	Undernutrition	Undernutrition, obesity and related diseases
Nutrient issues	Micronutrients	Micronutrients and fat, sugar and salt
Food-insecure	Peasants	Urban and rural poor, women, refugees, ethnic minorities, indigenous communities
Main sources of national food shocks	Poor rainfall and other production shocks; protracted crises	International price rises, and other trade-related problems; protracted crises
Main source of household food shocks	Poor rainfall and other production shocks; family circumstances (death, disease)	Collapse of livelihoods, Income shocks causing food poverty; family circumstances (death, disease)

(*Continued*)

Table 1.2 ***Cont'd***

Theme	*Food policy old*	*Food policy new*
Remedies for household food shortages	Safety nets, national/international food aid	Social protection, income transfers, national/international food aid
Food and health	Diseases associated with undernutrition	Diseases associated with undernutrition; diseases associated with overnutrition; antibiotic resistance; diseases crossing species barriers (HIV, Ebola, SARS, avian flu – all recent examples thought to be of this type)
Role of the State	Domestic food production; securing domestic food security	Reduced role for the state in many countries; food production for export markets
Petroleum inputs	Limited	Extensive
Fora for food policy	Ministries of agriculture, relief/ rehabilitation, health	Ministries of trade and industry, consumer affairs, ministers of Health and environment, food activist groups, NGOs, INGOs (Oxfam, Cafod, etc.)
Focus of food policy	Agricultural technology, land reform, parastatal reform, supplementary feeding, food for work	Corporate control, economic pressure on primary producers, food speculation, land reform and access, health education, food safety, competition and rent seeking in the value chain, structure in the retail sector, waste management, advertising
Key international institutions	FAO, WFP, UNICEF, WHO, CGIAR	FAO, UNIDO, ILO, WHO, WTO. IMF, WB

Source: Maxwell and Slater (2004: 4)

transformed the nature of the debates about food security and hunger. His work since 1981 has been extended to include analyses of human capacity, development and poverty (Dreze and Sen, 1989; Sen, 1999). For our purposes, his work on entitlement is central:

> Starvation is the characteristic of some people not having enough food to eat. It is not the characteristic of there not being enough food to eat. While the latter can be the cause of the former, it is but one of the many possible causes.
>
> (Sen, 1981: 1)

Sen's work established that hunger and sufficient food may exist together, in the same place at the same time; clearly, things are not simply about food availability. In other words, enough food being available does not guarantee that no one will be hungry. This is obviously the case at present when the world produces more than enough food to feed everyone adequately but available food is not distributed equitably. Millions of poor people simply do not have the capacity to access the food in their midst. Suffering from malnutrition is not about there not being enough food; it is about people not being able to acquire available food, that is, about food distribution.

Understanding people's entitlements shifts the analytical focus from food production and availability to understanding distributive mechanisms; that is, what determines how available food is distributed and how politics, economics and ideology influence distribution. As is detailed in later chapters, processes operate at every scale of analysis that influence how food is distributed and how entitlements are constructed. The ability of some people to command (acquire) food may reflect their political, economic, social, military or inherited position within the international system and its national and sub-national elements. The term 'command' is employed because it suggests that an individual's or group's ability to acquire food is correlated to their access to power; it reflects their political status. Table 1.3 suggests the variety of factors, historical, political and economic, at every level of analysis that helps construct entitlements.

As employed here, the term 'entitlement' refers to the power of any individual or group to acquire a decent diet. In the United Kingdom, any person's entitlement comprises of the capital they might possess, inherit or accumulate, income they earn, supplements from the state and assistance from family. In rural areas of the developing world, many peasants have entitlements based on a small plot of land

Table 1.3 ***The complex construction of entitlements***

Level of analysis	*Historical*	*Economic*	*Political/ideological*	*Social/cultural*
Global	Emergence of merchant capitalism in Europe and consequent European colonial projects; nature of peripheral integration into the capitalist world economy since the fifteenth century; development of colonies and settler states as sources of tropical foods, meat and cereals; since 1980s the role of globalization	The evolution of colonial patterns of production and trade relations; preferential trading arrangements for colonial possessions; since the 1980s, the dominance of economic liberalism; institutions of global economic governance (WB, IMF and WTO); the emergence of industrial food policies and dominance of TNCs; massive, volatile capital flows now characterize the global economy	Uneven development of world capitalist economy with a core and peripheries; asymmetrical geopolitical relations with changing hegemony (Spain/Portugal, France/Dutch, the United Kingdom and the United States); development philosophies reflecting dominant economic philosophies reflected in national development priorities; intense global competition for FDI; erosion of some national decision-making capacity; ideological dominance of consumerist paradigm	Mass migrations of European populations through colonial and allied systems; demographic transformations in most colonial possessions, collapse or enslavement of indigenous populations and initiation of various migrations including slave trade and the coolie system of labour transfer; increased role of marketing in transforming the concept of wants and needs; economic restructuring has strengthened some local identities and weakened others; globalization of consumerist paradigm; rise of individualism
National	Processes of economic and political integration into the global economy; character and capacity of the state; property relations; access and control over resources	Uneven economic development; colonial economic legacies, trading relations, etc.; development policies	Emergence of political elites allied to global accumulation strategies; development strategies; public policy; gender relations; differential access to power by class, race, ethnicity and region	Social differentiation and polarization; urbanization and transformed patterns of mobility and cultural affinity
Household	Socially sanctioned familial systems	Rural/urban; contrasts by class, race, ethnic variables	Gender relations; public policy; access to power and resources	Changing gender relations; changing family structures; knowledge of nutrition; religion, taboos and social customs

(inherited assets), from which they may produce food for consumption or sale, occasional earnings from selling their labour, earnings from any food sales and assistance from other family members. Urban dwellers in the south often depend on very insecure entitlements; earnings from casual labour, crops grown on common land, public-funded food programmes. Generally, in the south, there is limited public assistance available. Entitlement packages are dynamic and alter as social change differentiates individuals and groups by class, ethnicity, gender, age and region. People with limited entitlements will endeavour to improve them by a variety of strategies, sometimes violently, sometimes less dramatically by squatting on land, resisting land purchases, fighting for policy changes or voting for sympathetic political parties. By employing the entitlement concept, we are much better equipped to reveal the real factors that explain hunger, and therefore are better able to intervene to reduce its incidence. The problem requires us to consider livelihoods and their vulnerability to economic and social changes.

The term ‘entitlement’ may be used to understand differential power relations in international relations too. The United States and the countries in the European Union have diverse and generous aggregate entitlement packages based on their natural resource endowment and their inherited status as major world powers. This latter status grants them greater influence in modifying and directing contemporary world economy. Many of the states in sub-Saharan Africa are at the less privileged extreme on the international entitlement continuum; they have weak economies and little political influence to exert for the advancement of their national ambitions. Countries classified as newly industrializing (NICs), or emerging economies, fall in between these two extremes and are extending and diversifying their economies and political influence in international affairs. Human development indices (HDIs) rankings are a rough guide to any country’s entitlements, although it is important to stress that, as with all aggregate statistics, these conceal great differences among countries.

The entitlement concept has been the focus of very intense debates since its emergence in the 1980s; this discussion does not do justice to the complexity of the concept, but has focused on its specific relevance for the themes in this book. For a very helpful review of the strengths and weaknesses of the entitlement concept, see Devereux (1993: 66–85), and to consider its application to the Irish and Highland Famines of the 1840s, see Young (1996).

Food sovereignty

As the debates about the food crisis deepened, many analysts judged the concept of food security to be too limited. Food sovereignty is a more helpful concept because it draws attention to how and where decisions about the food system and its direction are implemented. Unlike food security, it confronts the question of power directly. It puts the question of power and politics at the centre of the debate; it emphasizes that consumers and producers should be in control of the food business, not corporations or unaccountable international institutions of regional or global governance. Food sovereignty is about resisting further corporate control of the food system; indeed, it argues that it is vital that the power of corporations be reduced, and that these powers be assigned to producers and consumers.

Food sovereignty is defined as 'the right of peoples to healthy and culturally appropriate food produced through ecologically sound and sustainable methods, and their right to define their own food and agriculture systems. It puts the aspirations and needs of those who produce, distribute and consume food at the heart of food systems and policies, rather than the demands of markets and corporations' (Source: www.foodsovereignty.org/FOOTER/Highlights.aspx; accessed 26 February 2011).

The core concept of food sovereignty is adopted by those who most seriously challenge corporate control of the food system. The Declaration at Nyelení, Mali in 2006 established the main objectives of the food sovereignty movement.

They are fighting for a world where:

- All peoples, nations and states are able to determine their own food-producing systems and policies that provide every one of us with good-quality, adequate, affordable, healthy, and culturally appropriate food
- There is recognition and respect for women's roles and rights in food production, and representation of women in all decision-making bodies
- People in all countries are able to live with dignity, earn a living wage for their labour and have the opportunity to remain in their homes
- Food sovereignty is considered a basic human right, recognized and implemented by communities, peoples, states and international bodies

- We are able to conserve and rehabilitate rural environments, fish stocks, landscapes and food traditions based on ecologically sustainable management of land, soils, water, seas, seeds, livestock and other biodiversity resources
- We value, recognize and respect our diversity of traditional knowledge, food, language and culture, and the way we organize and express ourselves
- There is genuine and integral agrarian reform that guarantees peasants full rights to land, defends and recovers the territories of indigenous peoples, ensures fishing communities' access and control over their fishing areas and ecosystems, honours access and control over pastoral lands and migratory routes, ensures decent jobs with fair remuneration and labour rights for all and a future for young people in the countryside
- Agrarian reform revitalizes interdependence between producers and consumers, ensures community survival, social and economic justice and ecological sustainability, and respect for local autonomy and governance with equal rights for women and men
- The right to territory and self-determination for our people is guaranteed
- We share our lands and territories peacefully and fairly among all people, be they peasants, indigenous peoples, artisanal fishers, pastoralists or others
- In the case of natural and human-created disasters and conflict-recovery situations, food sovereignty acts as a kind of 'insurance' that strengthens local recovery efforts and mitigates negative impacts; where we remember that affected communities are not helpless, and where strong local organization for self-help is the key to recovery
- Where peoples' power to make decisions about their material, natural and spiritual heritage is defended
- Where all people have the right to defend their territories from the actions of transnational corporations

(Source: Extract from the Declaration of Nyelení available at: www.landaction.org/spip/spip.php?article37; accessed 2 March 2011)

Reading through this list reflects the much wider remit of the concept of food sovereignty as compared to food security. Themes include: securing livelihoods; gender equity; environmental objectives; land and property rights and agrarian reform policies; local autonomy; and democratic systems of governance. And, it urges resistance to corporate control of the food system. The discussion of food sovereignty establishes politics at the heart of the food debate. The food crisis is about who controls the food system; it is about property rights and human rights, and its resolution will be political too. We resume the discussion about solutions in Chapter 9.

Food and politics in the past and present

Every stage of the modern food chain, from field to plate, is embedded in contentious economic and scientific debates which are essentially political. They are deemed political because they are about the role of markets and the state, about global trading regimes and the ability of governments to control their agricultural and food policies. They are also political because food security is the most basic expectation that citizens have of their governments; failure to provide a decent diet is associated with autocratic or corrupt regimes or conflict-ridden political situations. Finally, they are political because scarcity leads to food riots which have been associated with radical regime change in the past and present. Sen (1982) has argued that famines never occur in democratic countries, emphasizing the relationship between politics and food. Understanding the decline in mortality in Western Europe is another illuminating case study and is considered in Box 1.4.

A few other examples serve to confirm the political nature of food. In the past, Chinese emperors where held responsible for food shortages, and during food crises their power was precarious. Declines in food security were associated with challenges to central imperial power, and so ensuring food security was essential to maintaining the political legitimacy of the state. Another case is more notorious. Although most agree that Marie Antoinette never said 'let them eat cake', the fact that she was believed to have uttered this callous pronouncement had political import. It would be simplistic to 'explain' the French Revolution as a consequence of food shortages alone; it was influenced by Enlightenment philosophies and political ideas diffused after the American War of Independence, among other things. However, it would be wrong to dismiss the role of food shortages and rapid food price increases to the anger that erupted and the political transformations that ensued in 1789.

In Ireland too, the political ramifications of the 'Great Irish Famine' in the 1840s has had a long legacy that blighted British–Irish relations

Box 1.4

Mortality decline in Britain

Traditional interpretations of mortality decline in Britain stressed the role of technological advances. Technology was deemed important in several ways. Increases in food production were explained by improvements in agricultural technology in the eighteenth century. In addition, because of improvements in transport technology, food was more readily imported from further afield and was more easily distributed to areas suffering from food scarcity. Increased food production then helped improve the health of the population and contributed to the decline in mortality rates. Advances in medical technology were also considered important to explanations of mortality decline in Britain in the nineteenth century.

Recent interpretations have emphasized other variables, however. Chief among these is the role of the state in provisioning its population when famine threatens. Arnold argues that:

> in the European context the need to mitigate famine and provision the people was one of the most important factors behind the rise of the modern nation-state, just as neglect of this responsibility exposed regimes to some of their most serious challenges.
>
> (Arnold, 1988: 104)

Increasingly, technology advances offered the opportunity for those in power to intervene to alleviate hardship, but the crucial variable was the political will to intervene. As the British state became territorially integrated and expanded its military and administrative capacities, legislative and institutional changes evolved to ensure food security. Mortality rates declined in Britain, because it became politically rational for the political elite to intervene to relieve hunger and disease. Reductions in mortality then, as today, owed more to public health investments, especially investments in sanitation, than to another other factor. The role of medicine was important because professionally qualified people promoted public health reforms. A more enlightened elite was also alarmed at the poverty and squalor in which so many millions existed in the rapidly expanding urban centres; in addition to being a Christian obligation, it became economically important to improve the health of the industrial workforce.

until very recently; arguably, it was crucial to the Republican rhetoric which delivered Irish independence.

More recently, in the mid-1970s, food riots erupted across the developing world and were known as the 'IMF food riots' because many held that institution responsible for the policies that caused food price hikes. Apparently, in January 2011, presidents Obama and Sarkozy discussed how to cope with the political chaos they expect this spring if food prices increase as predicted by their security teams.

It is also obvious that food security has re-emerged on international and national political agendas after some years of neglect. At present, some analysts claim that political instability in North Africa is associated with food price rises. I think most will allow that food security has political leverage.

Food – its production, distribution and consumption – is more contentious than ever before. As affluent consumers agonize about ethical food consumption and their waistlines, millions still struggle to command a decent basic diet. This text addresses the nature and character of the global food system and focuses on its serious shortcomings with particular reference to its failure to deliver healthy diets to millions in the developing world. It employs the concepts of food security, entitlements and food sovereignty to expose fundamental contradictions at the heart of the globalization project. It examines both these concepts and evaluates how national strategies to promote them have been undermined by economic liberalization since the 1980s. It finds that, unless current policies are drastically reformed, traditional patterns of food insecurity will continue in the global south.

The diffusion of obesogenic environments is also allied to globalization and is responsible for the 'double burden' of undernutrition and obesity now challenging health delivery systems across the developing world. Food sovereignty must be reasserted to address the problem of obesity and undernutrition; national governments must challenge the inequities in food provisioning that corporate dominance of the global food system has caused. Imbalances in power throughout the food chain help explain food insecurity and the erosion of food sovereignty in the past and present. The book concludes that, as in the past, purposeful public intervention is required to promote healthy diets. This analysis is not about globalization *per se*, although it holds that the contradictions inherent in global economic liberalization are nowhere better exposed than through analyses of diets and food. Using a political economy approach to food crises exposes the political power struggle at its heart; such perspectives are vital, and educating students to challenge simplistic interpretations of hunger and obesity have never been so imperative.

Partial analyses abound. Will population growth, combined with the massive shift in food consumption habits (the diffusion of carnivorous diets), ultimately deliver a Malthusian crisis? Is Malthus going to be vindicated in 2011–2012? Certainly, such doom-laden warnings are re-emerging in the global media and are promulgated from corporate boardrooms; a main ambition of this book is to encourage students to be critical of such simplistic interpretations of global hunger and obesity. Simplistic conceptualizations of the problem are allied to equally simplistic solutions: produce more food to relieve undernutrition and employ medical interventions to reduce obesity. The analysis presented in this text requires us to appreciate the historical roots of these two contemporary problems and to evaluate how economic liberalization, or globalization, has exacerbated circumstances since the 1970s. Undernutrition and obesity are symptoms of a political problem and must be addressed by politically informed decisions at the national and international levels; they are not simply problems for health professionals or economists. How and what we eat are political decisions which have ethical implications, and understanding the politics of food is essential to delivering solutions to hunger and obesity.

Scales of experience

This book employs different scales of analysis to explore the nature and character of the global food system and the processes that created and continue to sustain it in its present form. The scales are not discrete; quite the opposite, processes operate across different scales and serve to integrate the food system. However, given the complexity of the food system, employing various scales for analysis helps illuminate some otherwise very complicated interactions. Processes which operate at the global scale are mediated by national processes, and processes at the national scale are modified by a variety of internal variables.

Table 1.4 suggests the myriad paths and processes that create our food environments, their historical roots and contemporary political, economic and social contexts. Not all these variable are considered in detail in this analysis, but most are mentioned and some are examined in depth. The examination begins by understanding processes operative at the macro (global) level and then attempts to understand what national-level processes mediate global processes to generate distinctive national geographies of food production and consumption. The limitations of national analyses are obvious, particularly when we attempt to understand conflict situations and how they constrain food production and consumption. However, most of us still consume our food in national contexts and, rightly or wrongly, still presume that national governments can intervene to improve food provisioning systems or the nature of our diets.

Table 1.4 ***Scales of experience: selected factors that explain under- and overnutrition***

Global level	*National level*	*Sub-national level*	*Local level*	*Individual level*
Historical factors:	Agricultural policy	Regional policy	Work environment	Gender
Development policies	Industrialization policy	Rural and urban policies	School environment	Age
Globalization and institutions of global governance (IMF, WB and WTO)	Food and nutritional policy	Local health care	Home environment	Ethnicity
Global media and cultural change	Health policy	Sanitation	Infections	Class
Gendered policies	Gender and national policy	Gendered relations	Gender relations	Literacy
FDI	Urbanization	Food distribution networks	Local environment	Status in household
TNCs	Social security policy	Local agricultural outputs	Kinship networks	Marriage status
Geopolitical factors: conflict, food aid	Media and cultural policy	Local livelihoods (rural and urban)	Gender relations	Access to land
Medical factors: pandemic diseases	Environmental policy			Access to employment
Environmental change	Educational policy			Access to welfare
Regional alliances	Transport policy			

Summary

- **Most analysts agree that there is a global food crisis which has intensified since the economic recession of 2008.**
- **Evidence of the food crisis is that more people are hungry today (2011) than ever before; most of these people live in Asia and sub-Saharan Africa, but pockets of hunger still exist in the affluent world.**
- **Paradoxically, more people are overweight and obese than ever before; these people live in the north and in the global south.**
- **Explanations of the global food crisis are contentious and political, and so are the proposed solutions.**

- **The food crisis may be examined with reference to proximate and structural factors.**
- **Sen's (1981) work on entitlements significantly altered the analyses of hunger.**
- **Food matters are political matters; there is nothing 'natural' about our food system. Food has proved a potent political issue in the past, and it continues to be politically salient today.**
- **The urgency of various food-related issues has intensified in recent years.**

Discussion questions

1. This chapter referred to proximate and structural variables to explain the contemporary food crisis. What do you understand by these terms?
2. Compare and contrast the concepts of food security and food sovereignty.
3. Assess the significance of Sen's (1981) entitlement concept to analyses of hunger.
4. With reference to contemporary and historical circumstances, explain why food is a political issue.
5. Debate the contention that food is just like any other product and hence a suitable commodity for financial speculators.

Further reading

Lang, T. and Heasman, M. (2004) *Food Wars. The Global Battle for Mouths, Minds and Markets*, London, Earthscan. An analysis of the battles being fought over food policy and its future direction; the authors recognize the political and ideological nature of the 'food wars'. They detail the rise of corporate control and the dominance of the productionist paradigm. They outline an alternative food paradigm that prioritizes healthy diets based on the production of goods in an environmentally sustainable fashion.

Magdoff, F. (2008) 'The World Food Crisis: Sources and Solutions', *Monthly Review* at: www.monthlyreview.org/080501magdoff.php (accessed 12 November 2009). This short article stresses that the food crisis is not limited to the global south, but that it has ramifications for people in the developed world too. Magdoff's analysis stresses the very unequal economic and political power relations within and between countries, which explains the contours of the contemporary crisis.

Buttel, F., Magdoff, F. and Bellamy Foster, J. (eds) *Hungry for Profit The Agribusiness Threat to Farmers, Food, and the Environment*, New York, Monthly Review Press. This analysis sets the current food crisis in its historical context, specifically the emergence of global corporate agriculture. The authors offer an incisive overview of the issues and debates surrounding the global commodification of agriculture. The authors also examine the extent to which our environmental, social and economic problems are intertwined with the structure of the contemporary food system. The emphasis is on the economic and political nature of our food system. An associated video is available at: www.newday.com/films/Hungry_for_Profit.html

Millstone, E. and Lang, T. (2003) *The Atlas of Food: Who Eats What, Where and Why*, London, Earthscan. A terrific introduction to all the main issues raised in this text. The Atlas is beautifully produced; its colourful maps and accompanying text make this a joy to use with students. You don't have to be a geographer to love this text!

Useful websites

www.foodfirst.org The Food First Institute's mission is to end hunger by addressing the injustices that perpetuate it across the globe. There is a great deal of material available at the website: briefings, fact sheets, books, etc. Read the introduction to a new book (2010) 'Food Sovereignty: Reconnecting food, nature and community', Wittman, H. Desmarais, A. and Wiebe, N. at: www.foodfirst.org/sites/www.foodfirst.org/files/pdf/origins%20&%20potential%20intro.pdf (accessed 8 July 2011).

http://www.ifad.org There is a great deal of useful material at this organization's website: www.ifad.org. Of relevance to this chapter, see the debate 'Food: Who pays the price' at: www.ifad.org/media/video/food

Chapter 1

Follow Up

1. Read about some of the problems associated with the global system identified by Oxfam at: www.oxfam.org.uk/get_involved/system/?intcmp=hp_column-1_system-join_080811

    - Select one of the issues (climate change, land grab, food price hikes or intensive farming) and summarize it as a presentation to your peers.
    - Then challenge the arguments by Oxfam and identify any alternative arguments that might be used to explain the problems they identify.

2. Floods in 2010 created a food crisis in Pakistan. Read about the livelihood crisis one year later at: www.oxfam.org.uk/resources/policy/conflict_disasters/downloads/pakistan-progress-report-floods-260711-en.pdf (accessed 18 August 2011)

    - Consider what proximate and structural causes help explain this crisis

3. Read an article that links food security to political stability at: www.guardian.co.uk/environment/2008/apr/09/food.unitednations (accessed 18 August 2011)

    - Select one of the countries that has experienced food riots in recent years and do some research to understand what proximate and structural factors may be involved.

4. Read more about food speculation at: www.guardian.co.uk/global-development/2011/jan/23/food-speculation-banks-hunger-poverty and at: www.foodfirst.org/en/node/3279

5. 'Financial speculation on food should be banned'. Organize a debate around this proposition.

2 The contemporary nature and geography of malnutrition

Learning outcomes

At the end of this chapter, the reader should be able to:

- **Detail the complex nature of malnutrition**
- **Understand the basic connections between food and health**
- **Describe the changing geography of malnutrition**
- **Describe epidemiological and nutritional transitions**
- **Appreciate the problems of measuring and mapping malnutrition**

Key concepts

Acute and chronic hunger; famine; dietary deficiency; secondary malnutrition; undernutrition; overnutrition; nutritional transition; epidemiological transition

Introduction

This chapter details the changing nature and extent of food-related mortality and morbidity (death and illness). It opens with a consideration of the terminology and statistics employed in analyses of malnutrition and considers their strengths and weaknesses. Some core concepts are evaluated and revealed to be less straightforward than expected. A major part of this discussion explores the complex connections between food and health and the difficulties that such connections cause for policy-makers and health professionals.

The discussion also describes the very rapid changes that have transformed food-related health problems in recent years. Traditionally, food-related health issues have been concerned about ensuring sufficient food for poor populations, and food security has been the main concern (see Chapter 1). Malnutrition was understood to be confined to the circumstance of not having adequate intake of food, or having food that was too little in terms of calories or nutrients. In the last two decades, the quality of food consumed by poor and affluent populations has emerged as an equally pressing issue (Young, 2004). Malnutrition must now be understood to include all those who (as before) have not got enough to eat, and the condition includes millions who have an inappropriate diet in terms of quality – people whose diets include too much sugar, fats and salts. Changes in diets are identified as being of central concern, and the concepts of 'nutritional transition' and 'health transition' are considered before the health consequences associated with obesity are assessed with reference to some parts of the developing world. Again, it is crucial to explore the complex linkages between nutrition and health – in this latter case, health burdens allied to obesity.

After reviewing the major terms and concepts, this chapter concludes by examining problems associated with mapping the incidence of food-related mortality and morbidity (incidence of ill health). Problems encountered with collecting and aggregating data are evaluated before questions about scale selection are considered. This discussion emphasizes on how statistics may be manipulated for political ends and introduces readers to the problems inherent in conducting research on malnutrition of various kinds. The case of Brazil is examined to illustrate the dangers of aggregate national-level statistics that often conceal more than they reveal.

Before mapping the patterns of contemporary malnutrition, it is important to understand the complexities of some related core terminology. One might think that hunger is a pretty straightforward concept, but as with many socially constructed conditions, hunger, or malnutrition, is not a simple concept at all; rather, it has a variety of manifestations, all of them admittedly serious.

Terminology

Malnutrition

This concept means 'bad diet'. Traditionally, it was assumed to refer to a diet lacking in calories, protein, minerals or micronutrients; in recent decades, it increasingly applies to a diet high in fats, sugar and salt. Hence, malnutrition may refer to having a poor diet – this may be having too little or too much.

Upon investigation, we find that exact definitions of what constitutes 'malnourished' or 'undernourished' and 'overnourished' are fraught with problems. However, while we must accept that absolute definitions are bound to elude us, it is crucial that an attempt be made to standardize definitions. That such definitions vary is illustrated by the range of estimates available from international and national agencies faced with measuring malnourishment globally and nationally.

As of 2009, the FAO estimates that 1.02 billion people are undernourished worldwide. This is the highest number since 1970, the earliest year for which comparable statistics are available (FAO, 2009). These malnourished people may suffer from any one of the following types of malnutrition: dietary deficiency; secondary malnutrition or undernutrition. These distinctions are important because they suggest diverse policy interventions, and are considered in the following text.

Undernutrition

This form of malnutrition is the one most people think of when you mention malnutrition; it is undernutrition. Undernutrition occurs when an individual's diet is short of calories and/or protein necessary for normal growth, body maintenance, and the energy required for normal activity. This type of hunger is most common among the poorest populations in the developing world but is not absent either in wealthier countries (Everington, 2010). There are a number of symptoms of undernutrition, ranging from the moderate to the severe. All undernourished populations are more vulnerable to infection and disease than those that enjoy a diet that is adequate in calories, proteins and nutrients. Several physiological symptoms suggest that an individual is, or has been, undernourished. These are expressed as a comparison to a reference population.

Two clear indications of undernutrition are low birth weights and high infant mortality rates (IMRs). IMR is the number of deaths of infants under 1 year of age per 1,000 live births. Compare the IMR of the United Kingdom (4.8) with that of Sierra Leone (160.3). Remember that these figures are means (averages) and conceal a great deal of domestic divergence from the mean. Even in developed countries such as the United Kingdom, IMRs vary by class and ethnicity (see http://news.bbc.co.uk/1/hi/health/7471874.stm).

Three other indications of undernutrition are often used by researchers in surveys to establish its extent. Low height for age may be associated with undernutrition in the past even if the diet is adequate at present. Tables are computed to establish whether an individual is significantly small for his or her age; such calculations allow for normal height variations. Then, if an individual has a low weight for his or her height, this probably indicates undernourishment. A low weight for age may indicate past or present undernutrition. A useful indicator of nutrition in young females is the age when they get their first period. A delayed age of menarche (first menstruation) indicates low levels of calorie intake. The female sex hormone, oestrogen, is produced from cholesterol, so reduced calorie intakes are associated with both late menarche and infrequent periods.

The most extreme forms are exemplified by nutritional conditions known as kwashiorkor and marasmus. These are fatal if not countered with quick, intensive medical intervention. The word 'kwashiorkor' comes from Ghana and actually means 'the evil spirit that infects the first child when the second is born'. It is clear how this name arose. When a mother becomes pregnant with a second child, she weans her first child, and its diet of protein-rich mother's milk is replaced by a protein-deficient diet. Children suffering from kwashiorkor have swollen bellies from oedema, and some or all of the symptoms of undernourishment: stunted growth, loss of hair colour, patchy and scaly skin, ulcers and open sores. They sicken easily and are weak, fretful and apathetic. When the body has insufficient proteins available to meet all its needs, it reduces body maintenance and prioritizes the most vital functions. Hence, the hair and skin are neglected in favour of maintaining the heart, lungs and brain tissue. Many of the antibodies are also degraded in the body's attempt to build the vital organs. This precipitates a downward spiral because, as antibodies are reduced, the body becomes vulnerable to infection and readily contracts dysentery, a disease of the digestive tract. Dysentery causes diarrhoea, leading to the rapid loss of nutrients, which worsens the protein deficiency, thereby increasing the probability of a second or third attack of dysentery (see the following text for a fuller consideration of secondary malnutrition).

Marasmus, a wasting disease, is another extreme condition associated with severe undernutrition. When the body does not get enough calories (energy), it breaks down protein to use as energy, so many children with marasmus get protein-deficiency problems too. Marasmus may occur in adults or children, but if it occurs within the first two years of life, brain development is impaired.

The symptoms of marasmus are shocking because children look aged and lack all the normal interest and energy of infants. Children are usually sick because their resistance to disease and infection is low and their muscles are wasted; this may include the most vital of organs, their hearts. Their metabolism is slow, and they have very little fat to keep them warm. Marasmus is most likely to occur in populations suffering from extreme poverty, whose access to calories is very inadequate.

Dietary deficiencies

Deficiencies in any of the micronutrients necessary for good health, e.g., iron, iodine, vitamin A, the other vitamins, major and minor trace elements, are manifest in various ways and are known as dietary deficiency malnutrition. Because this form of malnourishment is insidious, it is sometimes known as 'hidden hunger'. Although it is less visible than protein-energy undernutrition, if a diet lacks one or more essential nutrients then the consequences for health can be very serious. Since the early 1990s, considerable effort has been devoted by major international donors to addressing this problem. Solving the problem of dietary deficiencies is economically cheap and politically uncontentious; however, many people still suffer from dietary-deficiency and related diseases. Some of the most serious such problems are described in the following text.

Iodine deficiency is a very serious form of dietary deficiency and is the world's biggest cause of preventable brain damage. Unlike many other forms of malnutrition, iodine deficiency can be solved very simply by adding iodine to common salt. Because salt is used in cooking and to preserve foods, sufficient amounts are absorbed by the general population. The World Health Organization (WHO, 2010) has had considerable success in reducing iodine deficiencies in recent decades, but success has been limited in some regions. Most recent estimates suggest that 266 million children and 2 billion people worldwide are still at risk of iodine deficiency, and almost one-third of school-age children (228 million) do not have access to iodized salt.

A second common and serious deficiency is of vitamin A (VAD). The United Nations International Children's Emergency Fund (UNICEF) estimates that 33 per cent (190 million) of pre-school-age children and 15 per cent (19 million) of pregnant women do not have enough vitamin A in their daily diet, and can be classified as vitamin A deficient. Furthermore, another 5.2 million of pre-school age suffer from clinical VAD. The highest prevalence and numbers are found among countries of sub-Saharan Africa and Southeast Asia. Death rates in these children are commonly 20–30 per cent higher than in children whose vitamin A intake is adequate. Vitamin A deficiency is the leading cause of preventable childhood blindness, but it has other less well-known health impacts too. The links between diet and disease (see secondary malnutrition in the following text) are exemplified by vitamin A deficiencies because when suffering from this dietary deficiency, people, especially children, are very vulnerable to other diseases – diarrhoea, malaria and measles, for example. Improving vitamin A in the diet will reduce, not eliminate, the toll from these three childhood killer diseases. Again, the remedy is not complicated or politically contentious. A small intake of leafy green vegetables every day, or a tablet which costs very little, taken three times a year, or vitamin A added to sugar or cooking oil, would eliminate this problem. Improving vitamin A intake is one of several obvious, powerful, low-cost strategies with the potential to reduce illness, blindness and death among the children of the developing world. Other small-scale initiatives, such as supporting aquaculture in poor communities, can have quick, positive results at low costs (Box 2.1).

Box 2.1

Aquaculture and food security

Fish contributes to national food self-sufficiency through direct consumption and through trade and exports. In traditional fish-eating countries in Asia and Oceania, per capita consumption is often above 25 kg. In some island countries in the Pacific, the per capita consumption is above 50 kg per year, or even as high as 190 kg in the Maldives. The extreme importance of fish to food security and nutrition may be illustrated by assessments on the situation in Africa. FAO estimates that fish provides 22 per cent of the protein intake in sub-Saharan Africa. This share, however, can exceed 50 per cent in the poorest countries (especially where other sources of animal protein are scarce or expensive). In West African coastal countries, for instance, where fish has been a central element in local economies for many centuries, the proportion of dietary protein that comes from fish is extremely high: 47 per cent in Senegal, 62 per cent in Gambia and 63 per cent in Sierra Leone and Ghana.

In general terms, aquaculture can benefit the livelihoods of the poor either through an improved food supply and/or through employment and increased income. However, at present, little or no hard statistical information exists concerning the scale and extent of rural or small-scale aquaculture development within most developing countries and low-income, food-deficit countries (LIFDCs), nor concerning the direct/indirect impact of these and the more commercial-scale farming activities and assistance projects on food security and poverty alleviation. Despite the lack of information concerning the role of rural aquaculture, there is one sure benefit of consuming fish, and that is the nutritional and health benefit to be gained from its valuable nutritional content. Food fish has a nutrient profile superior to all terrestrial meats. It is an excellent source of high-quality animal protein and highly digestible energy, as well as an extremely rich source of omega-3 polyunsaturated fatty acids (PUFAs), fat-soluble vitamins (A, D and E), water-soluble vitamins (B complex) and minerals (calcium, phosphorus, iron, iodine and selenium). In fact, if there is a single food that could be used to address all of the different aspects of world malnutrition, it is fish – the staple animal protein source of traditional fishers.

Source: Subasinghe, R. (2005) 'The role of aquaculture in food security' at: www.fao.org/fishery/topic/14886/en (accessed 21 November 2010)

Iron deficiency leads to iron anaemia and is particularly common in women of menstruating age. Even small amounts of iron deficiency are associated with impaired work performance, damaged learning ability and dysphasia. Anaemia increases susceptibility to illness, pregnancy complications and maternal death. Modifying or supplementing diets only slightly could markedly reduce this form of deficiency (Box 2.2).

Plate 2.1a,b Fresh fish and seafood are important sources of protein in many countries: examples from China (Courtesy: Rosie Duncan).

Plate 2.1a,b Cont'd

Box 2.2

Women and anaemia in the developing world

> Anaemia is the world's second leading cause of disability and thus one of the most serious global public health problems. Anaemia affects over half of pre-school children and pregnant women in developing countries and at least 30–40 per cent in industrialized countries. In poorer malaria endemic countries anaemia is one of the commonest preventable causes of death in children under 5 years and in pregnant women.
>
> (Source: www.who.int/medical_devices/initiatives/anaemia_control/en/index.html)

Anaemia is a particularly serious problem for women in the developing world, because women need more iron than men to lead healthy, productive lives.

Anaemia may be caused by a lack of iron in the diet but it is often exacerbated by the presence of hookworm and diseases such as malaria and HIV/AIDS. An anaemic woman is tired and cannot be as productive as others; she is also more prone to disease. Both have serious implications for herself, her children and the community.

Anaemia is a particularly good example of the interactions between diet and disease. Unless her diet is adequate in iron, each pregnancy depletes her supply of iron, which means that she is more vulnerable to infections, disease and death, and less able to deal with the demands of breastfeeding and the next pregnancy. An anaemic woman is likely to have low-birth-weight babies who are liable to become undernourished children prone to infection. Malaria has a dangerous association with anaemia. Its incidence is higher among young pregnant women. Both can be identified early and treated, so it is vital that young women be screened for both. Although anaemia has been recognized as a public health problem for many years, there has been little progress on its eradication, and the global prevalence of anaemia remains unacceptably high.

A variety of social factors have negative implications for women's nutritional status and health: poverty, low social status, discrimination against girls, none/limited family planning services, restricted access to education, poor primary health provision and early pregnancy. At different stages of her life cycle, poor women in many developing countries is at risk from different nutritionally related health problems (see Figure B2.2).

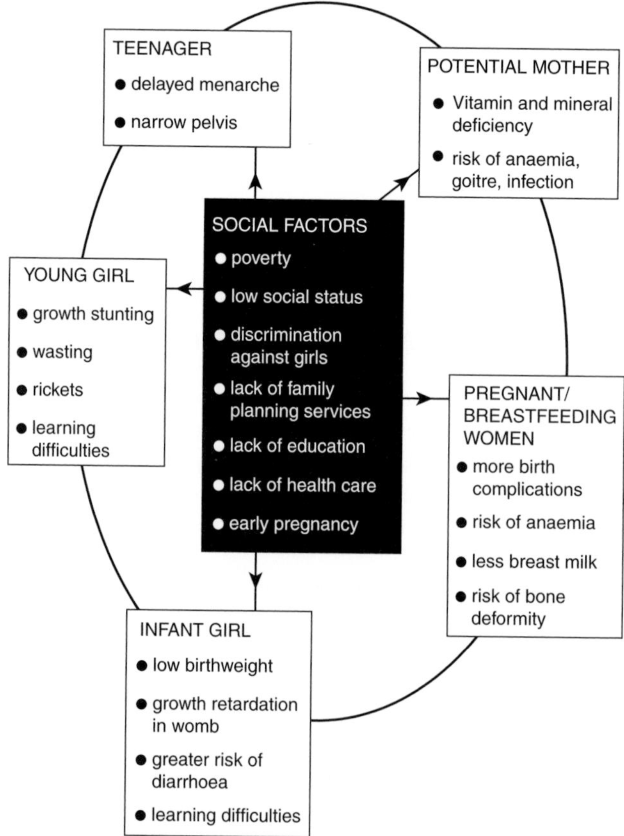

Figure B2.2 Risks at different stages of a woman's life cycle in developing countries.

Source: Smyke, P. (1991) *Women and Health,* London, Zed Books. p. 18.

Infant girls born to poorly nourished mothers are low in birth weight because of growth retardation in the womb, is at greater risk from diarrhoea and more likely to suffer from learning difficulties. As a teenager and young woman, this child will continue to have learning difficulties and stunted growth, and will have a narrow pelvis and may have delayed menarche. As a potential mother, she is at risk from vitamin and mineral deficiencies and their attendant health problems, and especially now she will be liable to anaemia. As a poorly nourished young mother, she may have anaemia, less breast milk and more birth complications, and she risks bone deformity.

Source: Smyke, 1991. For a review of the global incidence of anaemia, see: http://whqlibdoc.who.int/publications/2008/9789241596657_eng.pdf

Secondary malnutrition

Another type of malnutrition is known as secondary malnutrition and is associated with the complex interactions between diet and disease or illness. Some diseases are particularly virulent in undernourished populations. The main killer of undernourished children is diarrhoea, but other major killers associated with undernourishment are pneumonia, influenza, bronchitis, whooping cough, measles and, in recent decades, the HIV/AIDS epidemic (Box 2.3). Some conditions mean that the body cannot adequately exploit the food that is available;

Box 2.3

Awful synergies: HIV/AIDS and nutrition

'Sowing the Seeds of Hunger', a film produced by the FAO and the Television Trust for the Environment (TVE), a London-based independent film company, explored the terrible connections between nutrition, poverty and HIV/AIDS and the vicious cycle that exacerbates their awful impacts in sub-Saharan Africa. The situation is, however, very similar across the developing world where the disease often strikes the poor and vulnerable, exploiting their disadvantage.

The film illustrates how the crisis directly affects millions of people living with the disease – many of who are agricultural workers. Since HIV/AIDS typically strikes during the most productive years – ages 15–49 – many fields across southern Africa now lie fallow. As a result, families are not only losing food and cash crops, but also valuable resources, such as livestock and tools.

Unfortunately, health services in this part of the world are not able to meet the needs of those affected by the epidemic. In such a situation, good nutrition is the key. 'When people are malnourished, they do not have the strength to withstand infection and AIDS develops that much quicker', says Karel Callens, an FAO nutrition officer working in the region who provided technical support for the documentary. 'Food may not cure HIV/AIDS, but it can help people live longer, more productive lives'.

Food can be prevention

When a person dies from AIDS, hardships often intensify for the family members they leave behind, particularly women and children. In some communities, for instance, a woman may lose access to land and other assets when her husband passes away. Since food production is frequently a female responsibility, such inheritance practices can affect the entire household. Family members may move away in search of food or work, increasing their chances of contracting HIV – and bringing it back home. For others, commercial sex may be the only option for survival.

In the film, 19-year-old Mercy stands along a busy highway corridor, trying to help support her two younger brothers through prostitution. On a good night, she says, she will have sex with 10 or more men – and for the right price; she will do so without a condom. 'What I need is the money', says Mercy, defiantly.

This film is a poignant illustration of the deadly connections between nutrition and disease and the importance of recognizing these when designing national or international interventions.

Source: www.fao.org/english/newsroom/focus/2003/aids.htm (accessed 6 January 2011)

this results in secondary malnutrition. The most common causes of secondary malnutrition are diarrhoea, respiratory illnesses, measles, AIDS/HIV and intestinal parasites. These conditions are associated with some, and occasionally all, of the following symptoms: loss of appetite, poor nutrient absorption and diversion of nutrients to parasites. Deaths due directly to undernutrition are limited, the bulk being from diseases associated with undernutrition. Undernourished people are more vulnerable to diseases, so while their deaths may be attributed to measles, for example, the real culprit may be poverty and associated undernutrition. This suggests one reason why health statistics may be problematic – what is the cause of death, diarrhoea or poverty? It also confirms that purely medical interventions are doomed to be partial as long as whole communities suffer from poverty and environmental pollution.

Diarrhoea and pneumonia remain two of the most serious causes of death among children in the developing world; each of these claims millions of young lives per year. They are both cause and consequence of malnutrition and illustrate the complexities of the interactions between unhealthy living conditions and malnutrition. Sometimes, they also reflect the marketing success of major multinationals. Consider the breastfeeding debate reviewed in Box 2.4.

Box 2.4

The breastfeeding debate

Imagine a new 'dream product' to feed and immunize everyone born on Earth. This product requires no storage facilities or delivery and helps mothers plan their families and reduce their risk of cancer. What is this product? Human breast milk, available to all at birth, and yet breastfeeding is declining in both rich and poor countries. Despite over-whelming scientific evidence that human breast milk is superior to any of the infant formulas, even if mixed with clean water and given from sterilized bottles, formulas continue to usurp breast milk as the primary food in the first months of the newborn's life. Technological advances in packaged foods during the 1950s made it possible to offer breast milk substitutes to infants in the industrialized world. Aggressive marketing, free samples and intensive promotion through hospitals and health centres help explain why infant formulas rapidly usurped breastfeeding in the West. Potential markets in the developing world were targeted and exploited next. As the use of formulas was seen to be 'modern' and promoted by health professionals, its success in the developing world was assured.

As poor households in the developing world are not equipped to sterilize bottles and do not have access to safe water to dilute the formulas, babies are exposed to a greater likelihood of malnutrition and diarrhoea. A global campaign by health organizations and citizens' groups led to the adoption of the International Code of Marketing of Breast-milk Substitutes by the World Health Assembly in 1981. However, the 1980s saw bottle-feeding continue to increase and breastfeeding continue to decline. More recent declarations from the United Nations and governments have stressed on the unparalleled benefits of breastfeeding. The Convention on the Rights of the Child, entered into in 1990, made it a legal obligation of states to provide mothers and families with the knowledge and support needed for breastfeeding. The Innocenti Declaration, signed by 32 governments and 10 United Nations agencies in 1990, recognized the need for global support for breastfeeding. A consortium of major international non-governmental organizations (NGOs) formed the World Alliance for Breastfeeding Action (WABA) in February 1991. In 1999, the International Association of Infant Food Manufacturers promised to stop supplying free and low-cost breast milk substitutes to hospitals and maternity centres in the developing world. A Report (2005) recognized considerable progress but concluded that in the 15 years since the adoption of the original declaration in 1990 'remarkable progress has been made in improving infant and young child feeding practices worldwide. Nevertheless, inappropriate feeding practices – sub-optimal or no breastfeeding and inadequate complementary feeding – remain the greatest threat to child health and survival globally' (source: http://innocenti15.net/declaration.pdf.pdf).

Nestle has been the major target of the Baby Milk Action Campaign which promotes breastfeeding and campaigns to expose corporate misdeeds. Nestlé is targeted with the boycott because monitoring conducted by the International Baby Food Action Network (IBFAN) finds it to be responsible for more violations of the World Health Assembly marketing requirements for baby foods than any other company. You can follow this campaign at: www.babymilkaction.org/pages/boycott.html

Where people are living without access to clean water and adequate sanitation, illnesses gradually erode their health and ability to utilize the food they do eat. To eradicate malnutrition, then, people must be granted access to both clean water and safe sanitation. The International Drinking Water Supply and Sanitation Decade (1981–1990, organized by the UN) saw the proportion of families with access to safe drinking water rise from 38 to 66 per cent in Southeast Asia. However, although improvements were achieved in Africa, access remains low. Water affects every aspect of our lives, yet a recent estimate states that nearly 1 billion people around the world don't have clean drinking water, and 2.6 billion still lack basic sanitation (World Water Day, March 2010). These people are effectively condemned to lives plagued by disease and premature death. Review the issue in more depth at: www.worldwaterday.net. Freshwater is vitally related to good health and nutrition, which explains why some major international non-governmental organizations devote much of their efforts to improving water provisioning across poor regions of the developing world.

The catastrophic, complicated connections between nutrition and health are exemplified by the links between HIV/AIDS and nutrition which have engulfed whole communities in the developing world in recent decades. Initially, AIDS has a devastating effect on a person's nutritional well-being:

- Nutrient absorption is reduced
- Appetite and metabolism are disrupted

- Muscles, organs and other tissues waste away
- Secondary infections and other stresses increase demands for energy and nutrients

A deadly cycle is established where illness exacerbates undernutrition as many of the symptoms have a direct impact on the body's ability to utilize food properly (see Box 2.3). Even before symptoms are obvious, the illness lowers the immune system and then the symptoms – diarrhoea, weight loss, sore mouth and throat, nausea and vomiting – all mean that nutrition is compromised. Of course, the impacts are not limited to the individual with the illness; there are ramifications for whole communities and countries; some of the worst examples are from regions where peasant farming predominates – for example, in much of sub-Saharan Africa.

The disease often attacks the most productive members of the community first. As a consequence, they cannot work so effectively in the fields, and food production declines. In these circumstances the family and/or community then suffer a rapid decline in food security

Plate 2.2 AIDS pandemic: message in Kolkata.

and become more susceptible to a myriad of opportunistic diseases. In sub-Saharan Africa, for example, malaria, bacterial infections and tuberculosis (TB) have been identified as the leading causes of HIV-related deaths. All these diseases are manifestations of malnutrition, but they may appear in quite different places and in different ways. The next discussion describes another important distinction, that is, the difference between acute and chronic hunger.

Acute and chronic hunger

Most people are vague about the distinctions between acute and chronic hunger, and the boundary is indeed very problematic; however, it is important to establish the difference. Ask students to mark on a map the famine-prone regions of the world, and most will identify Ethiopia, Sudan, Bangladesh and perhaps Somalia. Asked to identify where most of the world's hungry live, they fare less well. Why? One explanation is that while the great majority of hungry populations suffer from chronic hunger, the world's media grant the problem of acute hunger, that is, famine, more attention and bigger headlines.

The conference 'The F-Word: Hunger in the Media' was held in London on 2 June, 2010, to examine how the media covers hunger. Barrow (2010) points to a persistent problem: 'Sometimes it seems that unless newspapers can use the word "famine" in their headlines, they are not interested in reporting on hunger. But real famines are quite rare and aid organizations like the World Food Programme (WFP) are doing their best to get rid of them entirely'. Unlike everyday chronic malnutrition, famines make a better story and their images are more lasting and moving; they are very visible tragedies. The day-to-day debilitations associated with chronic hunger are less amenable to headlines but have been described as 'the insidious sabotage wrought' on millions of children in the developing world. Famines, unfortunately, still claim some lives, although the absolute numbers affected are much less than those who are chronically hungry, who suffer debilitation from malnourishment. A recent estimate suggested that less than 10 per cent of all hunger-related deaths are in famine situations, while the other 90 per cent are caused by chronic, persistent hunger. Chronic malnutrition breaks down resistance to even the mildest diseases like the common cold and diarrhoea. Although it may almost be time to call famine 'history' (O'Grada, 2010), it is instructive to consider their nature.

Acute hunger: famine

Trying to define exactly what constitutes a famine is more complicated than would first appear. Arriving at a universally accepted definition is impossible, but because it is a concept around which institutional responses are geared, and because it is, at some stage, distinct from generalized situations of chronic hunger, it is essential to review attempts at its definition. This task is also useful because it helps indicate the difference a word makes and how complex apparently simple concepts may turn out to be upon investigation. Devereux and Arnold suggest some of the difficulties that complicate the concept of famine:

> Famine is like insanity, hard to define, but glaring enough when recognized. (quoted in Devereux, 1993)
>
> Famine is one of the most powerful, pervasive, and arguably one of the most emotive, words in our historical vocabulary, and that in itself makes it all the more difficult to isolate its meaning and wider significance.
>
> (Arnold, 1988: 5)

A famine is an exceptional event. In different societies, at different times, they may be very rare or more frequent, but they are never the normal state of affairs. Famines certainly happen and victims and observers recognize them as specific 'events', but commonly their beginnings and endings are problematic because they rarely happen without warning and their consequences continue when more normal circumstances are re-established. One way of identifying famine is by the associated increases in mortality. Sometimes, however, mortality increases are not as drastic as might be expected, because populations often have famine prevention strategies, and domestic or international relief efforts may avert a massive mortality crisis.

Famines are always associated with horror and confusion. They have been interpreted as signs from God, as divine interventions in response to human wickedness, and they sometimes precipitate a change in government. In famine circumstances, conventions are disregarded and systems of law and order collapse; they are correlated with awful social dislocation, migration, family break-ups and the breaking of taboos. Everything considered 'normal' is threatened.

Every famine, then, has its own character and an exact definition is impossible, but the following is a useful working definition:

> Famine refers to a shortage of food or purchasing power that leads directly to excess mortality from starvation or hunger-induced diseases.
>
> (O'Grada, 2010: 4)

In a recent book describing and explaining the history of famines, O'Grada (2010) describes how the incidence and severity of famines has fallen in recent decades, and predicts that this trend will continue:

> Even in Africa, the most vulnerable of the seven continents, the famines of the past decade or so have been, by historical standards, 'small' famines. In 2002, despite warnings from the United Nations World Food Programme and non-governmental relief agencies of a disaster that could affect millions, the excess mortality during a much-publicized crisis in Malawi was probably in the hundreds rather than the thousands. As for the 2005 famine in Niger, which also attracted global attention, experts now argue that it does not qualify as a famine by standard criteria. Mortality there was high in 2005, but apparently no higher than normal in that impoverished country.
>
> (O'Grada, 2010: 1)

Let us hope O'Grada is correct, and that the incidence and severity of famines will continue to decline, but his optimism is not shared by everyone. Sadly, during the summer of 2011, tragic headlines and news reports are coming again from the Horn of Africa, where 16 million are at risk from famine if international interventions are not immediate and effective (http://news.bbc.co.uk/1/hi/world/africa/696803.stm). Devereux (2007) is less sanguine than O'Grada and describes how, since 1984, approximately 1.4 million people have died from famines. He suggests the character of these 'new famines' is different from that experienced before the most recent economic and political changes associated with globalization. He categorizes these famines as either 'acts of commission', where they were deliberately caused by ruthless or unaccountable governments, or 'acts of omission' where governments simply failed to intervene to reduce deaths. Such situations are almost always associated with power struggles and conflict.

It is interesting to note that some of the most catastrophic famines occurred in the twentieth century, and that some of these were wrought upon populations in Europe. A detailed evaluation of these twentieth-century famines by Devereux (2000) is available to download, and his text on the theories of famine (1993) remains exemplary, as is Arnold's consideration of famine, social crisis and historical change (1988). Two very infamous famines in the twentieth century occurred in the former Soviet Union and China; these are considered in Box 2.5.

Box 2.5

Politics and famine in the twentieth century: case studies of the USSR and China

> It is a great irony that the most deadly of famines of the last century – including the worst ever in terms of sheer numbers – occurred under regimes committed, at least on paper, to the eradication of poverty.
>
> (O'Grada, 2009: 232)

The most catastrophic famines of the twentieth century, in the Soviet Union in the 1920s and 1930s, China between 1958 and 1962, Ethiopia in 1984 and North Korea in the 1990s, were largely the result of 'the malevolent or incompetent exercise of state power by authoritarian or unaccountable state regimes' (Devereux, 2007: 4).

Famines wrecked havoc in the initial years of the Soviet Union. The First World War had left an awful legacy of economic disruption and hardship and then, after the October Revolution in 1917, civil war exacerbated the political unrest and economic chaos. The Soviet Regime was forced to request international assistance in early 1921. The American Relief Administration was the major international contributor, and estimates suggest that they mounted a programme of relief 'unmatched in the history of famines' and by May 1922 they were feeding 6 million people across the Soviet Union (O'Grada, 2009: 234). Despite such efforts, the death toll from this famine is estimated to be over 6 million; as in most food crises, these deaths were largely from the ravages of diseases, especially typhus and fevers.

A second devastating famine occurred in the Soviet Union during 1932–1933. This example exemplifies two themes that surround almost all famines; the causes are complex and politically contentious and their associated statistics also subject to debate. Unlike the famine of 1921–1922, news about this famine in the early 1930s was largely suppressed by the Soviet government and its international sympathizers. Its causes and consequences remain contentious; some argue that it was designed to eliminate political resistance to forced collectivization in the Ukrainian countryside by effectively starving the opposition into submission. Devereux (2007: 4) writes of Stalin's genocidal policies while O'Grada (2010) suggests that, while its horrors were undeniable in the Ukraine and a region stretching from the Caucasus's to the Urals, its occurrence owes more to the regime's ruthless instigation of collectivization and industrialization than to genocide.

Establishing the facts in food crises situations is always problematic. The exact conditions in the countryside are difficult to evaluate because there were contradictory pressures on various bureaucrats when estimating grain yields around these years, it being in the interests of some to herald good harvests and in the interests of others to exaggerate the harvest failures, indeed; '[F]ear and terror distorted information flows' (O'Grada, 2009: 238). Problems associated with recording or assessing death rates in famine conditions are also well illustrated in this example as estimates range between 4 to 6 million excess deaths due to famine and associated diseases. More famines occurred in the Soviet Union: one in 1941 when Leningrad was blockaded by the Nazis' and some three-quarters of a million people died, and another at the end of the Second World War in 1946–1947 when approximately 1–1.5 million died because of poor harvests and poor political judgments.

Famines had plagued China throughout its history and their frequency and death tolls suggest that it was justifiably known as the 'land of famine'. The following statistics illustrate the country's crisis-prone nature in the nineteenth century; the famine in 1849 may have claimed 13 million, famines between 1851 and 1864 allegedly killed another 20 million, while that of 1878–1879 resulted in 9–13 million deaths (O'Grada, 2010: 245). However, these figures are dwarfed by the famine that has been labelled the 'worst man-made famine in human history' which occurred in China during 1959–1961. Analyses of this famine, like those of the Soviet famines considered earlier, are coloured by politics, specifically the Cold War, which dominated international relations between 1947 and 1991. At the time, its impacts and awful death toll were denied nationally and internationally. Estimates of excess famine mortality are again much contested and range from approximately 18 million to as high as 60 million (O'Grada, 2010: 95). While some part may have been played by poor weather, most now understand this human tragedy as a result of the misguided economic and political policies ruthlessly pursued by Chairman Mao, euphuistically known as '[T]he Great Leap Forward'. These policies included the forced diversion of labour and resources from the agricultural to the industrial sector, agricultural policies associated with forced collectivization, and the generalized economic, political and social chaos that ensued. In addition, the central government continued to demand excessive food procurements from regions suffering from acute food shortages. These examples illustrate that, contrary to popular opinion, there is nothing 'natural' about famines.

Overnutrition

> An estimated 300 million people around the world are obese (Body Mass Index >30). The task force's conservative estimates suggest that obesity levels will continue to rise in the early 21st century – with severe health consequences – unless urgent action is taken now.
>
> (International Obesity Task Force, 2010)

Once considered only a problem for small numbers of people in affluent societies, the issue of overnutrition, in its most extreme manifestation, obesity, has now reached panic proportions. Alarming headlines appear regularly in media outlets across the globe; there is no escape from the 'obesity pandemic'. So, what is obesity, how is it measured and should we be alarmed if it is becoming more prevalent?

Overweight and obesity are defined as abnormal or excessive fat accumulation that presents a risk to health. A crude population measure of obesity is the body mass index (BMI), a person's weight (in kilograms) divided by the square of his or her height (in metres). A person with a BMI of 30 or more is generally considered obese. A person with a BMI equal to or more than 25 is considered overweight (World Health Organization, 2010).

Simple, perhaps not? The use of BMI as a measure of obesity is not universally accepted and is criticized by some in the academic community as being simplistic at best and seriously negative at worst. These voices urge caution when categorizing large numbers of people in ways that can cause anxiety and discrimination, especially if such populations are not ill or suffering any serious medical complications from being heavier than the 'norm'. In addition, while mainstream medical opinion considers obesity to be highly correlated with a number of health problems, the most serious of which are heart disease, some cancers, and type 2 diabetes (WHO, 2010), some also challenge this association (Colls and Evans, 2010). While the mainstream opinion is accepted in the following review, that the BMI serves as an adequate if not totally satisfactory measure of obesity and that obesity is associated with serious health risks, it is important to be aware of these sceptical perspectives. Certainly, reading about the various perspectives on obesity reminds us that scientific opinion is always both contested and dynamic.

With that caveat, trends in obesity rates have caused public health officials across the world to raise the alarm, and international and national investments are being made to counter what are viewed by most medical experts to be serious increases in the incidence of overweight populations, however measured. The following section of this chapter maps the changing incidence of all the food-related heath issues detailed in the preceding text.

Changes identified in the following quote are thought to be the major proximate (immediate) factors that explain the emerging obesity pandemic:

> Changes in the world food economy have contributed to shifting dietary patterns, for example increased consumption of an energy-dense diet high in fat, particularly saturated fat, and low in carbohydrates. This combines with a decline in energy expenditure that is associated with a sedentary lifestyle, with motorized transport, and labour-saving devices at home and at work largely replacing physically demanding manual tasks, and leisure time often being dominated by physically undemanding pastimes.
>
> (WHO, 2002a: 1)

The relative contribution of changes in diets and changes in physical activity to obesity trends are debated, but we can certainly detail the dietary changes known in the academic literature as the 'nutritional transition' which has transformed many of the diets in the developing world in the last two decades of the twentieth century (Popkin, 2004).

The nutritional transition describes a shift from a traditional diet high in cereals and fibre to a 'Western' diet high in saturated fat and added sugar as well as increases in meat and dairy foods; such changes are also associated with increases in processed foods and alcohol consumption. The 'nutritional transition' helps explain the health transition or 'epidemiological transition'.

The epidemiological transition is a shift in the patterns of death and diseases (mortality and morbidity). In traditional agrarian societies, most deaths and diseases resulted from undernutrition and associated infectious diseases (smallpox, pneumonia, tuberculosis, cholera, malaria, etc.). Increasingly in developing countries, people are suffering and dying from chronic non-communicable diseases, such as diabetes, heart disease and some cancers, the main killers in the affluent world. The transition, therefore, is from a high incidence of infectious diseases and deaths to a decline in the incidence of these and an increase in non-communicable diseases traditionally associated with affluence. See Box 2.6 for examples from Southeast Asia.

The contemporary extent of malnutrition

The preceding discussion describes some of the difficulties which surround the terminology of malnutrition. This section reviews the

Box 2.6

The epidemiological transition: case studies from Southeast Asia

In Hong Kong, infectious diseases and parasitic disease (mostly diphtheria, enteric fever and dysentery) accounted for nearly a quarter of deaths in 1951, a proportion that had fallen to 3 per cent by 1988. The pattern of cancer deaths was the reverse: in 1988, these accounted for 30 per cent of deaths (and in 2005, just over 31 per cent), compared to 4 per cent in 1951. Rates of heart disease and stroke have also increased (29 per cent in 1988, similar in 2005). This is evidence of very rapid epidemiological transition. Thailand shows a similar pattern: the crude death rate from tuberculosis dropped from 47 to 12 per 100,000 between 1970 and 1983, while the crude death rate for all cancers rose from 13 to 27 per 100,000. Malaysia and Singapore offer a similar picture. Note, however, that many of these figures pre-date the emergence of HIV/AIDS as a global health problem; this coupled with the re-emergence of diseases such as tuberculosis means that we cannot point (even in the developed world) to an end stage of the epidemiological transition in which only diseases of 'affluence' remain (Smallman-Raynor and Phillips, 1999).

Source: Gatrell and Elliott (2009: 89)

type and quality of the statistics employed to map malnutrition before describing its distribution in 2010. The political context of accounting needs a mention first. Different interest groups at the international, national or regional level may be best served by manipulating statistics to exaggerate or diminish the incidence of malnutrition. Consider the following examples.

Classifying a situation as 'famine' may result in the provision of international relief, which may be welcomed by national governments, or the use of the term may be eschewed because of the implications it has for national policy. After recognizing the positive contributions of numerous international non-governmental organizations (NGOs) to famine relief and global awareness of poverty and underdevelopment, O'Grada discusses some examples of where they may overstate or 'hype' the nature of food crises to generate more funds (O'Grada, 2009: 218–225). Similarly, governments, careful of their reputations domestically and in the international arena, often deny the seriousness of malnutrition: frequently, the well-being of specific minority groups is politically charged.

If a country has suffered serious declines in well-being, then its government may want to suppress the statistics. Certainly President Mugabe will not be celebrating the health statistics for Zimbabwe since 1990: life expectancy has collapsed between 1990 and 2005, from 60 years to 37.5 years; infant mortality has climbed from 53 to 81 in the same period and maternal mortality has risen from 610 in 1995 to 1,100 in 2000 (Gatrell and Elliott, 2004: 90). In Zimbabwe, the ongoing food crisis has affected all social segments, even the rich: '[F]or the first time, malnutrition became serious among the richest urban residents, affecting 25% of that population, compared to 29% of the poorest urban population' (State of World Cities, 2010: 103).

The World Food program (WFP) was delivering food aid to over a million people in Zimbabwe in 2009–2010, and they explain:

> In Zimbabwe, hyper-inflation, acute shortages of basic supplies and a series of very poor harvests led to serious food shortages and acute food insecurity in recent years. Along with these factors, the collapsing economy, very high unemployment (estimated at over 80%), a rapidly devaluating currency and a high HIV/AIDS prevalence rate (15.3%) – all contributed to increasing levels of vulnerability. This situation necessitated large-scale humanitarian food assistance operations in the country.
>
> (Source: www.wfp.org/countries/zimbabwe; accessed 28 July 2010)

The crisis, as described here, sounds almost inevitable, although we know that the situation has its roots in the country's colonial history and recent political turmoil. This example illustrates that circumstances in many countries are very dynamic, and that it cannot be presumed that they will always improve; nutritional and health conditions may suffer reversals for diverse reasons and be manifest in a multitude of ways. It is also important to remember that these statistics are aggregate, and that there will be communities in Zimbabwe that will not have suffered such declines; they may well be community or ethnically specific.

At the opposite end of the spectrum, there are great profits to be made from a world where being overweight is deemed unacceptable and where body image is so closely correlated to self-esteem and social acceptability. Emphasizing the health problems allied to being overweight can increase sales figures for things as varied as diet foods and gastric surgery to gym memberships. The 'life-changing surgery for a whole new you' business and 'beautiful beings' business is

Plate 2.3 Health foods are big business: sales of porridge soar when it is marketed as a health food. (Source: Wikimedia Commons)

Plate 2.4 Oprah: lifestyle guru. (Source: Photo by Alan Light, Wikimedia Commons, http://en.wikipedia.org/wiki/File:Oprah_Winfrey_(2004).jpg)

global in nature, both in the origin of its clients as well as the location of its clinics.

Measuring malnutrition

What statistics are available to analysts of malnutrition? How may we determine the incidence of undernutrition and overnutrition? The Food and Agriculture Organization (FAO) is the main source of data on world hunger, and the World Health Organization is the best place to source statistics on related health problems. The FAO's most recent estimates of undernourished populations are available online (www.fao.org/publications/sofi/en).

Measuring malnutrition is fraught with problems, and estimates of the number of hungry people may vary significantly if a few basic parameters are adjusted. Recently, a new Global Hunger Index has been developed to help refine analyses (Box 2.7). However, for the purpose of this discussion, the general patterns illustrated on the following map (Figure 2.1) are adequate.

Box 2.7

2009 Global Hunger Index

The Global Hunger Index is a composite statistic; that is, it combines several statistics to produce one that is more helpful than any one alone. The 2009 Global Hunger Index (GHI) ranks 84 developing and transitional countries using three equally weighted indicators and combines them into one score. The three indicators are:

1. The proportion of people who are calorie deficient, or undernourished, which is a key indicator of hunger.
2. The prevalence of children under the age of five being underweight, which is a measure of childhood malnutrition—children being the most vulnerable to hunger.
3. The under-five mortality rate, which measures the proportion of child deaths that are mainly caused by malnutrition and disease.

By using these three indicators, the GHI captures various aspects of hunger and undernutrition and takes into account the special vulnerability of children to nutritional deprivation. In the 2009 report, the GHI is calculated for 1990 and for the most recent period for which data are available to measure progress over time.

Countries are ranked on a 100-point scale, with 0 being the best score and 100 being the worst possible score, respectively. Some countries with severe hunger, including Afghanistan, Iraq and Somalia, are not included in the GHI because sufficient data are not available to calculate their scores.

Global and regional trends

Globally, hunger has decreased by less than one quarter (24 per cent) since 1990, with overall scores improving from 'alarming' (20.0) in 1990 to 'serious' (15.2) in the 2009 GHI. Since 1990, a handful of countries have been able to reduce their GHI scores by half or more, and about one-third have reduced their scores between 25 and 49.9 per cent. As regions, South Asia and sub-Saharan Africa performed the worst, with 'alarming' levels of hunger represented by scores of 23.0 and 22.1, respectively.

Source: International Food Policy Research Institute (IFPRI) at: www.ifpri.org/sites/default/files/publications/ghi09keyfacts.pdf (accessed 10 August 2010); visit the interactive map and explore the variety of global hunger indices statistics for various countries at: www.ifpri.org/publication/2009-global-hunger-index.

'India below sub-Saharan Africa in hunger index'

When this statistic is applied to regions within a country, some interesting comparisons emerge. Below is an article that reviews the results of the GHI for India in 2008 and illustrates the great diversity of results within India, by state.

NEW DELHI: Despite robust economic growth in recent years, India's record on hunger is worse than that of nearly 25 sub-Saharan African countries and all of South Asia, except Bangladesh. The International Food Policy Research Institute (IFPRI)'s 2008 Global Hunger Index says that with over 200 million people insecure about their daily bread, Indian scenario is 'alarming' in terms of hunger and malnutrition.

The first ever India Hunger Index found that not a single state in India fell in the 'low hunger' or 'moderate hunger' categories. Madhya Pradesh had the most severe level of hunger in the country, followed by Jharkhand and Bihar. Punjab and Kerala scored the best on the index.

When Indian states are stacked against countries in the 2008 Global Hunger Index, Madhya Pradesh ranks between Ethiopia and Chad. Punjab, the best-performing state, ranks below Gabon, Honduras, and Vietnam. India's high levels of child malnutrition and calorie insufficiency are the primary reasons for its poor performance. India's rates of child malnutrition are higher than most countries in sub-Saharan Africa.

Source: *The Economic Times*, 15 October, 2008, available at: http://economictimes.indiatimes.com/News/PoliticsNation/India_below_sub-Saharan_Africa_in_hunger_index/articleshow/3596612.cms (accessed 10 August 2010)

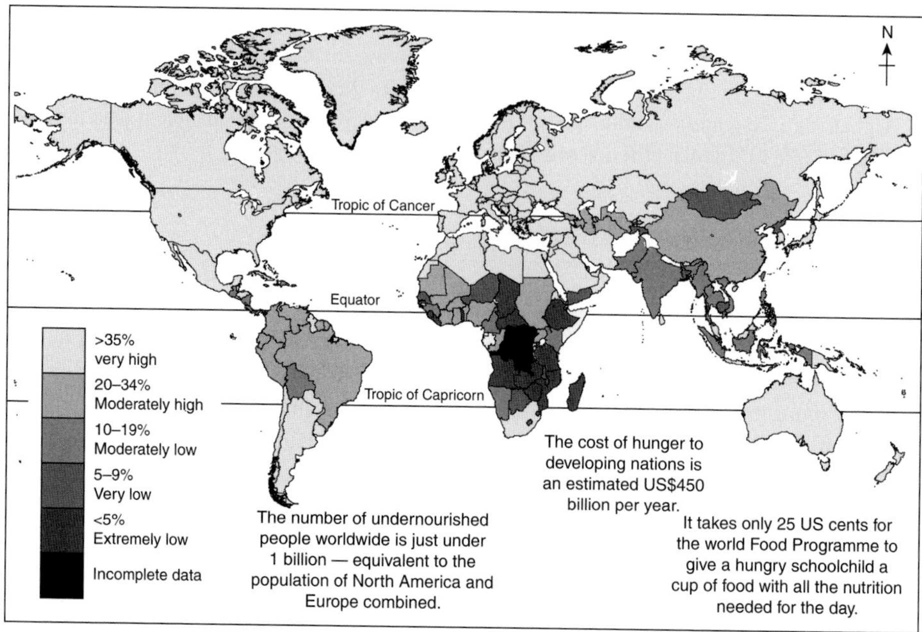

Figure 2.1 The geography of world hunger (adapted from: Food and Agricultural Organization, 2009. For a basic overview of hunger statistics and definitions, visit: www.fao.org/hunger/hunger-home/en and www.wfp.org/hunger/stats).

The geography of global undernutrition is complex, and at all places is mediated by sub-national variables such as region, class, ethnicity, gender and age (see Chapter 5). Essentially, patterns of hunger are correlated with patterns of power; those without it tend to suffer first and those who enjoy access to power seldom experience hunger (Young, 1997). Patterns are also dynamic, and show substantial seasonal and annual shifts; high global food prices in 2007–2008, for example, were estimated to increase the number of hungry people by 75 million over the 2006 estimate (FAO, 2008). However, allowing for these complexities, it is important to appreciate that the majority of chronically undernourished people live in Asia, where China, India, Pakistan and Bangladesh together account for most of the total. The prevalence of Asia contradicts popular perceptions which assume that hunger is most serious in sub-Saharan Africa (SSA). Hunger is very serious in parts of SSA where the incidence of hunger is most extreme, but because the total population of the region is much smaller, absolute numbers are much less than those for Asia. The following table illustrates broad regional totals.

Table 2.1 ***World hunger: regional totals***

Numbers in millions	*Region*
642	Asia and Pacific
265	Sub-Saharan Africa
53	Latin America and the Caribbean
42	Near East and North Africa
15	Developed countries
Total 1,010 million	

Source: adapted from FAO, 2009.

However, 15 out of 16 countries where the incidence of hunger exceeds 35 per cent are in SSA, and at present severe hunger threatens to overwhelm relief organizations in East Africa and the Horn of Africa. These areas are suffering from acute hunger, which, in contrast to chronic hunger, implies a real threat of death if emergency relief is not available. In October 2009, Ethiopia was in the headlines

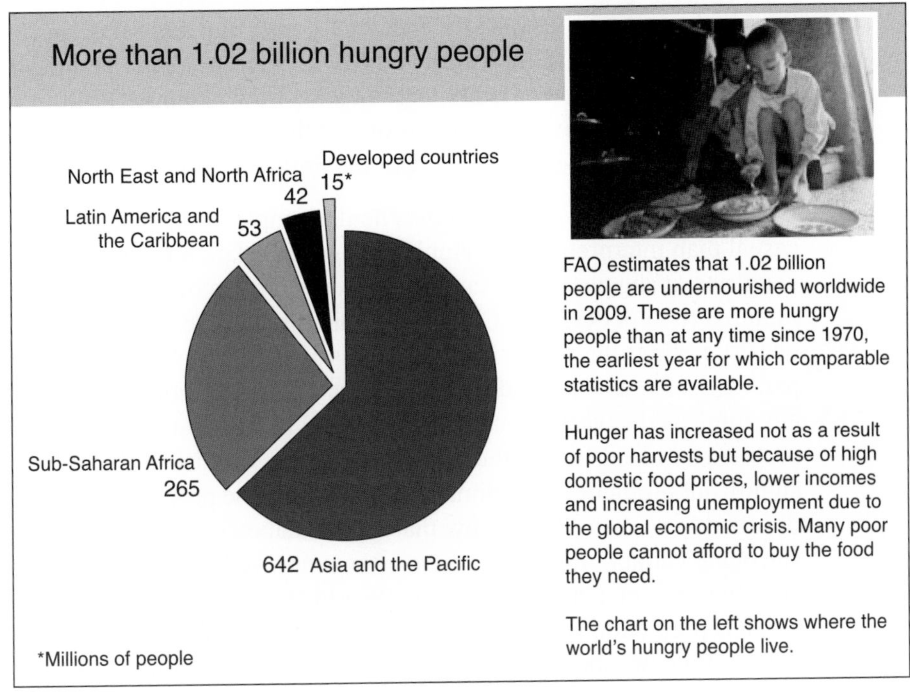

Figure 2.2 Geographical incidence of hunger by world region (adapted from: FAO, 2009).

again as the World Food Programme warned that famine was imminent. So, 25 years after the famous Food Aid concerts, launched by Sir Bob Geldof during the emergency food crisis in 1984–1985, this country is currently suffering another food crisis. Although drought may be the proximate (immediate) cause of the current crisis, as in the 1984 example, conflict and political mismanagement are also implicated, and unravelling the various threads of causation is difficult. In Kenya, too, approximately 4 million people are facing severe food shortages and have become dependent on food aid, while in Uganda the picture is even bleaker. All food emergencies are the end result of a toxic mix of economic, political and environmental factors (hence the term 'complex emergencies'), and the specific contribution of each factor will vary in nature and significance. Fuller coverage and daily updates on these and other acute food emergencies are available at the FAO and World Food Programme (WFP) websites.

Public health debates about obesity in recent years help explain the availability of statistics and information about overnutrition, its incidence and trends. Both developed and developing nations are paying a high price for malnutrition (Halweil and Gardner, 2000). Simple distinctions between the incidence, costs and nature of malnutrition in the affluent north and less affluent south are no longer satisfactory. The World Health Organization now identifies health problems linked to overnutrition as one of the most serious public health issues facing countries in *both* the north and south (2002a).

Beware, reading this map is an excellent example of the dangers of casual map interpretation. It employs aggregate data for very large countries; it suggests, for example, that the problem of obesity is minimal (0–5 per cent) in China and India. In fact, the problem of obesity is alarming public health officials across Asia. Sadly and paradoxically, as with traditional patterns of malnutrition, Asia is emerging as the world centre of obesity and its associated health problems. Indeed, across the developing world, obesity is already a significant problem, and estimates indicate that the problem is destined to worsen. If we allow that type 2 diabetes is a good surrogate variable for obesity (it is often employed when obesity statistics are not available), then the World Health Organization's (WHO) prediction is frightening, it argues that 'three-fourths of the 350 million, or more, diabetics projected in 2025 will inhabit the third world, and that diabetes deaths will increase by more than 50 per cent

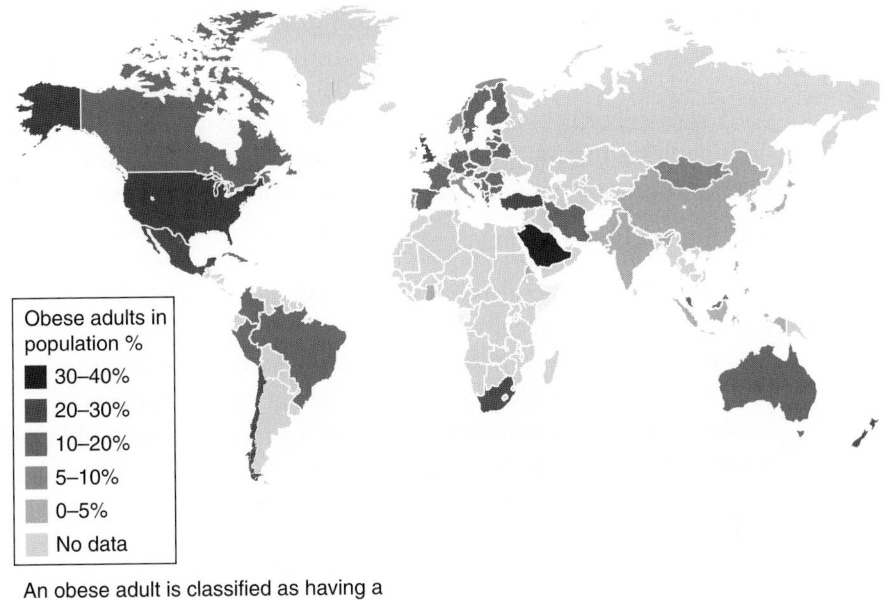

Figure 2.3 The global incidence of obesity (WHO, 2005).

worldwide by 2016' (WHO, 2009). All the health problems associated with obesity are expensive to treat and are generating dilemmas for public health officials trying to balance budgets between treating traditional killers, such as infectious diseases, and the more recent diseases associated with non-communicative diseases. Countries across the developing world are facing this 'double burden' of disease and its multiple social and economic impacts (WHO, 2009; www.who.int/mediacentre/factsheets/fs311/en/index.html).

Conclusion

This chapter has introduced readers to the complexity of the concept of malnutrition and its diverse manifestations across the world. While traditional patterns of malnutrition (undernutrition) continue to take its toll on millions of lives every day, the role of famine and acute hunger has declined. The impact of chronic hunger remains serious across the global south and is not eliminated in the affluent north either. A double burden now exists in many countries that have experienced economic growth in the last two decades. In these communities, traditional malnutrition is still present, but now overnutrition is emerging as an equally serious threat to public health.

The following is a summary of this chapter.

Conceptualizing and measuring malnutrition is not straightforward; the diverse types and categories are sometimes problematic. Malnutrition may be present even if the individual suffering is unaware of their problem, and it is not simply about not having 'enough' to eat. Chronic malnutrition is the most serious problem in terms of numbers affected and may be present amidst plenty and is not always reported in the media. Some cheap and easy medical interventions are available to reduce malnutrition but are not always employed. Examples of acute hunger, associated with famines and food crises, have been less frequent in recent decades but may still emerge, given certain circumstances; they are especially allied to conflict situations and political unaccountability. Many acute crisis situations are difficult to solve because of their complex character; these are considered in more detail in Chapter 8.

There is a 'geography of malnutrition' and, at every scale of analysis, it is dynamic. World regional categorizations help us begin to map the patterns of world malnutrition. At the global level, while hunger persists in the 'rich' world amidst some of the wealthiest populations, the great majority of the world's hungry live in the developing world, specifically Asia. However, the countries in the category 'developing world' are now so diverse that its use is increasingly problematical. Once thought of as an American problem, obesity now occurs across much of the developing world; it is more prevalent in countries that have experienced rapid industrialization and urbanization and is a largely urban phenomenon. Some countries have seen their economies stagnate or decline in the last two decades, while others have experienced rapid economic expansion; these are usually reflected in their nutrition trends. These contrasting experiences are largely the result of the differential impacts of globalization and have different implications for malnutrition, as is analysed in the next chapter.

At the sub-national level, some regions are particularly associated with hunger: the Sahel region of West Africa, the Horn of Africa and the northeast of Brazil are notorious examples. Intra-national contrasts exist in all countries and are related to patterns of power and status established in the past or present; age, race, gender, ethnicity and class are all important variables that help explain the incidence of all forms of malnutrition (see Chapter 7). Describing the international and intra-national contrasts in patterns of malnutrition is the first task, and explaining them is more challenging and contested; this is what is attempted in the next chapter.

Summary

- **Although famines still claim too many lives, the majority of the world's deaths due to hunger are from chronic malnutrition and associated diseases.**
- **There are approximately 1.02 billion hungry people in the world; the majority of these people live in Asia.**
- **There are four forms of malnutrition; undernutrition, micronutrient deficiencies, secondary malnutrition and overnutrition. Each of these types of malnutrition is addressed by different policy interventions. The easiest to eliminate is micronutrient deficiencies.**
- **There are millions of people in the global south who eat good, diverse and healthy diets.**
- **There are stark contrasts in diets within countries as well as between them.**
- **There are hungry populations in the world's wealthiest countries, and in all countries in the global south, many people enjoy a good diet.**
- **Obesity is a health problem in all of the emergent economies that have experienced rapid industrialization and urbanization allied to globalization in the last two decades.**
- **The nutritional transition describes a shift in consumption from traditional foods to a diet consisting of more processed foods and animal-based products; such shifts are correlated to increases in obesity.**
- **The nutritional transition is associated with an epidemiological transition where non-communicable diseases begin to replace infectious diseases as the major cause of death and disease.**
- **Within countries, extreme contrasts in food intake may exist between regions, classes, genders and ethnic/age groups.**
- **Infant mortality rates (IMRs) are a useful indication of well-being, including nutrition. When IMRs fall below 50 per 1,000, persistent hunger is considered to have been eliminated.**

Discussion questions

1. What problems complicate the definition and measurement of malnutrition?
2. Describe the various manifestations of malnutrition observed across the globe.
3. Evaluate the geography of global malnutrition and its recent trends.
4. What is the nutritional transition?
5. Evaluate some of the connections between nutrition and disease.
6. What is the epidemiological transition, and how is it associated with the nutritional transition?

Further reading

Arnold, D. (1988) *Famine; Social Crisis and Historical Change*, Oxford, Blackwell. This remains a short and accessible investigation between social change and famine. Arnold's discussion is rich and wide-ranging.

Devereux, S. (1993) *Theories of Famine*, London, Harvester Wheatsheaf. Again, although published in the earlier 1990s, this remains an excellent discussion of famine. It is accessible, and explanations reviewed are sadly still relevant to famines in the headlines in the early twenty-first century.

Useful websites

www.fao.org An arm of the United Nations, the Food and Agricultural Organization is the best place to find statistics about nutrition and global trends in malnutrition. Among their numerous publications, 'The State of Food Insecurity', published annually and available online, remains the best source for statistics about malnutrition. See the 2010 publications at www.fao.org/publications/sofi/en. They have useful country briefs and profiles as well as video resources.

http://www.ifad.org The International Fund for Agriculture and Development was established in the late 1970s and concentrates of promoting rural development in the developing world. Their Rural Poverty Report (2010) is available online, and they make a great deal of material available through their rural poverty portal at www.ruralpovertyportal.org/web/guest/home. IFAD also makes available numerous videos and case studies from the field.

www.wfp.org Another part of the United Nations group, this programme is dedicated to emergency relief and promoting food security among some of the most vulnerable communities in the world, refugees and in conflict situations. A good basic introduction to the topic of hunger is available at: www.wfp.org/hunger

www.who.org This is an essential source for information about global health. Their annual 'World Health Report' selects different themes every year. The site also has helpful fact sheets about diseases and country profiles. They have various campaigns and special-themed publications, recent activities included support for World Breastfeeding Week and a report on tobacco consumption.

Government departments often provide information about poverty and malnutrition; for example, the USDA in the United States produced the Report of Household Food Security in the USA in 2008, available at: www.ers.usda.gov/Publications/ERR83/ERR83.pdf

National Newspapers usually provide updates on themes such as malnutrition and poverty. The Guardian recently published statistics about malnutrition in the United Kingdom and India; see, respectively:

www.guardian.co.uk/society/2009/feb/10/malnutrition-health-social-exclusion

www.guardian.co.uk/world/2010/dec/07/india-hunger-malnutrition-ration-shops

Chapter 2

Follow Up

- Learn about the global hunger index (GHI) and read a summary of trends at: www.guardian.co.uk/global-development/datablog/2010/oct/11/global-hunger-index
- Visit this FAO web page and read the summary flyer about world hunger in 2009. Also listen to the presentation about hunger trends and causes at this address: www.fao.org/publications/sofi/en
- Visit this address for some vital statistics about contemporary hunger: www.wfp.org/hunger/stats
- Read about vitamin A deficiency and examine its incidence globally at: www.childinfo.org/vitamina.html. Write a short summary about its global incidence. What simple interventions would help eliminate vitamin A deficiencies and the associated ill health?
- Watch the video called 'Formula for Disaster' about the effort to promote breastfeeding in San Roque, a poor district of Quezon City in the Philippines. However, it is also important to appreciate that the promotion of milk formula is not limited to poor communities; see the article at: www.guardian.co.uk/media/2009/jul/22/cow-and-gate-ads-banned
- Visit this page for a fuller analysis of the links between HIV/AIDS and nutrition: www.fao.org/ag/agn/nutrition/household_hivaids_en.stm

- Study the global incidence of iodine deficiency at: www.who.int/vmnis/iodine/status/summary/median_ui_2007_color.pdf
- Study the incidence of night blindness in pre-school children as a consequence of vitamin A deficiency at: www.who.int/vmnis/vitamina/prevalence/vita_fig1a.pdf
- Read about the global incidence of anaemia among pregnant women at: www.who.int/vmnis/anaemia/prevalence/summary/PW_anaemia.pdf
- Review a video about a water and nutrition project financed by Action Against Hunger at: www.actionagainsthunger.org/resources/video/acf-brings-clean-water-communities-after-war-northern-uganda
- The World Food Programme has a review of the causes and consequences of the crisis in the Horn of Africa at: www.wfp.org/crisis-horn-africa?gclid=CPPGvrfW6KICFQT92AodZEA0uw

3 Theoretical perspectives: understanding the patterns

Learning outcomes

At the end of this chapter, the reader should be able to:

- **Appreciate the contested nature of theories of food security**
- **Appreciate the difference between proximate and structural determinants of food security**
- **Evaluate some important proximate explanations for malnutrition**
- **Evaluate some of the structural determinants of malnutrition**

Key concepts

Food security; proximate causes of malnutrition; structural determinants of malnutrition; Malthusian ideology; political and economic theories of malnutrition; development philosophies; neoliberalism; globalization

Introduction

The preceding chapter described the changing global patterns of malnutrition; it noted the continued prevalence of undernutrition and detailed the emergence of a global obesity pandemic. How might we explain this apparently paradoxical situation, where approximately two million people have such deadly diets; one million people have too little to consume, and are ill or die as a consequence, and one million have too much to eat and are ill or die as a consequence? Describing a situation is a relatively unproblematic task but understanding or explaining the patterns is much more complex.

Theories or explanations that purport to explain why some people in some places are still undernourished have been controversial since

Biblical times (God's wrath or punishment?), and they remain so. This terrain is a battleground because theories suggest policies and very diverse political and/or economic interventions; they also often apportion blame and urge action. Analyses of malnutrition are politicized because they consider the role of markets in society (and how to regulate them), appropriate roles for the state (how much, if at all, should governments intervene in markets or attempt to modify society) and how much public funding should be devoted to addressing undernutrition or overnutrition. Even if there is a consensus that government interventions are necessary, what policies will be most effective? Hence, debates continue in every corner of the globe, whether in China about obesity or in Malawi about undernutrition; what should corporations do or abstain from doing or what might governments do or what must they avoid doing? How are individuals implicated in their under- or overnutrition condition? This chapter tackles these questions.

The first task is to review a core term employed in the literature of food and introduced in Chapter 1, namely, food security. Food security is a complex concept, and a useful definition is:

> ... food security, at the individual, household, national, regional and global levels, is achieved when all people, at all times, have physical and economic access to sufficient, safe and nutritious food to meet their dietary needs and food preferences for an active and healthy life.
>
> (World Food Summit, 1996)

Examining this definition with reference to the statistics detailed in Chapter 2, it is obvious that we have failed miserably to deliver sufficient and safe food to millions of people worldwide. The object of this chapter is to examine why so many millions of people do not enjoy food security, that is, access to a 'sufficient, safe and nutritious' diet. Why are some 'stuffed and some starved' (Patel, 2007)?

Analyses of both these manifestations of malnutrition must be examined with reference to the political economy of international trade and development. Similar historical and contemporary economic and political processes, associated with power and profit, explain these contrasting types of malnutrition. Obesity has joined undernutrition as a global problem, and both are symptomatic of the dysfunctional workings of the international political economy. My analysis emphasizes the crucial role of corporate players and international ideologies and institutions while allowing that complex interactions at the national and sub-national levels mediate these global processes. The following discussion explains that food systems are both socially generated and dynamic; they are neither natural nor inevitable but, as social constructions, may be modified by politically informed choices. It concludes that change is not simply advisable but essential for a just and sustainable future.

A 'political–economic' perspective informs this analysis and emphasizes the social production of malnutrition which holds that food crises and hunger 'are rooted in the social relations of agricultural production and the political, cultural and economic relations of food access, distribution and consumption' (Jarosz, 2009: 2066). Such perspectives have long been employed to understand hunger; this analysis argues that they are equally fruitful for analyses of obesity, a newer manifestation of these asymmetrical global social relations. It examines how politics and economics create structures within which individuals, nations and organizations operate; these structures create opportunities and constraints which are temporally and spatially dynamic. A political–economic perspective argues that, by understanding how problems are created, we may also understand how they might be addressed; in other words, by correctly theorizing problems we may appreciate what political interventions may solve them. Such a perspective holds that there is a crucial link between theory (how we understand or conceptualize a problem), and practice (what can we do to solve it). The analysis presented here is not simply academic, it hopes to initiate changes in the way food is produced and consumed in the contemporary world, so that it is more socially just, ethical and sustainable.

Proximate and structural causes of malnutrition

A useful first step is to differentiate between proximate and structural causes of malnutrition. Simplistic, popular analyses of malnutrition abound, and must be challenged to expose the structural connections that explain the simultaneous existence of hunger and plenty.
As illustrated in World Hunger (Young, 1997), the hungry are not hungry because there is not enough food, and neither are the obese simply victims of their own greed. The following discussion examines some popular explanations of undernutrition to expose their weaknesses before considering similarly unsatisfactory explanations of obesity.
The terms 'proximate' and 'structural' were introduced in Chapter 1.

Proximate causes of malnutrition, sometimes referred to as 'trigger factors', are those which can be identified immediately. Hence, obviously, undernutrition is a consequence of not having access to a decent diet, and obesity is 'caused' by eating too much and/or not having enough physical exercise. However, a comprehensive analysis of the patterns of global hunger and obesity reviewed in Chapter 2 must analyse factors from the global to the local and appreciate their historical and ideological contexts. Structural analysis thus seeks to expose the historical and contemporary context of poverty and vulnerability. Such analyses consider the structural determinants of any community's location, in a historically constructed system of social relations (political, economic and cultural) from the global to the local.

Similarly, the probability of any individual becoming obese varies according to his or her social and economic characteristics, rather than any inherent biological variables. If we consider the incidence of obesity, we find that it too illustrates patterns that transcend the individual; it is obviously unhelpful to assert that some individuals are greedy while others are more able to modify their eating and exercise to keep healthy. This is not to deny that individuals have some control over their habits and health but that the degree to which they are able to exert control is modified by variables beyond the specific individual's capacity. When we map the contours of the obesity pandemic, patterns emerge which require explanations that consider why some individuals in some communities in some places are more susceptible to obesity than others. As with the incidence of undernutrition, patterns of obesity are usually correlated to poverty, race, class, educational attainment and gender – factors that are considered in more detail in later chapters. Hence, again, 'explaining' obesity requires a structural analysis that examines the individuals 'at risk' as being part of a specific community, within a specific region, within a specific country which has a specific location within a historically created global political economy. Even within the United Kingdom, we might wonder why obesity is higher among women in Scotland than women in England. Clearly, such a condition cannot be 'explained' by simply suggesting that women in Scotland are inherently or biologically more prone to obesity than either Scottish men or their English equivalents; there must be deeper structural determinates that explain such patterns.

It is crucial to appreciate the argument being made here, and the difference between proximate (trigger) and structural (underlying) causations and their implications for theory. If we accept individualistic explanations of malnutrition, under or over, then interventions will target individual behaviour and attitudes and favour short-term remedial action; these may be medical, technical or educational, for example. To address undernutrition, these might include targeted food aid or education campaigns for mothers. However, if on the other hand, we identify broad structural determinates that explain these patterns, then interventions must be much more radical and target underlying problems associated with economic deprivation and political marginalization – political interventions that challenge the status quo, rather than limited medical or technical interventions that are fated to be partially effective, if at all.

So too, with obesity, if our theories of causation emphasize individual behaviour and attitudes, then interventions might include healthy eating campaigns or improving food labelling to help individuals select healthier foods. Local governments might invest in walking-to-school projects or target overweight children for specific health campaigns. These may marginally improve matters, but will they seriously challenge the roots of the problem? The following discussion considers some common explanations of malnutrition with reference to proximate variables and reveals their limitations.

Proximate explanations of undernutrition

Most people, when asked to rank the causes of world hunger, prioritize natural causes over human ones; floods, droughts and poor soils are most popular. When the human dimension is acknowledged, the 'problem of population' is most frequently offered, followed by war. Several other assumptions are exposed through discussions with students. Among the most relevant are the following:

- That hunger exists only in the developing world
- That famine is the main problem
- That the problem of hunger is most serious on the African continent
- That population growth causes hunger
- That people are hungry because there is 'not enough food' and that increased food production is imperative

- That hunger occurs in places that lack numerous essentials; most frequently cited are the lack of good soils, inefficient agriculture and therefore inadequate food production, poor infrastructure and food-storing facilities and poor weather
- That hunger is caused by natural disasters, floods, droughts, hurricanes, etc.
- That hunger in the developing world is explicable with reference to the internal characteristics of those countries alone; that is, they are ignorant of the historical and international dimensions of the problem

Some of these assumptions have already been considered and rejected in Chapter 2; namely, we have established that hunger does exist in the developed world, that chronic hunger is more pervasive than famine and that, while millions are indeed hungry in sub-Saharan Africa, more hungry people live in Asia. The following discussion challenges the last five assumptions; that population growth 'causes' hunger; that people are hungry because there is not enough food in the world; that malnutrition is caused by absence of some essential characteristics in that region; that natural disasters are a main cause of malnutrition; and that we can understand the problem of malnutrition with reference to the internal characteristics of the country in which it occurs.

Since the middle of the nineteenth century, the most popular assumption has been that hunger is caused by population growth or 'overpopulation', that population growth proceeds at a faster pace than food production and that poverty and malnutrition occurs if population growth is not reduced. This is testament to the persistence of Malthusian interpretations of hunger, which are adequately critiqued elsewhere (Arnold, 1988; Devereux, 1993, 2007; Lappé and Collins, 1986; Ross, 1998). Its limitations are only summarized here.

Malthusianism is fascinating to consider because, although associated with Thomas Malthus who wrote at the end of the eighteenth century, it still has many adherents, and elements of his thesis inform that 'there is not enough food' position (considered in the following text). It is also an excellent example of the relationship between theory and policy. If, as Malthus argued, population growth caused malnutrition, then policies to reduce population or population growth would reduce malnutrition. A great deal has been written about Malthus and his

'Essay on the Principles of Population' (1798), which was to become very influential. Malthus postulated that population increases geometrically (2, 4, 8, 16, etc.), and that food production increases arithmetically (2, 4, 6, 8, 10, 12, etc.). Population expansion, he argued, would eventually outpace food production, and starvation would result.

Malthus appreciated that population was expanding and also that poverty was increasing, and because they occurred together, he concluded that one caused the other, namely that population expansion caused poverty and consequent malnutrition. He argued that population growth caused poverty. This remains a very popular fiction in all sorts of places, but the causes of poverty are very much more complex than this theory allows. One of the greatest causes of poverty is the hierarchies that exist between places and people, which mean that the world's goods and privileges are very unevenly distributed. In other words, the way the political and economic system is organized helps explain why some people enjoy more goods than others; these are considered in more detail in the following as structural determinates of undernutrition.

Malthus' theory attempts to 'naturalize' poverty, to construct it as a natural condition of human society rather than as the consequence of political and economically informed decisions and policies. The theory asserts that it is 'natural' that as population expands poverty increases, when in fact there is no such natural rule at all. Population expansion may be associated with increased affluence. and it often has been in the past. The way goods are distributed in any society (national, local or global) is governed by socially constructed networks; that is, how we produce and distribute goods is designed by humans, through it often appears very inhumane! It is worth remembering that, in the eighteenth century, many great minds accepted that the hierarchies between people were part of God's Grand Design, and were natural. Few people accept this today, but some continue to feel that poverty is something that 'is always with us' or is 'natural' for some groups of people. The danger in Malthus's theory, in the past and present, is that it blames the poor for their condition rather than understanding their poverty as a consequence of the way society is organized. Many people still think that poverty in much of Africa is because it is 'overpopulated', or that hunger there is because there are too many people; neither is an accurate assessment of the situation.

Malthus, writing in England at the end of the eighteenth century, was living in a country in the middle of a societal transformation; it was changing from a predominantly rural economy and society to an industrialized urban one, and everything was in turmoil. The agricultural revolution had occurred, and millions of poor agricultural workers were migrating to the emerging industrial cities. The population of the United Kingdom was indeed increasing, and so too was urban poverty; Malthus made a simple judgement and concluded that, because population was expanding and poverty was increasing, that the first condition 'caused' the second. This presumption ignored the many and diverse alternative reasons why poverty was increasing; the most important of these was that over the preceding decades the rural poor had lost their access to common property resources, so vital to their survival before the agricultural revolution and the privatization of rural land. Major structural changes happened to the economic, political and social conditions in the United Kingdom in the eighteenth and nineteenth centuries, and these have to be explored to understand why poverty was intensifying, and urban poverty in particular appeared to be increasing. The case of the Irish Famine, considered by contemporaries as a classic Malthusian crisis, is discussed Box 3.1, and its implications for other explanations of malnutrition are considered in the text that follows.

Box 3.1

The Great Irish Famine

Famines, or food crises, occurred in Europe until the nineteenth century, and the majority of people were poorly nourished; diets for all but the privileged few were uninteresting and inadequate, and until the last decade of the nineteenth century, food and drink were liable to be adulterated. The last famine in Western Europe accompanied by massive increases in mortality was the Great Irish Famine, which was at its most serious during the winter of 1846–1847. The Irish population in 1841 was approximately 8.2 million. After the potato failures in 1845, 1846 and 1847 and the deaths and migrations which followed in its wake, the population in 1901 was approximately 4.5 million. During the famine, over a million men, women and children died, and millions more emigrated. The proximate cause of the food crisis was a fungus, which devastated the potato crop that formed the bulk of the diet of the poor. The structural causes were more complex and are more contentious, but certainly a 'space for famine' had been created. That space was the result of the particular process of capitalist incorporation which had been integrating Ireland into the British and larger North Atlantic economy since the seventeenth century, the specific changes in capitalism that had occurred in the decades which preceded the Famine and, finally, but perhaps most crucial, the prevalence of ideologies about the nature of poverty and the source of funds to relieve it.

The capitalist incorporation of Ireland between the sixteenth and nineteenth centuries was aided by English and Scottish colonial plantations, which created a system of property relations that were hierarchical and unequal. There were a few privileged landowners who owned estates from the vast (over 10,000 acres) to the more modest (1,000–10,000 acres), but the great bulk of the population had farms of less than 30 acres or were squatters who had colonized common land or had no rights to land at all. While most of the largest landowners were from the settler populations, not all were, and certainly a great many of the settlers were not privileged but small tenant farmers too. Between the late eighteenth century and the 1840s, changes in economic circumstances meant that unemployment and underemployment intensified in Ireland. The rurally based proto-industrial (cottage) industries, especially the textile-related ones, which had been so vital in supplementing household incomes, collapsed in the face of competition from industrial producers in Belfast, Derry and industrial cities in Britain. The 1840s was a decade of economic hardship all over Western Europe (with food riots in England and serious food crises across Europe), so income from temporary migration to Britain also declined in this decade.

On the eve of the Famine, at least 40 per cent of the population had very limited entitlements and lived on a diet based on potato. When that crop failed for three years in succession, these populations were at the mercy of the public purse and charitable donations; neither source proved sufficient to alleviate the impact of the crop failure. It is increasingly realized that, in this case, as in other famine conditions, it was social dislocation allied to disease that caused the high death rates, not starvation. Typhus, typhoid and cholera were all present between 1846 and 1850 and killed the poor and some of the better-fed too (for detailed accounts, see Young, 1996; Daly, 1986; O'Grada, 2006).

The ghost of Malthus is evident in the allied theory that malnutrition persists because there is 'not enough food' in the world. This explanation is particularly popular with biotech companies like Monsanto and subsidized farmers in the developed world, especially the United States and European Union. Obviously, if this is an adequate explanation, it is indeed worthy to keep producing ever more food for the world's hungry populations and perhaps the genetic modification of food to increase production is a noble cause. However, does this interpretation of the problem stand up to scrutiny? The answer is no, but it is a further illustration of how theorizations of a problem determines solutions. Let us review some facts.

Is there not enough food produced in the world to feed its inhabitants adequately? We considered Sen's (1981) work in Chapter 1 and how it challenged simple correlations between food availability and hunger. The evidence contradicts this simple assertion. Food production has, in fact, been increasing more rapidly than population growth. While there are some serious deficits in some places, the main reason why undernutrition exists is because too many people remain desperately poor and their livelihoods are vulnerable to sudden collapse or erosion by forces beyond their control. A large part of their food insecurity is also because food and agricultural policies are not designed to deliver adequate diets to the world's inhabitants and often have the directly opposite effect; that is, there are structural explanations that explain poverty and food insecurity, but the idea that there is 'not enough food' persists.

The legacy of Malthus is obvious in the banner headlines on the Monsanto home page (www.monsanto.com; accessed on 10 August 2010); the banners are in a loop which links the following assertions:

- The world's population is growing
- To keep up with the population growth, farmers will need to produce more food
- More food in the next 50 years than in the last 10,000 years combined
- American farmers will meet the challenge

We see blatant use here of Malthusian theory, 'too many people and not enough food', couched in alarmist language, including a call to American farmers to help prevent catastrophe. The assumption is obvious: there is not enough food in the world, and Monsanto's technologies (including their genetically modified seeds) are an essential ingredient of any solution. Specifically, Monsanto hails its genetically modified technologies as the answer to this food-deficit dilemma.

The problems with this interpretation are myriad, and its foundations are certainly based on a false premise, that there is a world food deficit which explains why people are hungry. In a recent indictment of Monsanto, Lotter (2009) concludes that they have used their corporate power to promote genetically modified seeds and shape research, public opinion and regulatory agencies into a 'reckless acceptance of risky technologies and that scientists have looked the other way as they did so'. The role of transnational agri-businesses in the evolution of food production and distribution are considered in more detail in the next chapter, but this is such an excellent example of how theory guides policy and exemplifies how it is often allied to vested interests that it warrants inclusion here (see Box 3.2).

Jarosz (2009) describes the flawed theorizations of the contemporary food crises which commonly employ metaphors drawn from the natural world of 'perfect storms' and 'tsunamis'; such discourse implies that food crises are natural events rather than socially embedded ones. She then describes the ineffective policies adopted as a result. If the problem is understood as being a consequence of failed harvests due to natural events (disease, floods, earthquakes, droughts etc.), then food shortages are framed as 'failures in food supply'. Then again, as stated earlier, increasing supply is the answer – enter the Second Green Revolution and its rich commercial possibilities. She describes the reasoning:

> [When] Global institutions and world leaders identify the problem as one of supply and demand … then the international response urges a technological fix through increasing applications of synthetic fertilizers, pesticides and herbicides, enhanced soil and water management, plant breeding aimed at increasing yields, the development of genetically modified crops, the extension of credit and the development of free markets.
>
> (Jarosz, 2009: 2066)

Many students subscribe to the 'lack of' school of thought, usually associated with the 'cycles of poverty' school. This theory is based on the notion that where hunger persists it is because people lack everything from 'good weather' to 'modern technology', 'the contraceptive pill' and 'education and investment', and that all these

Box 3.2

Genetically modified organisms and world hunger: the solution?

The debate about genetically modified organisms (GMOs) has raged for many years and still excites fury and frustration in equal measure from both sides of the debate. This summary reviews some pros and cons in the debate and introduces the reader to some excellent websites where they may follow up the issue in more detail.

GMOs: the case for

Theory	*Solution*
Hunger exists because there is not enough food to feed the world's population.	It is imperative that more food be produced to avert widespread hunger and famine.
The world's population is growing very rapidly, and malnutrition will increase in the coming decades as population and incomes increase and the nutritional transition (see Chapter 2) diffuses across the globe.	Intensify all forms of agricultural output; cereals and animal-based food production must be increased by the application of advanced science and technology. Industrial farming systems are best suited to this end.
Economic globalization, the intensification of liberalization trade in agricultural and food commodities policies, will promote food security and food production.	Support the World Trade Organization, specifically its Agreement of Agriculture (AoA), which is designed to reduce all 'market distortions' in the food and agricultural sectors, namely the continued reduction in all quotas, tariffs or subsidies for food and agriculture.
Historical agricultural advances (in Europe in the eighteenth century, the United States in the twentieth century and the developing world with the Green Revolution of the 1970s) which have involved the application of science and technology offer a template for the future.	Technological advances, emanating from corporate research laboratories, should be adopted to improve agricultural production and efficiency. Governments across the globe should facilitate the diffusion of such technologies, for example: • GMOs • Animal cloning
Any reservations that some people have about these technologies are misguided at best and Luddite at worst.	Any reservations voiced about the benefits of these new technologies must be exposed as mistaken and vehemently opposed in the media whenever possible. Campaigns to promote such technologies should be well funded and promoted in all markets.

GMOs: the case against

Theory	*Solution*
Hunger exists because vested political and economic interests (at all levels of analysis, global and national) obstruct positive changes that could create an alternative, more ethical system of food provisioning across the globe.	Radically transform the nature of the food system by promoting more sustainable agriculture across the globe. Challenge the vested interests that obstruct positive alternative systems of food provisioning and support alternative agricultural models where they exist.
There is enough food produced in the world to feed the current and projected population numbers. Undernutrition persists because millions of poor people do not have the resources (money, land) to command an adequate diet (see entitlement concept). Dietary changes associated with the nutritional transition should be modified to reduce the energy-intensive production of meat and diary products	Promote poverty-reduction programmes and sustainable livelihoods in urban and rural regions of the developing world to ensure that all people at all times can access an adequate diet. Alter subsidies to ensure that meat and diary products reflect their 'real costs' of production, including their very high energy and environmental costs. Use fiscal interventions to encourage health diets, which are less dependent on processed foods rich in fats and sugars.
Economic globalization of the food and agricultural sectors in the last decades of the twentieth century has promoted unsustainable as well as unjust patterns of food production and distribution. The system depends on extensive and opaque commodity chains which, in many cases, are exploitative and environmentally damaging. Since the 1970s, economic globalization has reduced the role of national governments in agricultural and food policies. Rules and regulations have increasingly been established at the WTO where many developing countries have little influence. Policies are often designed to promote corporate interests, such as markets for GMOs, rather than domestic food security. The contemporary global food system has exacerbated food insecurity and been associated with continued undernutrition and the emergence of the obesity pandemic.	Shifts towards local provisioning systems which have short, more transparent commodity chains should be encouraged through modifications of the rules and regulations of the WTO. Efforts should continue to expose the social and environmental costs of some of the foods we consume. Externalities associated with petroleum-based industrial food systems should be reflected in the cost of food. Democratically accountable national governments should reassert, promote and protect domestic sustainable food production to ensure food security. Financial incentives should be increased to support local, environmentally sustainable farming and food systems.
A major problem with the current global food system is associated with the dominant position of transnational food corporations and their control over decision-making in the food and agricultural sectors. They can assert considerable pressure on governments to adopt technologies, despite their being unpopular with domestic consumers.	Promote more transparency when trade negotiations are undertaken at the WTO. Adjust subsidies, tariffs and quotas to promote food security and food sovereignty, rather than powerful vested interests at the national and international levels.

Science and associated agricultural technologies have a vital part to play in increasing food security. However, recent concerns about some technologies promoted by the biotech corporations are understandable (GMOs cloning).	Scientific and technological changes in agriculture should be monitored to ensure they promote positive change that promotes small farmer viability, food security and resilient sustainable farming systems.
Many are worried that such technologies will intensify corporate control over the food system (see Chapter 4).	
Our global food system has become increasingly concentrated in recent years and enormous influence is exercised by a few giant food corporations; many have well-founded fears about the degree of corporate control over our food system.	Every effort should be made to dilute the power of the global food conglomerates to ensure that they are answerable to public opinion and the public's concerns about the nature and methods of farming they promote.
The integration of some technologies, such as GMOs, will serve to strengthen corporate control over our food system.	

Source: The author based this study on a variety of sources, including: Shiva, V. (2000: 95–116); Potter, R. *et al.* (2004: 230–232); Altieri, M. (2004). To keep up to date with the latest news on the issue of genetically modified (GM) food and crops and find out about the deceptive PR campaigns being used to promote GM worldwide, visit: www.gmwatch.eu. Read about resistance to GM crops in Asia at: www.grain.org/articles/?id=67

reinforce each other and are exacerbated by 'overpopulation'. These mantras prove incredibly resilient in the popular imagination and media. It is important to challenge these assumptions and explanations.

While we may allow that the economies of many developing countries do indeed 'lack' elements that explain the economic success of some advanced economies, these weaknesses are symptoms rather than causes. The fact that the road system is inadequate or that technology or education is limited need to be understood as the result of that region or country's place in the international political economy. Again, there are structural explanations which explain why that place has been denied the funds needed to promote economic success. These 'weaknesses' are not reasons, but symptoms of a lack of power to influence development policy to deliver economic expansion and prosperity. What may be lacking, however, is an effective government that is democratically accountable to its citizens; the role of democracy in the decline of famine mortality is examined in the next chapter. In most cases where development has been arrested or hijacked, national politics and politicians are implicated.

Hence, if Malthusian explanations of 'overpopulation' are too simplistic (and have so far been proved wrong) and 'the lack of school' is limited, perhaps the 'natural causes' perspective proves more satisfactory? With reference to natural disasters, drought, flooding, late rains, and crop failures due to disease or pests are all often cited as causes of famine and hunger. A recent example is the emergency food crisis in North Korea during 2009–2010.

> The Democratic People's Republic of Korea continues to suffer food shortages due to a combination of factors – including a lack of arable land, poor soil management, insufficient water reservoirs to combat drought, shortages of fuel and fertilizer, outdated economic, transport and information infrastructure, and a general vulnerability to natural disasters.
>
> (Source: www.wfp.org/countries/korea-democratic-peoples-republic-dprk)

The World Food Programme (WFP) has been sending food aid to North Korea very often during the last 10 years as food crises recur with tragic regularity. At present (June 2010), the WFP estimates that approximately one-third of women and children are malnourished, and they are providing fortified food to prevent the situation declining into famine as experienced in the recent past (1990s). You can see a short video showing the work of the WFP at: www.wfp.org/videos/container-port-super-food.

This case illustrates the very political character of food policies established in Chapter 1. While the WFP is careful to stay politically neutral in this video, most experts agree that it is the economic and political nature of the regime that is ultimately responsible for the chaos in food production and provisioning in North Korea. The crucial phrase from the preceding quote is 'the general vulnerability to natural disasters'. Many places experience natural disasters; the question is why are some communities so fragile that their economy and social structures disintegrate when a natural disaster strikes them? While proximate variables trigger hunger or famine, these are only effective as triggers in specific 'spaces of vulnerability' (Watts and Bohle, 1993) that have emerged consequent upon historically created processes and ideologies which dictate access to power and govern economic policies. In the case of North Korea, the main reason why its people are hungry is that the country is ruled by a totalitarian dictatorship which ignores human rights and its basic responsibilities to the people under its jurisdiction.

There are ongoing economic and political structural variables that explain why so many people in North Korea face intermittent food shortages which often threaten to become famines if international assistance is withheld. This case should remind you of the problems in Zimbabwe reviewed earlier. In every example of undernutrition, poor governance emerges as a major explanatory variable. Food insecurity, not having access to a decent diet, is a political problem associated with historical circumstance and failed development policies or politics and/or corrupt or misguided policies (Devereux, 2007: 7–11). The problem of malnutrition in conflict zones and the relationship between natural and human-induced disasters are considered in Chapter 8.

Proximate causes of obesity

Proximate factors dominate discussions about the causes of obesity too, where behavioural and biomedical factors predominate. Some emphasize individual responsibility, exemplified by ‘it’s your own fat fault’ attitudes. These ‘blame the victim’ by asserting that fat people are lazy and greedy and are largely responsible for their poor health and fitness; this is certainly a very common attitude among my undergraduates, However, is it a satisfactory or fair judgment? Additionally, simple analyses stress changes in eating patterns (more fatty, sugary foods) and behaviour (eating out, less physical exercise). The ubiquity of ‘fast food’ outlets and cheap food and drink in supermarkets is cited as ‘causes’. The following discussion suggests that, at best, these are partial explanations, with limited analytical power.

These behaviours are often described with reference to the emergence of ‘obesogenic’ environments across the world in the last decades of the twentieth century. All populations in the developed world and increasing numbers of people in the emerging or developing world inhabit these environments that simultaneously constrain mobility and stimulate high energy intakes. Consider the main elements of an

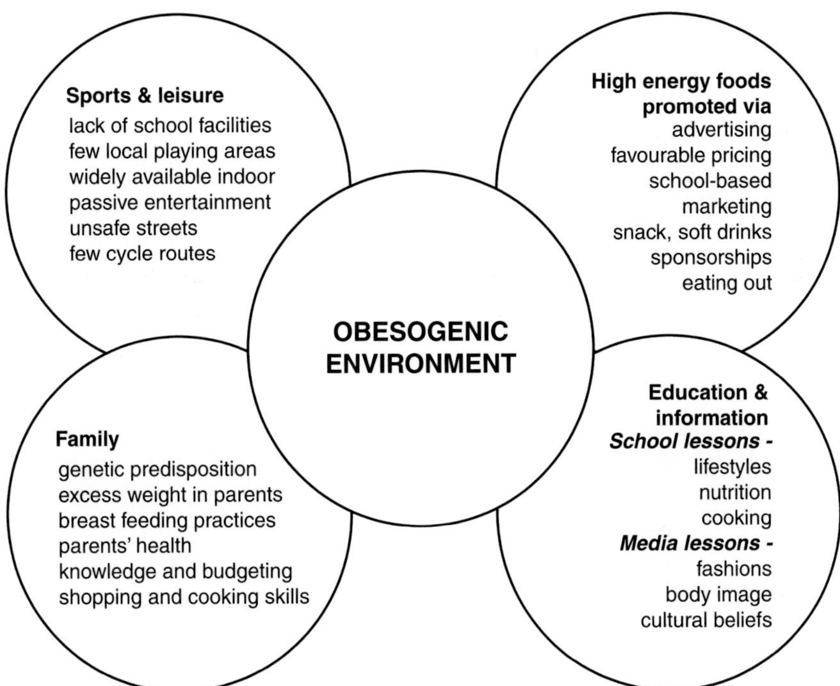

Figure 3.1 Obesogenic environments.

Source: European Association for the Study of Obesity 'Obesity in Europe' 2002.

obesogenic environment, as shown in Figure 3.1. Do you agree that these are all proximate in nature?

Understandably, the diagram focuses on eating and exercise, suggesting that changes in sports and leisure activity allied to the availability of high-energy foods has been vital to the emergence of obesity. However, even if we limit analysis to proximate variables, this diagram is very limited, and, of course, the specifics of all the changes identified will be very geographically varied; this diagram was based on the European situation. However, given this geographical variety, some changes, which the diagram neglects completely, have been very important.

Shifts in the nature of employment in the last 50 years have meant that a higher percentage of people work in the tertiary sector where physical activity is less than in many traditional manual jobs. This is as true for many developing countries as for the emerging and developed countries. It neglects to mention the massive rural–urban migrations which mean that, with a few important exceptions, most people today are urban dwellers. Most urban planning has been based on European and American models that prioritized motor transport and suburban sprawl, so that the majority of people walk less than in the past. The diagram fails to indicate the massive shifts in gender roles in the last 50 years in many countries, certainly one variable in the changing patterns of food consumed in the home. Allied changes in family composition also help explain changes to family eating practices.

It is interesting to consider why proximate variables dominate discourses of obesity, just as they do in hunger discourses. Such interpretations allow for simplistic policy interventions which prioritize individual efforts and responsibilities, effectively privatizing the problem. Hence, although we recognize that obesity is a major 'public health' problem, analyses eliminate the public from the debates. These explanations largely ignore the powerful corporate interests that both fuel the problem and exploit its commercial opportunities. Commercial opportunities are extensive and include everything from the functional foods, medical interventions and the diet and exercise business. When the problem is conceptualized as a personal problem, as opposed to a public one, then private solutions can capture the market and offer expensive private answers.

To illustrate some opportunities that this interpretation creates, consider the following. Sales in functional foods in the United Kingdom have risen from £134 million in 1998 to £1,720 million in 2007 (www.igd.com/CIR.asp?menuid=35&cirid=1145). These foods allow profits to be generated based on people's anxiety about diets, food and health. Private medical interventions to address obesity are also big business. *Bariatric surgery* is a procedure to reduce obesity. In addition to expanding commercial medical opportunities domestically, medical tourism is a growth industry that connects expertise in India, for example, to affluent markets in Europe and North America – a novel example of globalization and new market opportunities. Hence, the obesity pandemic certainly creates burdens on national health services, but for some it also represents an opportunity.

To summarize, proximate variables certainly help explain why more people are becoming overweight or obese, but this begs the question 'Why or what forces are causing our environments to alter?'. We now turn to consider structural determinates of malnutrition to appreciate the political economy of the changes identified.

Structural perspectives

Structural analyses recognize the historical, political and economic context of malnutrition, and my thesis is that both forms of malnutrition are explicable with reference to the changing nature of the world political economy. The role of these factors in undernutrition has a long history, but it is more novel to examine their role in the obesity pandemic (Young, 2004). It is important to reiterate a caveat here; by evaluating some significant structural variables that help illuminate patterns of global malnutrition, I am not suggesting that individual actions are uniquely determined by such structures, or that we are all puppets at the mercy of structures over which we have no control. This analysis accepts the dynamic nature of all such processes and their socially embedded natures which vary temporally and spatially. Understanding the respective roles of individuals and structures to explanations of human behaviour, the structure–agency debate, is central to all social sciences and differentiates them from the natural sciences where human agency is largely irrelevant.

That we are conscious subjects, not simple objects of material forces, animates all these debates and complicates analyses in the social sciences. However, for the purpose of this exposition, it is confidently asserted that to understand contemporary patterns of malnutrition, it is essential to consider the evolution of the world economy since the fifteenth century and its ramifications for access to power and privilege in the twenty-first century. By its nature, this broad brush review of the emergence of the modern world does an injustice to the very specific histories and geographies of places in its effort to simplify what is, of course, a phenomenally complex and contested story. However, there is some merit to appreciating these general trends within which the diversity of specific places may be contextualized and understood.

Pedagogically, it is helpful to simplify some very complicated processes to isolate some of the most relevant variables. Table 1.3 (Chapter 1) attempts to isolate some of these at different scales of analysis. The major concepts employed, historical, economic, political and social, are not discrete, but for our purpose here, they are useful categories. All these factors, at different times and places, in various degrees, help explain the contours of the contemporary food system and, therefore, the places and spaces within which some individuals are undernourished or overnourished. The rest of this book investigates in more detail malnutrition with reference to the historical, political, economic and social processes operative at the scales employed in Table 1.3. The following discussion is an introduction to their fuller examination in the later chapters.

Structural perspectives: the relevance of history

This analysis allows historical processes a central role to explanations of global malnutrition. No interpretation of the current international food system can neglect the importance of historical variables. Historical legacies are everywhere obvious, and sometimes they are very vital but not immediately obvious. Adopting a very long view, plants and animals have been sold, exchanged and transferred for thousands of years since the domestication of plants and animals during the first agricultural revolution around 9,000–7,000 BC in agricultural hearth areas. Two further agricultural revolutions are normally recognized as significant thresholds in the evolving global agricultural system and are significant developments that help explain the evolution of agriculture worldwide.

The 'second agricultural revolution' occurred in Western Europe, particularly the Netherlands and England, in the eighteenth century. The myriad changes introduced in the eighteenth and nineteenth centuries spelt first the decline and then the disappearance of subsistence, peasant farming. It was replaced by an innovative capitalist farming sector which was to provide the foundations of the better-known Industrial Revolution in nineteenth-century Britain. Changes wrought in the countryside provided people to work in the new industries, and also provided some of the required food surpluses. An essential element of this agricultural revolution was the enclosure of common property resources (land) and the consolidation of fields and farms which facilitated new farming technologies, new crops and crop rotations, improved animal breeding and husbandry, all accompanied by greatly extended markets for agricultural inputs and outputs. The landscapes of Britain were transformed, as were agrarian social relations. It is instructive to review these processes in the past because they help us understand similar processes of dispossession and transformation that are still occurring in some places today.

Rural areas lost millions of inhabitants as small holders lost access to land and livelihoods. Employing Sen's (1981, see Chapter 1) concept, they suffered a decline in their entitlements, access to land having traditionally been vital to their ability to produce staple food, especially potatoes and vegetables for domestic consumption. As with all social change, there were winners and losers in Britain in the eighteenth century. The winners were the capitalist farmers who prospered and expanded their holdings. The losers, peasant producers and small holders, became hired labourers or migrated to the cities in the United Kingdom. Millions more became part of the '[W]hite Plague' part of the mass migration that was the 'indispensible foundation of the [British] Empire' the 'Britannic exodus that changed the world' and' turned whole continents white' (Ferguson, 2004: 54). These migrants also helped the diffusion of these new agricultural technologies across the globe; agriculture, especially in the most fertile regions, changed forever, as a consequence.

This 'second agricultural revolution' was part and parcel of the intensification of capitalist farming, which was later transferred to America where it laid the foundations of the next major agricultural transformation, the emergence of the industrial food production system after the Second World War. This period of agricultural restructuring is also known as the 'productionist period', and is associated with the direct corporate involvement in every aspect of food production. These changes are hailed by some, as they have created unprecedented food surpluses and 'cheap' food for some affluent consumers, but they are also increasingly critiqued as being both socially and environmentally unsustainable as well as being a root cause of the cruel distortions in the contemporary global food system. The specific characteristics and geography of corporate control and its implications for global malnutrition are investigated in more detail later.

The preceding discussion establishes the dynamic nature of the evolving food system and some of its most significant conjunctures. An allied and equally vital factor that helps explain the contours of the contemporary food system are the colonial exploits of the Europeans between the sixteenth and twentieth centuries. During the early encounters between Europeans and indigenous peoples in the Americas, Africa and Asia, plants and animals were transplanted and traded. Some of the most important of these were the introduction of the potato from the Americas into Europe, where it had, by the middle of the nineteenth century, become the staple crop of the poor across Europe. In the other direction, Europeans introduced bananas and sugar into the Americas and initiated the infamous of plantation system of agriculture.

The nature of capitalism dominant at any particular period determined the character of transformations in the metropolitan regions (core regions) and their colonial possessions, effectively the periphery of the world economy during these centuries. A major part of the colonial project was the search for new markets and new resources and 'capitalist expansion entailed a forcible transformation of pre-capitalist societies whereby their economies were internally disarticulated and integrated externally with the world economy' (Knox, Agnew and McCarthy, 2003: 250).

Agricultural commodities and food became major elements of trade between the European colonial powers and their colonies across the globe, and here the historical and economic elements illustrated in Table 1.1 (Chapter 1) combine to explain past and even some present patterns of trade and resource use. Among the most important of the colonial commodities were sugar, tobacco and cotton. These were in great demand in Europe but could not be grown there, so the tropical parts of the colonial regions were devoted to their production. Great stretches of the Americas, Africa and Asia were appropriated for food production to meet growing demand in Europe, and the economic infrastructure in the colonies was constructed to facilitate its export. As these goods increased in economic significance, their production intensified across the colonial possessions. Patterns of property ownership and trade relationships established during these centuries are still obvious in today's food system and are examined in detail in Chapter 5. These examples illustrate how economic patterns of production established in the past prevail and continue to influence agricultural and systems today.

The place of development theory

Although this book is not about development theory, how 'development' is understood and promoted has had serious implications of agricultural change and food policy, and therefore nutrition across the global south (see Willis, 2005). Even when most of the former colonies in Asia, Africa and the Americas became formally politically independent, economic patterns established during the colonial age prevailed and left their legacies in internal property relations and infrastructure, as well as in international trading relations. There were other, equally important historical and economic legacies, because 'theories of development' promulgated in industrially developed countries, based largely on their experience, and became the template for economic development policies in the newly independent states. While these were varied and contested, modernization theory dominated policy and practice in many independent states in the decades between 1950 and 1970. This theory prioritized industrial development and related urban expansion, and neglected the agricultural sector and domestic food production.

Similar 'models of development' survived and were promoted through new policies in the 1980s and 1990s when globalization was deemed by Western economic theorists to be both inevitable and advantageous. As the debt crisis threatened to undermine the global economy, the World Bank (WB) and International Monetary Fund (IMF), through which global economic theory was instituted, encouraged or insisted that developing nations restructure their economies. Structural adjustment programs (SAPs) were the means through which these policies were implemented, and all those countries struggling to pay their debts were advised to reduce protection for domestic food producers. In many places, this resulted in large-scale 'dumping' of subsidized staples which undercut local rural producers. The livelihoods of domestic food producers consequently collapsed in these places (when one country subsidies agricultural production and then sells the goods in another country at below production costs, this is known as 'dumping').

While some countries and populations suffered from this forced liberalization, it also facilitated the emergence of new centres of food production in countries classified as 'new agricultural countries' (NACs). Among the most important of these are Argentina, Brazil, China, Kenya, Mexico, Thailand and Chile. These countries were encouraged to 'open up' and embrace foreign direct investment (FDI) to exploit access to cheap labour and land prices and a generous regulatory environment to produce food for export. Many of the produced goods were luxury, high-value foods for affluent consumers in Europe and North America.

Just as in the eighteenth and nineteenth centuries, these economies were developed to service external markets and affluent consumers in distant lands. This time, prawns, coffee, exotic fresh fruits and flowers were cultivated and delivered fresh or frozen to the supermarkets in milder climates. This helps explain the restructuring of agricultural production in many emerging or developing countries.

Integration into the global marketplace, advanced by globalization and consequent economic liberalization has consequences for consumption too. Hawkes (2006: 1) highlights three major processes that help explain agricultural and dietary transformation in the emergent market economies since the 1980s: the production and trade of agricultural produce; foreign direct investment in food processing and retailing; and global food advertising and promotion. All of these are examined in more detail in Chapters 4 and 5, but to establish a major structural determinant of dietary shifts – the nutritional transition (see Chapter 2), simply consider the last of these factors here.

Before liberalization, domestic food markets and retail sectors were protected from external competition, after it the markets in these countries were fair game for, the often ruthless, exploitation by foreign operators. The appearance of these actors helps explain the shift in food and drink preferences in the developing world and emerging markets. The most obvious concept to introduce here is 'coca colonization', a concept which links globalization to a specific American soft drinks corporation, but exemplifies a process which represents a transformation of lifestyles generally. This is a much contested concept, but with reference to changes in what people tend to eat and drink across the emerging markets, it has considerable purchase. It effectively suggests the means and implications of foreign direct investment by global corporations in the developing world and how they are implicated in the nutritional transition, and thereby, the obesity pandemic detailed in Chapter 2.

A major change is that people in the developing world are eating more animal-based products. 'Global consumption of livestock products has more than doubled in the last 30 years, driven mainly by substantial growth in meat and dairy consumption in developing nations' (Holmes, 2001: 1; also French, 2000: 3). The pattern in Japan, where meat consumption expanded by 360 per cent during 1960–1990, is being replicated in some of the largest developing countries; it is already well underway in China and Brazil (Holmes, 2001). Gandhi (1999) describes how meat consumption is often viewed as 'progressive' in India, and summarizes some of the negative consequences of the expansion of meat production for domestic and export markets.

The preceding discussion introduces some of the principal structural processes that are essential to explanations of contemporary malnutrition; all these are discussed in more detail in the chapters that follow. To summarize, to understand any individual entitlement to a decent diet requires us to explore their location in a historically situated, constantly evolving and complex global political and economic environment.

There is nothing inevitable about the persistence of malnutrition, either under or over. When the essential political character of malnutrition is appreciated, then it becomes possible to envisage a world where malnutrition is history. While the political character of the problem has long been appreciated by some academics, the 'problem of malnutrition', in popular consciousness and in some textbooks, continues to assume an apolitical character which denies the connections between feast in some regions and hunger in others. A fundamental shift in the discourse of malnutrition is required to appreciate the structural variables that explain both manifestations of malnutrition. Politics and profits are the key variables that explain both undernutrition and obesity, and only when this is accepted may appropriate public responses emerge.

Summary

- **Obesity has joined undernutrition as a global problem, and both are symptomatic of the dysfunctional workings of the international political economy.**
- **Theories employed to understand undernutrition and overnutrition are very contentious.**
- **All food systems are socially generated and dynamic; they are neither natural nor inevitable.**
- **Proximate and structural factors are usually implicated in both undernutrition and overnutrition.**
- **Malthusian explanations of undernutrition persist and are employed to suggest that 'overpopulation' causes hunger; such interpretations are simplistic and tend to ignore the structural causes of poverty and vulnerability.**
- **Historical legacies are often implicated in contemporary patterns of food insecurity.**
- **Theories and practices of development have helped shape economic and political patterns of undernutrition and overnutrition since the 1950s.**

Discussion questions

1. Evaluate why explanations of undernutrition and overnutrition continue to be very contentious.
2. Describe and explain the difference between proximate and structural causes of undernutrition and overnutrition, and assess the significance of this distinction.
3. Review the influence of Malthusian ideologies on theories of hunger.
4. Assess the relevance of historical perspectives to explanations of undernutrition with reference to a major world region or country.
5. Evaluate the various implications of development theory for undernutrition and overnutrition.

Further reading

Lappé, F. M. and Collins, J. (1986) *World Hunger: Twelve Myths*, New York, Grove Press. This remains a very good introduction to the relevance of ideologies to hunger analyses.

Ross, E. B. (1998) *The Malthus Factor: Population, Poverty, and Politics in Capitalist Development,* London, Zed Books. This remains one of the best investigations of Malthusian ideology and its relevance to poverty and hunger.

Potter, R. B., Binns, T., Elliott, J. A. and Smith, D. (2008) *Geographies of Development*, 3rd edition, London, Pearson, Prentice Hall. This is an excellent undergraduate introduction to the patterns and processes of development which reviews the legacies of colonialism as well as the diverse theories and practices of development since the 1950s.

Dreze, J. and Sen, A. (1989) *Hunger and Public Action,* Oxford: Oxford University Press. This is a more challenging read for students considering the politics of poverty and hunger in more depth.

Useful websites

www.odi.org.uk/default.asp The Overseas Development Institute provides research and reviews about all the major themes understood to be related to development.

www.foodfirst.org Another indispensible site for updates and discussion about food security.

www.fao.org Already mentioned (Chapter 2) as a useful website to read about contemporary food and agricultural issues; the material here is indispensable for research.

Chapter 3

Follow Up

1. Review some of the causes of hunger at: www.globalissues.org/article/7/causes-of-hunger-are-related-to-poverty
   - Select some of the less obvious causes identified here and assess the connections between the factors and food insecurity
2. Read a Briefing Paper (2001) published by the Overseas Development Institute (ODI) that summarizes some of Sen's (1981) work at: www.odi.org.uk/resources/download/1564.pdf
3. Read a critique by Devereux (2001) of Sen's entitlement theory at: www.sas.upenn.edu/~dludden/FamineMortality.pdf (accessed 18 August 2011). Discuss the issues raised by Devereux in seminar.
4. Describe your entitlement package and evaluate why you enjoy its various elements.
5. Read the article by Levine (2011) and summarize how he explains the current crisis in the Horn of Africa (think proximate and structural): http://blogs.odi.org.uk/blogs/main/archive/2011/07/06/horn_of_africa_famine_2011_humanitarian_system.aspx
6. Write an essay on why an historical perspective is essential to explanations of contemporary entitlements.
7. Read the Briefing Report about the global economic crisis by Willem te Velde (2009) at: www.odi.org.uk/resources/download/2822.pdf. Summarize his recommendations.
8. Read the article by Lappé and Schurman (2000) 'The Population Puzzle' online at: www.context.org/ICLIB/IC21/Lappe.htm (accessed 10 December 2011), and then answer the following questions:

   - Outline the traditional interpretation of the 'population problem' as detailed in this article.
   - Why it is rational for poor families to have numerous children?
   - The authors introduce an alternative theory 'the power structures perspective'. What are its central tenets?
   - Annotate a map of the world to illustrate some countries that have reduced fertility by progressive government interventions.

4 Globalization, development and malnutrition

Learning outcomes

At the end of this chapter, the reader should be able to:

- **Evaluate some definitions and contradictory theories of globalization**
- **Analyse the role of ideologies in the globalization process and the role of International Financial Institutions (IFIs) in the transformation of the global food system**
- **Understand some of the controversies that surround the globalization of agriculture and food**
- **Evaluate the role of corporate globalization in the creation of under- and overnutrition**

Key concepts

Globalization; generic globalization; corporate globalization; neo-liberalism; International Financial Institutions; Bretton Woods Institutions; fragmentation; export-led growth; ideology; alternative globalization

Introduction

> Globalization has been variously used in both popular and academic literature to describe a process, a condition, a system, a force and an age.
>
> (Steger, 2009: 8)

> Globalization refers to the expansion and intensification of social relations and consciousness across world-time and world-space.
>
> (Steger, 2009: 15)

This chapter begins by investigating the nature and extent of contemporary globalization before examining its implications for the global food system and, therefore, for agricultural development and malnutrition. It is important to appreciate the contested nature of the process and to establish its diverse geographical manifestations. It is neither the root of all evil nor the fountain of all that is good, as some of its critics or supporters assert. Vitally, it is a process that can be challenged and changed. The process is temporally and spatially dynamic and often contradictory (see Table 1.3). Everywhere, it is promoted and resisted with various degrees of urgency; sometimes celebrated and sometimes reviled at the same time at the same venue.

> Top politicians and business leaders have been converging on the Alpine resort (Davos) for five days of talks including boosting the recovery of the global market and tackling the thorny issue of government plans to tighten banking regulations.

While at the same venue:

> About 100 protesters demonstrated on Saturday against the annual meeting of the World Economic Forum in Davos in the Swiss Alps. Protesters tried to bring down a fence surrounding the venue hosting the meeting and threw snowballs at riot police. The protest, organized by anti-capitalist groups, the local Green Party and socialist movement, made its way to a hotel where some high-level guests were staying. Police used water cannons to try and disperse the demonstrators.

Since the emergence of the concept of 'globalization' in academic and popular discourses in the late twentieth century, it has generated an enormous literature and a great deal of fractious debates; every aspect of the concept has generated controversy (Goldblatt *et al.* 1999; Hirst and Thompson, 1996; Nederveen Pieterse, 2000; Scholte, 2000; Khor, 2001; Petras and Veltmeyer, 2001; Johnson, Taylor and Watts, 2002; Clark, 2003; Buckman, 2004; Wolf, 2004; Osterhammel and Petersson, 2005). It is not necessary to rehearse all of these debates here, but it is important to evaluate a few of the most salient and to clarify their relevance to the following discussion about globalization and food. The concept is notoriously problematic and open to myriad interpretations.

The discussion begins with a brief summary of the contentious character of contemporary globalization before examining its implications for agricultural development and global diets. It suggests that much of what is unethical about the contemporary world's food system may be explained with reference to global processes and actors; however, it concludes that alternative global processes and different actors may ultimately generate a more positive food future, but this is the subject of the final chapter of this text.

What is globalization? Some preliminary perspectives

To establish the nature and extent of globalization, we must first agree on a definition and some core characteristics, which is not easy. One of the most helpful definitions is:

> Globalisation is 'a process (or set of processes) which embodies a transformation in the spatial organisation of social relations and transactions assessed in terms of their extensity, intensity, velocity and impact-generating transcontinental or interregional flows and networks of activity, interaction, and the exercise of power'.
>
> (Held *et al.*, 1999: 16)

This definition has the merit of suggesting the myriad processes associated with the concept, it is not a single process but a 'set of processes' that may be economic, political, military, legal, cultural or social and, often, all of these combined in a variety of ways. It is vital to emphasize that globalization is not a singular condition or a linear process. There is nothing inevitable about its contemporary character or future contours, and in fact there is no consensus about its contemporary nature and import.

Three contrasting perspectives on the process were elaborated by Held *et al.* (1999) and are still useful because they are employed in popular and academic discourses and are associated with different political interventions; they also begin to suggest why the concept is contested. Crudely, they may be viewed as being on a spectrum, with at one end the hyperglobalists who understand the process to be the most important single factor shaping our contemporary realities, and at the other extreme, the sceptics who deny that it is transforming society in any significantly novel direction. The perspective that grants the process most transformative power and influence are the hyperglobalists. Interestingly, analysts who most vociferously oppose the process and those who celebrate it most share this perspective; they both grant it primacy in interpretations of how the contemporary world works. Its supporters emphasize its emancipatory potential, while its opponents see it as intrinsically oppressive.

The transformationalists inhabit a position somewhat between these two extremes. They accept that the process is significant and has deep structural roots, but they tend to see it as a more malleable process with diverse potential trajectories. While this text may occasionally be coloured by the anti-globalization hyperglobalist's worst fears, I hope it ultimately identifies with the more optimistic transformationalist's perspective. My position is that the process may be challenged and changed to deliver a global food system which is more socially and environmentally benign.

What is almost universally agreed, however, is that the process has generated new linkages between places and people across the globe (whether economic, political, military, social or cultural in character) that are deemed different from connections in the past by their intensity, extensity, velocity and impacts. It is helpful here to differentiate between what we might call 'generic globalization' and 'capitalist or corporate globalization' which is dominant at present (Evans, 2008; Sklair, 2010). Generic globalization is the technological framework through which the process is operationalized and embedded – the electronic revolution in communications and its attendant infrastructure as well as the creation of new transnational spaces. These technologies and networks of globalization may be moulded and developed by different ideologies to serve different interests. Globalization, as it is currently experienced, has been guided by neo-liberal economic ideologies which have been hegemonic since the 1980s. I contend that this ideology has privileged corporate interests and has created, in effect, 'corporate globalization'. As we discuss in Chapter 9, globalization does not necessarily have to satisfy these interests. Another type of globalization is theoretically possible which has different ends and interests. Critical analyses attempts to map and manufacture just such an alternative globalization (Applebaum and Robinson, 2005; El-Ojeili and Hayden, 2006; Evans, 2008).

Critical analysts of the process imagine an alternative globalization, more democratic in nature which prioritizes social and environmental justice rather than corporate profits. This alternative vision, or 'counter-hegemonic globalization' is based on different principals, and its objectives are to promote the equitable development of human capacities and environmental stewardship (Evans, 2008: 2). Realizing this vision and the politics which may generate its birth are the subject of debates among the diverse adherents of this ambition.

Globalization: a novel process?

The first controversial issue is to establish what's new about the process. Is contemporary globalization simply the most recent manifestation of capitalism which originated in Europe and diffused across the globe with colonization? This interpretation is convincing, and evidence from the developing world suggests that many of the similar processes of dispossession and exploitation that typified the colonial period are associated with patterns of development today. However, this discussion also holds that the nature of global interaction has indeed changed significantly since the 1970s. The interconnections have intensified and, some argue, have challenged the hegemony of the nation state system dominant since the Peace of Westphalia in 1648. The following discussion considers some evidence with particular relevance for the themes in this text.

Held *et al.* (1999) suggest a variety of 'indicators of interconnectedness' including: political–legal, military, economic, migration statistics, culture and environmental (see summary at the Global Transformations website, www.polity.co.uk/global/default.asp; accessed 12 July 2011). Various statistics may be employed to support the contention that globalization has intensified in recent decades, although this trajectory may be arrested with the economic crisis since 2008. Certainly, the global ramifications of the banking crisis and associated economic meltdown of 2008 was evidence of the integration of the world's financial markets; the collapse of Lehman Brothers damaged financial markets across the globe, and the implications are still being evaluated. Many would also agree that this global economic crisis resulted from at least two decades of global deregulation of economic activity, of specific relevance here, to aspects of trade allied to the agricultural and food sectors. All sorts of measures that were traditionally employed by nation states to guide, direct and control their financial and trading patterns were eroded. A remarkable shift in the world economy is captured in this assertion '[F]lows of capital rather than trade now drive the world economy' (Knox, Agnew and McCarthy, 2003: 87).

A statistic which indicates the growing financial connectivity between places is the growth of foreign direct investment (FDI) flows. These exceeded US$1.2 trillion in 2010 (UNCTAD, 2010). The agricultural and food business have accounted for some of these flows as northern agri-business interests, and increasingly southern-based interests, invest in agricultural opportunities in the global south. FDI in agriculture increased from US$2 billion in 2000 to US$6 billion in 2006–2008. FDI takes a multitude of forms, of course; the question is what form best promotes food security and development and what type jeopardizes the same (Panitchpakdi, 2010). Food prices rises are correlated to increased FDI in agriculture and land, so it is vital to ensure that such investments are responsible; evidence to date suggests that this is the exception rather than the rule. The emergence of responsible agricultural investment initiatives (RAIs) may help direct these capital flows towards more developmental and food security directions but, based on past experience, this is unlikely (see the following discussion about land grabs).

Changes in diets in the global south are also related to increases in FDI. Investments by major supermarkets, food companies, fast food chains and food advertising budgets account for a significant share of the total FDI. These investments, the availability of fast foods and processed foods as well their very effective promotion, help explain the rapid nutritional transition that so typifies dietary changes in much of the global south in the last two decades.

FDI is, however, not the most dramatic symptom of global economic integration; daily currency trading claims this title. In September 2010, the daily turnover in the world's foreign-exchange markets soared to US$4 trillion (Bank for International Settlements, 2010). Such massive daily currency shifts help explain why the global economy has proved so volatile in recent years and why the term 'casino capitalism' has been added to the lexicon of economic analysis. The events of 2008 unleashed a global recession of a magnitude not seen since the Stock Market Crash of 1929, which confirmed the 'casino' character of the prevailing system. The 'casino' terminology is a perfect description of dealings so devoid of any material basis as the sub-prime mortgage markets which were the air in the bubble that preceded the 2008 crash; speculation had run out of control, although it generated phenomenal profits for investment bankers. The role of speculation on food prices was mentioned in Chapter 1, and it continues to concern most analysts of the food crisis, the only issue in dispute in the magnitude of its impact. The rollercoaster illustrated in Chapter 1 (Figure 1.1) has disastrous

Plate 4.1a China: traditional fast food.

Plate 4.1b China: recent fast food outlet.

implications for the poor in the global south eking out an existence on meagre incomes.

One of the most dramatic statistics that illustrates increased globalization and which is relevant to entitlements across the global south is that of remittances. 'From the 1990s up to now (2008) flows to developing countries have witnessed a staggering increase, more than five-fold between 1990 and 2006, reaching US$177 billion' (United Nations Conference for Trade and Development Report, 2010: 34). These are a graphic illustration of the interdependence of the global north and south and, after FDI, these monies represent the most important source of external finance to developing countries. The volume of these capital transfers are also a measure of labour migration, itself a very obvious indicator of globalization. In the context of this text, these monies are vital to sustaining households in the developing world and their access to food. They form one of the principal entitlements of these communities. Economic recession in the global north will significantly impact entitlements in the global south as employment opportunities decline or collapse. To understand more about the place of remittances in the global south and their trends visit: www.ifad.org/remittances/pub/gateway_flyer.pdf (accessed 13 July 2011).

An obvious indication of globalization has been the expansion of the total value of world trade in recent decades, from US$57 billion in 1947 (arguably an atypical low rate following the Second World War) to US$12.6 trillion in 2005 (Steger, 2009: 42). However, the international economy has changed significantly in every sphere of activity; the primary, secondary, tertiary and quaternary sectors all reflect the impact of globalization (Dicken, 2011).

The specific way in which globalization is transforming agriculture and diets are explored in more detail in the following text and in the next chapter, but first we turn to the second core question about corporate globalization and consider what drives it and what its objectives are. Before commencing, it is important to emphasize, as mentioned at the beginning of this chapter, that the following discussion describes and explains aspects of globalization as it operates currently. This is not to suggest that it is immutable or inevitable. My thesis argues quite the contrary; it is neither of these things. Rather, it is socially constructed and the outcome of political and ideological contestation, and it can be transformed to deliver very different outcomes.

What drives contemporary globalization, and what are its objectives?

Certainly, as mentioned earlier, the role of this technology cannot be emphasized enough, but it is best understood as an enabling element rather than as a fundamental explanatory variable. The very same technology could be employed to deliver a very different type of globalization with a more sustainable and equitable food system. To appreciate why contemporary globalization has been so effectively embedded across the globe, we must turn to its main agents and appreciate the role of neoliberal economic ideologies. The dominance of neo-liberal economic policies since the 1980s helps explain the specific character of contemporary globalization and its main beneficiaries.

Proponents of contemporary globalization argue that a principal ambition is to spread wealth and prosperity through the expansion of capitalism. Its supporters adhere to a free trade philosophy and believe that the expansion of capitalism and the integration of world markets will deliver 'health, wealth and happiness'. Globalization in its most recent incarnation is justified with reference to liberal economic theory or free trade ideologies. One of the ways in which ideologies affect nations and individuals is that they become enshrined in institutions which make policy. Neoliberalism has enjoyed global hegemony since the fall of the Berlin Wall (1989). Until the Asian crisis of 1997 and the more recent economic crisis, it reigned almost unchallenged in international political and institutional circles. Although some of its advocates have become slightly more cautious, the core principles of the economic philosophy have survived the crisis so far.

Unregulated capitalism has been effectively critiqued and there is much talk about creating capitalism with a human face (UNDP, 2010). This usually implies state intervention or some form of international regulation to reduce the worst impacts of economic change. However, the global political economy is based on capitalist principles and market penetration, and expansion has been both rapid and ruthless since the 1980s. The next part of this chapter reviews how neoliberal ideologies have driven contemporary globalization through its institutionalization in the global economic architecture.

The role of global international financial institutions

Free trade theory was vital to the expansion of capitalism and is still the guiding principle behind all IFIs. The Bretton Woods Institutions

(BWIs), namely, the World Trade Organization (WTO), the International Monetary Fund (IMF) and the World Bank (WB), are the most influential institutions that guide development policy and investment decisions in the global south. As three of the most important institutions that govern world economy, they form the basic architecture of the world economic system. Free trade is also promoted by other alliances, notably between some of the world's wealthiest countries, regional development banks and transnational corporations. These various actors and institutions envision the world as a free market where everyone is 'free' to trade with everyone else, although, as we discover in the following text, some are freer than others. The theory holds that such a system is more efficient and produces cheaper goods so that the consumer enjoys lower prices. Theory of course is never as simple in practice, as we see in the following text.

It is vital to appreciate how very important this doctrine has been in promoting global economic expansion and integration in recent decades and subsequent resource exploitation and production decisions. Many of these decisions have implications for food production and distribution and diets. Institutionalized in the BWIs, economic liberalism has guided their most significant policies. Following the international debt crisis of the early 1980s, many developing countries were forced to negotiate loans from the IMF. These loans were subject to strict conditions, specifically, the borrowing countries had to restructure their economies by implementing structural adjustment policies (SAPs) and were required to alter their development programmes and investment decisions.

SAPs became the means whereby liberalization was effectively implemented across the global south and help explain the massive increase in global trade throughout the 1980s and 1990s. Both the IMF and the WB extended their roles in the economic order by becoming important agents of liberalization. Countries who borrowed from either institution were required to 'open up' their domestic markets to foreign imports by reducing protectionist policies, especially domestic subsidies and tariff restrictions, while at the same time they were instructed to start exporting goods to the world market by eliminating export licenses and export duties. Therefore, liberalization was implemented on both the domestic economy and external trade. However, IFIs have a wider remit, which is also pertinent to mention here. Tsikata (2010: 8) notes 'their role in guaranteeing projects of trans-national corporations against financial loss as well as political risk management'.

The rhetoric of free trade sounds convincing until you begin to examine how it actually works. What is touted as a level playing field, where northern industrial nations and developing nations operate on equal terms, is found to be, upon investigation, very unequal (see Chapter 5). As Chang (2007) illustrates, free trade philosophies have been very partially implemented and reflect northern biases, and these policies can seriously prejudice the efforts of nations in the global south to industrialize. Equally important, as detailed in the preceding discussion, is that the 'free' nature of international trade has always been a fiction; trade rules too often benefit the already powerful and marginalize the less powerful. A study published in 2002 dramatically contrasts subsidies to agriculture in the OECD countries in 1999 (US$360 billion) to total agricultural exports from all developing countries in the same year, a mere US$170 billion. They conclude that the notion of 'competition' in this context is farcical (Sreenivasan and Grinspun, 2002).

Wade (2003) discusses another impact of 'proliferating regulations formulated and implemented by international organisations', particularly the WTO (p. 621). He argues that the majority of regulations imposed on developing countries in the 1990s seriously reduced their 'development space'. By this, he means the policy options that they might employ to protect or support domestic industrial or agricultural expansion. Market liberalization has, he argues, consolidated the dominance of the already advanced economies and condemned those still developing to an inferior status in the global political economy. He details how government policies (often protectionist in nature) so successfully employed by Britain, Germany, the United States, Japan and, more recently, Korea and Taiwan have become illegal, strictly off limits to developing countries today. Many would call this an example of neo-colonialism, as the advanced Western nations (particularly the United States and European Union) used their privileged positions in the institutions of global governance to enshrine policies that favoured their economic objectives rather than policies that might advance the development of relatively less powerful nations. Such regulations effectively reduce the self-determination of developing countries by restricting their policy options. Nowhere is this more blatant than in agricultural sector, described in the following text. For those concerned about promoting sustainable or equitable development, the ideology of free trade is a serious obstacle. By emphasizing the 'free' issue, its visionaries neglect other elements of trade which are more important; that is, its social and environmental impacts.

Contemporary globalization: agriculture, food and nutrition

To further investigate who or what benefits and suffers from the nature of contemporary economic globalization, we consider examples from the food and agricultural sectors.

Without doubt, the agro-food industry has become a battleground with several fronts 'between producers and producers, between producers and consumers, between producers and governments (not least because agro-food is one of the most heavily regulated industries), and between governments' (Dicken, 2007: 348).

This is just the start of the list of conflicting interests ranged across the food business today. What explains its location as a 'battleground'? Several of the most effective exposés of the global food regime have been by activists who are eager to reveal its underbelly; the facts and figures that are often ignored by its champions (Box 4.1). Serious accusations are launched against it from a great many different directions: social justice advocates, environmental campaigners, animal rights activists and others. The discussion now considers why the global food system has generated so much opposition from such diverse directions. Central to this question is who or what benefits and who or what suffers from the costs of the liberalization of food and agriculture?

Industrial farming

During the twentieth century, prevailing models of capitalist farming were transformed and industrial food production emerged. This had unsavoury roots, in colonialism (plantation production systems), and later in the American chemical and military industries that developed during the 1930s (Ross, 1998). Friedmann (1990) traces the origins of the food system and some of its ecological and social correlates since the sixteenth century and suggests that a significant threshold was the coming of industrial farming in the twentieth century. Roberts too delivers a fascinating overview of the emergence of modern farming (2008: 3–28).

What is industrial food production and how does it contrast to previous models of food production? One of the main characteristics

Box 4.1

The exploitation of labour: its prevalence in the agricultural and food sectors

A report published by the US Department of Labor found that 122 goods were produced with forced labour, child labour, or both, in 58 countries (because of methodological problems of reporting, definitions, etc., this may well be an underestimate). The research uncovered more goods made with child labour than with forced labour. We are reluctant to draw conclusions from this, as it could be the result of the greater availability of data on child labour, or the possibility that forced labour is better concealed by perpetrators. When grouped by sector, agricultural crops comprise the largest category. There are 60 agricultural goods on the list, 38 manufactured goods and 23 mined or quarried goods. Production of pornography was a separate category; compelling evidence was found of this egregious labour abuse in six countries, with the likelihood that it occurs in many more. A number of goods were found to be produced with child labour or forced labour in numerous countries. Examples include cotton (15 countries), sugarcane (14 countries), tobacco (13 countries), coffee (12 countries), rice (8 countries), and cocoa (5 countries) in agriculture; bricks (15 countries), garments (6 countries), carpets (5 countries), and footwear (5 countries) in manufacturing; and gold (17 countries) and coal (6 countries) in mined or quarried goods.

Source: www.dol.gov/ilab/programs/ocft/pdf/2009tvpra.pdf (accessed 26 January 2011). For more information about contemporary labour exploitation, go to the Anti-Slavery Organisation web page at: www.antislavery.org/english/what_you_can_do/free_campaigns_resources.aspx, and for child labour and food production: www.antislavery.org/includes/documents/cm_docs/2009/c/child_labour.pdf (accessed 26 January 2011)

of the industrial food production system is its reliance on the heavy application of products from the petro-chemical industry, namely pesticides, fertilizers and herbicides. Their price is allied to world petroleum prices, which ensure that their prices are inherently volatile; hence, food prices too are vulnerable to wide price fluctuations. One of the most serious problems associated with casino capitalism is that the price of food fluctuates randomly with drastic consequences for the poorest peoples' access to a decent diet. The volatility of food prices, now so closely correlated to international oil prices, exposes millions of the poor in the global south to food insecurity.

This is not the only connection between food prices and oil prices. Most work on industrial farms is now completely dependent on petrol, machinery, irrigation systems, heating, drying of cereals, etc. However, the petroleum connection does not end on the farm. In a global food system, food chains are very long and are powered by petroleum; whether it be by road, train, sea or air, food miles account for a great deal of food's carbon footprint. The industrialization of farming helps explain how the corporate sector has emerged as perhaps the most significant beneficiary. All these inputs are produced and sold by global corporations, and their control of the market is already significant and is increasing as globalization intensifies. Across the world, this model of agriculture is critically transforming rural life and traditional patterns of food production.

This model of food production, initiated in the United States, diffused rapidly after the Second World War and has been vigorously marketed since then. Its emergence and dominance may only be explained with reference to powerful agro-business interests largely

Plate 4.2 Dung cakes in Indian market (Courtesy: Dr Louise Bonner).

based in the United States and the European Union. The BWIs effectively urged developing countries to ignore food security and to 'focus on leveraging their "comparative advantages" to earn hard currency through foreign trade' (Philpott, 2010). This development strategy suffers from two major weaknesses. First, millions of small peasant farmers lose their entitlements as they are forced off the land; secondly, earnings from export crops are subject to massive price fluctuations. So, effectively, this policy has eroded the entitlements of millions of rural people and exposed them to greater food insecurity as a consequence. To follow food price trends and their implications for food security, visit: www.ifad.org/operations/food/index.htm (accessed 13 July 2011).

Trade policies

Free trade ideologies also explain the major changes implemented to trade policies in the same decades, this time the main institution was the controversial World Trade Organization (WTO), established in 1995. This institution has become the unacceptable face of global capitalism to anti-globalization analysts and lobbyists. Indeed, it has continued to lose credibility across the global south, and the latest round of negotiations, the Doha Round, collapsed in Cancun, Mexico, in 2003. Critics of the WTO have argued that it suffers from several fatal flaws. The most important of these is that it is viewed as an undemocratic institution which, since 2003, has sought to extend its empire to include many issues deemed outside traditional trade negotiations. One of the most relevant of these to mention here is its role in the patenting process (Box 4.2).

Box 4.2

Trade-related intellectual property rights: TRIPs and the WTO

The TRIPs agreement came into force of 1 January, 1995, and was designed to protect intellectual property rights on a global scale by means of patents, copyrights and breeders' rights. Traditionally, patents were granted to persons over the 'creations of the mind', like inventions, works or art or designs, and the object being patented had to be 'novel, innovative or useful'. While legislation was established in 1995, not all countries were required to comply until 2006. This legislation can be understood to be a reaction to the perception in the developed world of widespread foreign intellectual piracy, but it has caused an understandable vociferous response.

> For many hundreds of civil society groups and NGOs around the world, TRIPS represents one of the most damaging aspects of the WTO. The legitimacy of the WTO is closely linked to that of TRIPS. TRIPS has, in fact, given the multilateral trade system a bad name. Contrary to the so-called free trade and trade liberalisation principles of the WTO, TRIPS is being used as a protectionist instrument to promote corporate monopolies over technologies, seeds, genes and medicines. Through TRIPS, large corporations use intellectual property rights to protect their markets, and to prevent competition. Excessively high levels of intellectual property protection required by TRIPS have shifted the balance away from the public interest, towards the monopolistic privileges of IPR holders. This undermines sustainable development objectives, including eradicating poverty, meeting public health needs, conserving biodiversity, protecting the environment and the realisation of economic, social and cultural rights.
>
> (Khor, 2001).

After the scientific and technological revolution associated with genetic manipulation, the opportunities and profits from new plants became even more attractive. Most of this scientific advance was funded by major transnational corporations who wanted to make sure any new plant varieties were protected by patent, so the TRIPS agreement emerged to protect their innovations. This was the first time that life forms were patented, and it was this aspect that proved very controversial. Before long, serious problems emerged with the TRIPS agreement, and it still causes tensions, largely between the northern countries and their corporations and southern countries and their farmers, often small-scale farmers with limited means.

The problem is that these 'new' plants (covered by patents) from the scientific laboratories in the north were modifications of plants gathered from across the globe, and there had been no sense of 'ownership' to that material. Rather, all these 'original' plants were freely available and accessible to all. In essence then, farmers and southern countries (so rich in genetic biodiversity) felt cheated at having to pay for plants/seeds that they had 'owned' originally and 'given away' free.

At the heart of debates surrounding the patenting of life and its adverse effects on food security, farmers' livelihoods, local communities' rights, sustainable resource use and access to genetic resources are the requirements of patent protection for life forms and natural processes in Article 27.3(b) of TRIPS. Patents on seeds and genetic resources for food and agriculture threaten sustainable farming practices, farmers' livelihoods and food security. Farmers using patented seeds are deprived of their right to use, save, plant and sell their seeds.

Increasing consolidation of multinational corporations in the seed, agro-chemical and food processing industries has further concentrated the control over seeds, seed choices and, ultimately, food security into the hands of a few corporations, and out of the hands of the farming communities. The patent system is also facilitating the theft of biological resources and traditional knowledge. The imposition of patent rights over biological resources and traditional knowledge unfairly deprives communities of their rights over, and access to, the same resources they have nurtured and conserved for generations. This contradicts the key principles and provisions of the Convention on Biological Diversity (CBD). 'The race to patent genes, cells, DNA sequences and other naturally occurring life forms has blurred the crucial distinction between discoveries and basic scientific information, which should be freely exchanged, and truly invented products or processes meriting patent protection' (Khor, 2001). Stiglitz (2002: 8) identified the main problem with TRIPS when he asserted that it 'overwhelmingly reflected the interests and perspectives of the producers'. Downes finished his helpful article by hoping that the review of TRIPS would generate 'a more equitable agreement that reflects the contribution of farmers and indigenous communities to the conservation and propagation of diversity in agriculture. … [N]ot only in the interests of farmers in developing countries but also to safeguard public research in the industrialised world' (Downes, 2004: 376).

Source: This overview is largely adapted from material by Downes (2004) and Khor (2001). For further reading, see Khor, M. (2001) at: www.twnside.org.sg/title2/par/mk002.doc (accessed 7 October 2010), Shiva, V. (2001), Madeley, J (2000)

Many are understandably suspicious of its neutrality and see in many of its judgements a bias towards Western corporate interests. An additional problem is that negotiations lack transparency, and countries with well-funded delegations have a higher success rate than those with fewer resources. Jensen and Gibbon recently concluded about the Doha 'Development Round' that 'the Round's agenda became dominated by topics and proposals of little relevance (to developing countries) and at times threatening for some groups of developing countries, especially those in sub-Saharan Africa' (2007: 5). It is not surprising that:

> Rural workers, peasants, women's organisations, environmentalists, trade unions, faith-based organisations, indigenous movements and organisations of small produces have been salient forces in the development of novel repertoires of action to resist the exclusionary and undemocratic nature of recent trade initiatives.
>
> (Icaza, Newell and Saguier, 2009: 8)

One of the most blatant examples of the biased character of the WTO policies is exemplified by reference to trade in agricultural goods.

While purporting to advance free trade across the globe and being ruthless on occasion by insisting that developing countries reduce all barriers to trade and any protectionist policies, they have countenanced some of the flagrant violations of this principle by the United States and the European Union. The agricultural sectors in both these regions enjoy very high subsidies; the sums are astonishing and, despite some severe international and domestic criticism, remain in place. This issue helps explain why the Doha Round of negotiations failed.

Other countries, including many developing countries that have been required to eliminate their agricultural support systems, now find their markets flooded by below-cost (subsidized) US and EU agricultural goods (Box 4.3). This practice is called 'dumping', the international selling of goods at below cost price; these subsidized

Box 4.3

Guatemala pays a high price for failures of the global food system

Domingo Tamupsis works as a harvester on a Guatemalan sugar plantation for a firm that exports bioethanol to fill the fuel tanks of cars in the United States. He works 10–12 hours a day, six days a week, in a country that is a leading producer of *food* for global markets. His settlement in the fertile Pacific coastal area is surrounded by industrial farms, but he earns so little that his family cannot afford to eat every day. His wife, Marina, is 23, but so slight that she might be mistaken for a girl. She has two daughters, Yeimi, aged six, and Jessica, aged two. Jessica is the size of the average European one-year-old, her distended stomach a sign of chronic malnutrition. When she smiles, hollow creases form in her cheeks, betraying her semi-permanent state of hunger. This family is typical of some of the very poorest people who are exploited in the global food system.

Guatemala exemplifies many of the worst aspects of the global food system. Half of all the nation's children under five are malnourished – one the highest rates of malnutrition in the world. Yet, the country has food in abundance. It is the fifth-largest exporter of sugar, coffee and bananas.

Its rural areas are witnessing a palm oil rush as international traders seek to cash in on demand for biofuels created by US and EU mandates and subsidies. But, despite being a leading agro-exporter, half of Guatemala's 14 million people live in extreme poverty, on less than US$2 a day. And the indicators are getting worse. The money to be made from the food chain here, as in most poor countries, has been captured by elites and transnational corporations, leaving half the population excluded.

Aida Pesquera, Oxfam director for Guatemala, says: 'The food is here but the main problem is distribution. Land is concentrated in very few hands. The big companies pay very little tax. Labour conditions on plantations are appalling. It's a classic case of how a very productive country with high rates of exclusion, especially among the indigenous population, cannot feed its own people'.

Source: Extracts from Lawrence, F. (2008) 'Guatemala pays a high price for global food system failings' at: www.guardian.co.uk/global-development/poverty-matters/2011/may/31/global-food-crisis-guatemala-system-failure/print

goods undermine domestic producers. Such trade policies can devastate local small farmer viability and are the root cause behind the collapse of domestic production in a variety of countries. It is worth exploring these staggering subsidies in some detail because they are a key element of the injustices of the contemporary food system (visit the CAIRNS Group website to access some perspectives on the global agricultural trading regime and development at: www.cairnsgroup.org/Pages/Factsheets.aspx (accessed 10 August 2011).

The European Union devotes €55 billion annually to the Common Agricultural Policy (CAP), subsidies are not necessarily wrong if they have beneficial social or environmental outcomes, but these effectively institutionalize inequalities and inefficiencies and have negative social and environmental impacts. This sum represents approximately 45 per cent of the entire EU budget, down from 70 per cent in 1970s! Farmers receive about 70 per cent of the money, but approximately 10 per cent is devoted to export support subsidies which end up in the strangest pockets, major food companies. Such policies have also been adopted in the United States, Japan and South Korea and are known as price support agreements. One estimate suggests that agricultural subsidies in the United States in 2008 amounted to US$8.1 billion and concluded that 'these trade-distorting subsidies push the global burden of adjustment onto commodity producers in Africa, Asia, and Latin America' (Gamberoni, 2009). Most analysts also decry the fact that the bulk of these farm subsidies are enjoyed by the wealthiest farmers in the United States and European Union and by corporate interests, not the struggling small holders in these places. Efforts to reform the CAP and make it more palatable to taxpayers by supporting more sustainable farming and environmental initiatives have had limited effectiveness, and big business interests have prevailed (Box 4.4).

Box 4.4

French farmers and EU agricultural subsidies

France receives around €11 billion each year from the EU in agricultural support, but very little of it actually goes to those who do the farming. There are a little over 500,000 recipients of EU farming subsidies in France, but over 80 per cent of the funds actually go to large industrial food processing businesses and charitable organizations.

In the list of beneficiaries for 2008, recently released by the French government, no independent farmer is listed among the top 20 recipients. The largest recipient is the chicken production conglomerate Doux, who received a whopping €62.8 million in aid between October 2007 and October 2008. In the year 2008, the group had a turnover of nearly €2 billion.

According to the radical small famers group, the *Confédération paysanne et exploitant agricole*, this turnover has been achieved mainly through the acquisition of smaller competitors, all who would have been in receipt of the subsidies, which are now paid to the holding group. Most of the subsidies were paid to the group to maintain their competitive position in the export market, by compensating the company for the lower prices at which they are obliged to sell into this market.

France is under strong competitive pressure from countries outside of Europe in the export of chickens, and this subsidy is considered to be important for the strategic development of agriculture in France and Europe. The same principle applies to the subsidies received by a number of large cereal and sugar producers and wholesale traders on the list.

France is the leading producer of cereal crops and sugar in Europe, and is also the largest recipient of agricultural support.

Other large industrial groups who head up the table are the dairy products group Lactilis, the luxury brands group LVMH and the sugar producers Tereos and Sucrière de la Réunion.

The common agricultural policy is due for a major overhaul in 2013, and France has been busy trying to ensure that, as far as possible, a strong system of agricultural support is maintained after this date.

Source: www.french-property.com/news/french_life/eu_farming_subsidies (accessed 7 October 2010)

Lawrence summarizes the history of farm subsidies in the United States and their role in promoting global corporate interests and asserts that '[J]ust five crops accounted for 90 per cent of the money – corn, rice, wheat, soya beans and cotton' (2008: 26). As in the European Union, these subsidies are enjoyed by the biggest farms and help subsidize the costs of raw materials for the processing corporations. She describes how this aspect of agricultural policy is 'designed to promote the export of US agricultural surpluses while bringing the money from value-added markets back home. In this they mirror patterns of trade established between previous empires and their colonies' (Lawrence, 2008: 27).

Two American critics of such subsidies draw attention to another consequence – their role in the obesity pandemic. They assert that cheap corn and soy (subsidized by US and EU taxpayers) are one of the causes of obesity; they point out that high-fructose corn syrup (from corn) and hydrogenated vegetable oils (from soy) 'did not even exist a few generations ago but now are hard to avoid. They have proliferated due to artificially cheap corn and soybeans' (Schoonover and Muller, 2006: 1). Derivatives of both these crops are used to flavour and sweeten all manner of food and drinks and, even if we don't always eat them, the animals we consume eat them first. The stories of corn syrup and soy are both fascinating (Patel, 2007; Roberts, 2008). Their ubiquity in the global food system exemplifies how we have surrendered control over what we consume every day (Patel, 2007: 166).

If the ethics of farm subsidies are problematic domestically, they are even less defensible at the international scale. The most serious indictment is that these policies devastate the livelihoods of small-scale producers in the developing world, whether of cotton, wheat, dairy products, corn, cattle, sugar, etc. Dumping of subsidized commodities therefore erodes or demolishes small holder entitlements and their access to income and a decent diet. These inconvenient truths are well hidden in both contexts, and most taxpayers in the European Union and the United States are unaware of just how much of their money is spent and its social costs. Critics quite rightly claim that these subsidies are responsible for some of the hunger and malnutrition suffered by communities in the developing world. Malloch Brown (2002) summarized the problem:

> 'Every cow in Europe today is subsidised two dollars a day. That is twice as much as the per capita income of a half of Africa. It is the extraordinary distortion of global trade, where the West spends US$360 billion a year on protecting its agriculture with a network of subsidies and tariffs that costs developing countries about US$50 billion in potential lost agricultural exports' See here for more details: http://content.undp.org/go/newsroom/2002/november/mmb-uganda.en;jsessionid=axbWzt8vXD9 (accessed 12 July 2011)

It is unsurprising that some experts argue that phasing out of market support for agricultural producers in developed countries is necessary as a first step in the fight against obesity, poverty and hunger worldwide. Yet, the WTO asserts that trade policy is devoid of any social or environmental implications. The collapse of the Doha Round reflects the defection of the so called G21 countries, a group that includes most of the biggest southern economies, who concluded that their interests were being ignored in favour of policies designed to further vested interests in the developed countries. It is important to note that the WTO's Agreement on Agriculture (AoA) has proved to be the most divisive of all. Just why this has become such a bone of contention between the 'rich' and the 'rest' is indicated in the preceding text, but there are additional problems.

Corporate connections

There is a consensus that if the IFIs in general, and BWIs in particular, are the institutions through which liberalization is promoted, then global agri-business corporations are among its greatest beneficiaries. The global agri-business giants have certainly benefitted from their unprecedented access to resources, markets and profits as liberalization policies were implemented across the globe. The global food business is highly concentrated, and a few of its biggest corporations are forming ever closer alliances and penetrating all world regions. The food sector has experienced a high degree of vertical and horizontal integration in recent years; corporations are merging or forming alliances with businesses across, upstream and downstream in the commodity chain. Such processes help consolidate and concentrate power and influence.

> From Seattle to Shanghai, from Brussels to Brisbane, we are *Nourishing Lives* around the clock with brands that are known for quality (General Mills, Inc., http://www.generalmills.com/Brands.aspx)

The power of these corporations over such a vital commodity as the food we all eat is, probably, the most contentious of all the issues that surround the globalization of the food system and our diets. Damning indictments are available which detail the myriad ways in which the transnational agricultural corporations use their power in the marketplace to exploit people and places in the food chain; it is called a scandal by some. In 2008, the President of the General Assembly of the United Nations condemned corporate profiteering:

> The essential purpose of food, which is to nourish people, has been subordinated to the economic aims of a handful of multinational corporations that monopolize all aspects of food production, from seeds to major distribution chains, and they have been the prime beneficiaries of the world crisis. A look at the figures for 2007, when the world food crisis began, shows that corporations such as Monsanto and Cargill, which control the cereals market, saw their profits increase by 45 and 60 per cent, respectively.
>
> (Source: www.gmwatch.org/gm-firms-mobile/10595-monsanto-a-history; accessed 11 October 2010)

To appreciate the power of these global corporations, it is helpful to conceptualize international food or 'agrifood' chains as consisting of several elements. Figure 4.1 details some of its most important component parts. It is obvious from examining this diagram how very complex food chains have become. It also suggests the variety of sites where legislation may be exercised and profits extracted. Over the last 30 years, power over every part of this 'agrifood' chain has become concentrated in the hands of a few corporations.

As globalization and corporate concentration has intensified in the food and agricultural sectors, the gap between producer prices and

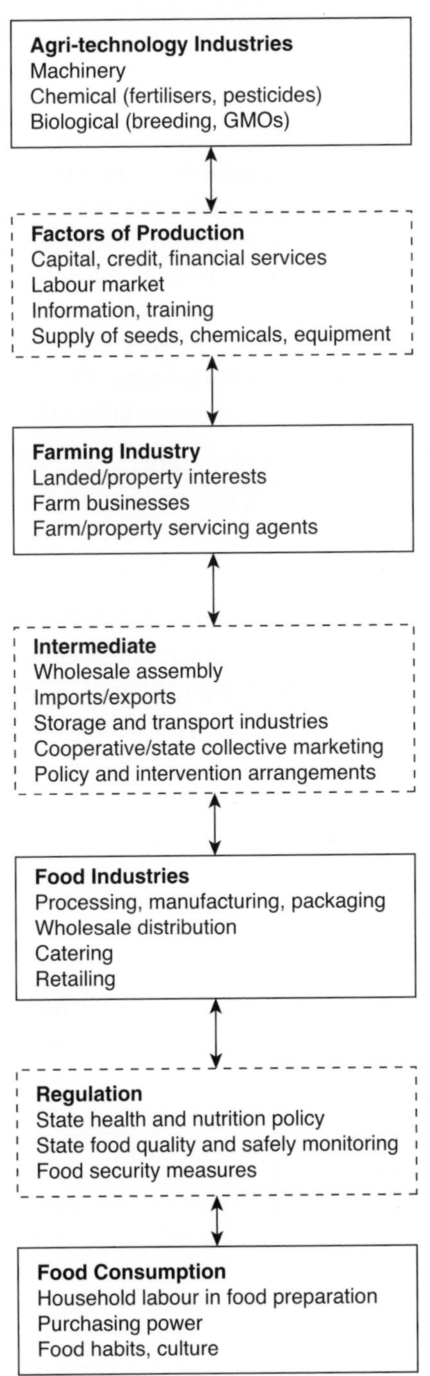

Figure 4.1 The global food chain: some components.

retail prices has grown. In other words, most of the money paid by consumers is enjoyed by various actors in the food chain, not the producers. In some cases, this is very extreme. Much of this divergence between producer price and retail price can be understood to reflect monopoly conditions; that is, a few major players can manipulate prices to ensure that they extract the lion's share of the profits. Their power is indeed exceptional; consider the statistics about corporate control over global seed sales in Table 4.1. Approximately 67 per cent of all seed sales are made by just 10 global corporations; this is part of the concentration at the pre-production end of the food chain (Goodman, 2009). This 67 per cent is the estimated share of all commercially sold seeds.

In industrial food systems, after seeds have been planted, the fields are treated with pesticides and herbicides to reduce crop loss due to insects and competition with weeds. Commercial farmers now buy 82 per cent of all pesticides from just 10 corporations; see Table 4.2.

Note that several names repeat in both tables. Pesticides are just one of the main inputs to industrial food production; this farming model requires the intensive application of other inputs from the chemical industry, notably herbicides and fertilizers. These are produced by the

Table 4.1 ***Top 10 seed companies***

Company	*Home base*	*Percentage of global seed market*	*Sales in 2002 (millions of US$)*	*Sales in 2007 (millions of US$)*
Monsanto	United States	23	1,600	4,964
DuPont	United States	15	2,000	3,300
Syngenta	Switzerland	9	937	2,018
Groupe Limagrain	France	6	433	1,226
Land O'Lakes	United States	4	NA	917
KWS AG	Germany	3	391	702
Bayer	Germany	2	250	524
Sakata	Japan	<2	376	396
DLF-Trifolium	Denmark	<2	NA	4,391
Takii	Japan	<2	NA	347
Top ten corporations		67	7,133 (includes some not included here)	7,652

Source: ETC Group at: www.gmwatch.org/latest-listing/1-news-items/10558-the-worlds-top-ten-seed-companies-who-owns-nature. NA: information not available.

Table 4.2 ***Pesticide corporations and percentage share of global market***

Corporation	*Home*	*Percentage of global share*
Syngenta	Switzerland	18
Bayer	Germany	17
Monsanto	United States	10
BASF	Germany	9
Dow Agro Sciences	United States	9
DuPont	United States	5
Makhteshim Agan	Israel	4
Nufarm	Australia	4
Sumitomo Chemical	Japan	4
Arysta LifeScience	Japan	2
Others	Various	18

same global corporations and exhibit similar patterns of market penetration.

After the food is harvested in the industrial system by massive machines manufactured by a few giant producers such as John Deere or AGCO, it is then traded and processed.

> Largely invisible to the general public but controlling a great part of the international food supply … are the behemoth trading and primary processing companies, mostly, but not exclusively, US based.
>
> (Lawrence, 2008: 31)

These trading and primary processing corporations have prospered as US farm subsidies guarantee their access to below-cost grains, which even after processing can be sold in most developing countries at below the costs of domestically procured grains. This 'dumping' has been condemned at numerous WTO meetings but has proved very resilient to reform (see the preceding sections). The dominance of US corporations is suggested by the following information, based on Lawrence (2008).

Four global giants dominate the trading and primary processing landscape: Cargill, Archer Daniels Midland (ADM), Bunge and Louis Dreyfus – the first three are based in the United States and the last one in France. Cargill is the largest privately owned corporation in the world, and controls approximately 45 per cent of the global

Box 4.5

The Bunge story: a corporate history of consolidation, expansion and diversification

This is a case study of how a relatively small commodities trader became a global agri-business corporation with investments in every corner of the globe and at every link in the extensive food chains that connect producers and consumers across the world in the twenty-first century. Visit the timeline at: www.bunge.com/about-bunge/timeline.html. It is fascinating to chart the expansion of their activities and their global presence since their beginnings in Amsterdam in 1818. They were very alert to new business opportunities in the past, and they became involved with agricultural expansion in North America and then Brazil and Argentina in the nineteenth and early twentieth centuries as the agricultural potential of these regions was becoming obvious. More recently, since 2000, they have been quick to invest in infrastructural and processing facilities in 'the traditional breadbaskets' of Eastern Europe and Russia as well as in the booming agricultural markets of China and India. They appear to have facilities in every world region. Diversification has also been their habit, most recently foraying into biofuels and ethanol production. Visit this page to explore the diverse activities of the company by world region: www.bunge.com/about-bunge/abr.html

grain trade. It is also the world's largest crusher of oilseeds such as soya and rapeseed. Its revenues for 2007 were US$88 billion. Archer Daniels Midland (ADM), another US-based corporation, controls about 30 per cent of global grain trade, and is one of the world's largest processors of soya beans, corn, wheat and cocoa. In addition, it has a huge portfolio of interests from sweeteners and food processing ingredients to energy and animal feed production. The history of Bunge Corporation is detailed in Box 4.5 and it illustrates the global status of these businesses. The Louis Dreyfus Corporation celebrates its diversity and global presence on its website:

> Founded and based in Europe, LD Commodities enjoys strong regional presence in North and South America, Europe, Asia, the Middle-East and Africa. … The Group's diversification and expansion strategy has enabled it to post outstanding year-on-year growth; it is the number one cotton and rice

merchant worldwide and ranks in the top three in orange juice, wheat, corn and sugar and in the top five in oilseeds and coffee. In addition, LD Commodities is the second largest sugarcane crushing and renewable energy group in the world.

(Source: www.ldcommodities.com/-About-us-.html accessed 27 September 2010)

Once traded and processed, the food is then manufactured. The dominance of the world's major food manufactures is illustrated in Table 4.3.

Their power, originally based in the developed world markets, has grown as globalization intensified, and their presence in many emerging markets is already significant, facilitated by trade liberalization since the 1980s. The comments, gleaned from their web pages, have been selected to indicate their corporate strength and ambitions. The top ten corporations' share of food production is estimated to be 28 per cent of the global total. It is important to note that their market share is very varied. In some countries, their share is over 50 per cent; in others, it comprises only a small percentage. However, their power is increasing; in both its penetration and global reach, concentration in this sector has more than doubled since 2001 (Mulle and Ruppanner, 2010). The role of food processors in the context of this book has more sinister implications because they are increasingly changing the character of what we eat, considered later. There are some new corporations, based in emerging markets, that are beginning to expand into northern markets too (Table 4.4).

The penultimate link in the chain from farm to plate is the retail sector. Here, the story is remarkably similar to that in production, trading and manufacture; that is, a few global market leaders are increasing their dominance. Their power has received a great deal of criticism in the United Kingdom and United States, but across the world they are enjoying increased presence and profits. The 'supermarketization' of the globe is already well advanced and has been effected in a remarkably short time frame. The 30 largest retail chains accounted for 29 per cent of all food sales in 1999, but this figure had grown to 33 per cent by 2002 (Vorley, 2003: 28). These names are more familiar than the grain traders and processors. Among the most powerful and famous, or infamous, ranked by size in 2002, are: Wal-Mart (United States); Carrefour (France); Ahold (Netherlands); Kroger (United States); Metro Group (Germany);

Table 4.3 ***The world's largest food processing corporations***

Name	*Food sales in millions (2007)*	*Some facts/brands*	*Selected comments from web pages*
Tyson Foods	US$25,246,000	One of the world's largest processors and marketers of chicken, beef and pork.	Any'tizers snacks are the perfect way to satisfy hunger, anytime.
PepsiCo Inc.	US$24,474,000	Pepsi Walkers Crisps Lipton Tea	Meet the four major divisions of the PepsiCo family: PepsiCo Americas Beverages, PepsiCo Americas Foods, PepsiCo Europe, PepsiCo Middle-East and Africa.
Kraft Foods	US$23,939,000	Terry's Chocolates Kenco	Today, we are turning the brands that consumers have lived with for years into brands they can't live without. Millions of times a day in more than 150 countries around the world, consumers reach for their favourite Kraft Foods brands.
Nestlé	US$23,3000,000	Smarties Nestlé has (September 2010) announced the establishment of an R&D Centre in India.	Nestlé has a strategic focus and leadership in emerging markets where the company expects sales to reach 45 per cent of total by 2020. See their role in the controversy about formula milk in Chapter 2.
Anheuser-Busch, Inc.	US$14,159,000	Budweiser Michelob Irish Red Ale Bacardi Silver Strawberry	Anheuser-Busch also has interests in malt production, rice milling, real estate development, turf farming, metalized and paper label printing, bottle production and transportation services. As the world's leading brewer, we are committed to ensuring our sponsorships, advertisements, packaging, promotions and other marketing communications are carried out responsibly.

Dean Foods Co.	US$11,821,000	Dean Foods is one of the leading food and beverage companies in the United States and a European leader in branded soy foods and beverages.	Alpro is the pan-European leader in branded soy food products.
General Mills, Inc.	US$11,093,000	All Betty Crocker products	From Seattle to Shanghai, from Brussels to Brisbane, we are *Nourishing Lives* around the clock with brands that are known for quality. In 1945, Fortune magazine named Betty Crocker the second-most popular American woman. Eleanor Roosevelt was named first.
Smithfield Foods, Inc.	US$9,749,000	Our core products include canned, continental, chilled and fresh meat.	In recent years, cheap flights have made people more interested in sampling more 'exotic foods'. This group of consumers has helped to grow the Spanish market.... Hardly a day goes by without a celebrity chef or cooking programme using these products and they, together with tapas, are also becoming more and more popular at gastropubs and restaurants across the United Kingdom.
Con Agra Inc.	US$8,864,000	See some recipe videos at: www.conagrafoods.com/consumer/recipes/recipetvvideo.do?id=173.	Nourish Today, Flourish Tomorrow Our consumer foods are found in 97 per cent of America's households, and 24 of them are ranked first or second in their category.
Kellogg Co.	US$7,786,000	Special K	Add sweet variety to afternoon snacks with *Austin*® Zoo Animal Crackers. Kids love the fun animal shapes. Moms love that they're a low-fat treat.

Source: www.fmi.org/docs/facts_figs/top_man_process.pdf and selected quotes from corporate web pages

Table 4.4 ***Ten southern agri-business TNCs***

Name	*Description*
Sime Darby (Malaysia)	World's largest producer of palm oil
CP Foods (Thailand)	Asia's largest meat producer expanding into Europe, Africa and Middle-East
Wilmar (Singapore)	Major palm oil and sugar producer. This company is allied to ADM which has a minority stake. A good example of the increasingly frequent joint venture between northern- and southern-based corporations.
Olam (Singapore)	Major agricultural commodity trader, with a presence in Asia, Latin America and Africa. Recently expanding into the production of staple foods.
JBS (Brazil)	World's largest meat company with a focus on beef. Recently expanded into North America and Australia and into poultry.
Karuthuri (India)	One of the largest producers of cut flowers in the world; it has recently moved into the production of food crops too.
Savola (Saudi Arabia)	The largest food company in the Gulf region, it is involved in the production and processing of foods as well as in retailing through its ownership of the Panda supermarket chain.
COFCO (China)	Still a state-owned conglomerate, this is China's largest food processor and trader.
COSAN (Brazil)	Fourth-largest sugar producer in the world. It recently entered into a major ethanol joint venture with Shell Oil, in another north–south joint venture.
New Hope (China)	A privately owned conglomerate that is China's largest producer of feed and one of its largest producers of pork, poultry and dairy. The company has recently launched operations in Vietnam, the Philippines, Bangladesh, Indonesia and Cambodia.

Source: adapted from www.grain.org/seedling/?id=693 (accessed 27 September 2010)

Target (United States); Tesco (United Kingdom); and Costco (United States).

The sector is very competitive, which may be good for the consumer initially, but this advantage is eroded as concentration accelerates. Intensifying the downward pressure on price is the emergence in recent years of new 'hard discounters' such as Costco and Aldi. The race to reduce prices of food has, of course, consequences for all components of the food chain. Producers are faced with stiff competition and in some cases are forced to sell below cost price. They require greater 'productivity' from the workers in the system and the animals they rear. Workers on the farms and in the

Plate 4.3 Traditional street vendors replaced by supermarkets (Courtesy: Dr Louise Bonner).

manufacturing and retail parts of the food chain are usually not unionized, while animals in the industrial food system are manipulated and managed in some very unsavoury ways to ensure 'cheap food'(Nierenberg, 2005).

The impact of the downward pressure on food prices in complex. Clearly, some benefit from access to cheaper food. Farmers, small farmers in particular, are usually at a disadvantage and are often squeezed out of production, so these populations lose their entitlements. In the global south, these people swell the volume of rural urban migrants (Patel, 2007: 173–214). Other workers in the food chain, in harvesting, manufacturing or retailing, also experience a squeeze on their entitlements as their wages remain low. Patel (2007) describes the prevalence of slavery in the agricultural sector in Brazil where between 30,000 and 50,000 people are thought to be exploited in the production of sugar, biofuels and soy. Barrientos (2011) has exposed how global production networks allow for labour contracting which 'involves a separation between the productive activity of labour and the formal "employer"', often through a network of intermediaries. These networks provide 'space for unscrupulous intermediaries to extract economic rent from vulnerable workers through coercive practices involving debt bondage and entrapment' (Barrientos, 2011: 4). Although exploitation may be most severe in the global south, it is important to note that it is also relevant for workers in the food chain in northern countries. For material about exploitation in the food chain in the United States, visit the Food Chain Workers site at: http://foodchainworkers.org/?page_id=38 (accessed 12 July 2011).

Supermarket dominance is due to intensify in the next decade and, if the experience of Latin America is replicated, it will be swift and include rural as well as urban consumers. In the last 10 years, supermarkets in Latin America have grown from accounting for 10–20 per cent of the market to between 50–60 per cent (Vorley, 2003: 31). Almost completely unknown in China in 1990, by 2002, supermarkets controlled approximately 7 per cent of sales. Carrefour first entered China in 1995 and, on the 6 July, 2007, it opened its hundredth store in Shaoxing, in the province of Zhejiang. The company planned to open 23 new stores during 2007, the largest number of hypermarkets it had ever opened in one country in a single year. More information is available at: www.carrefour.com/cdc/group/current-news/china-opening-of-the-100th-carrefour-hypermarket-.html (accessed 27 September 2010). Similar examples are available from across the global south. Global chains are now well represented in Southeast Asia where they are 'cannibalizing' traditional outlets (Vorley, 2003: 33), and in sub-Saharan Africa, where European and US operators are appearing as well as some from South Africa.

The appearance of supermarkets is swiftly followed by the transformation of supply chains, as producers have to deliver highly standardized products. As happened in the 1960s and 1970s in the United Kingdom, in Brazil the emergence of supermarkets has spelt the demise of small specialist retailers, local slaughterhouses, small trucker companies and small, locally embedded wholesalers and traders. Such changes extend the food chains that connect producers and consumers and all the elements of that chain, live or dead, which are incorporated in the global food web. Indeed, the distances that now separate producers from consumers are of concern as conditions of production, processing and delivery may be obscured from consumers. So, whether it is the human rights of labourers in the food system (farms or factories) or the suffering endured by live animals as they are delivered over huge distances from farm to slaughterhouses, changes in the food system generate ethical problems for all concerned.

The preceding discussion details some important changes in global food system since the 1970s. Many of these have direct impacts on entitlements and diets. The diffusion of industrial farming across the global south has marginalized many small producers who have lost their entitlements either to land or to income as their livelihoods have collapsed. The increased dominance of the corporate sector in food production, manufacture and retailing has had other implications for entitlements. They have helped secure low wages across the food chain.

Globalization and the 'land grab'

One further global trend in corporate and government behaviour requires consideration. There has been an 'explosion' of transnational commercial land transactions, popularly known as the 'global land grab' and variously described as 'astonishing', 'alarming' and 'criminal' (Matondi *et al.*, 2011). These transactions take a multitude of forms, which means their analysis is problematic (Box 4.6). Their myriad forms (bilateral government deals, TNCs and government deals, TNCs and private property deals, etc.), allied to the very different sources of capital (international, national, corporate, public), interacting with the complexities of local property relations (small peasant farmers or large private landowners, or public lands with unclear or ambiguous land ownership patterns, tribal communal systems, etc.) and political realities (authoritarian or democratic states, corrupt or transparent political regimes) mean that generalizations are difficult. To illustrate the diverse potential impacts, see Broadhurst (2011). His research in Tanzania established how different models of jatropha (biofuel crop) production (plantation, out grower-led and community-focused) had significantly different impacts for local food security, socio-economic development and the environment. However, there remain serious problems around the contemporary land grab phenomenon. In Africa, the issue is particularly controversial because in that continent food security is already seriously compromised. Figure 4.2 illustrates some important recent

Box 4.6

Billionaires and mega-corporations behind immense land grab in Africa

> 20+ African countries are selling or leasing land for intensive agriculture on a shocking scale in what may be the greatest change of ownership since the colonial era.
>
> (John Vidal, March 11, 2010, *Mail & Guardian*)

Awassa, Ethiopia: We turned off the main road to Awassa, talked our way past security guards and drove a mile across empty land before we found what will soon be Ethiopia's largest greenhouse. Nestling below an escarpment of the Rift Valley, the development is far from finished, but the plastic and steel structure already stretches over 50 acres – the size of 20 soccer fields.

The farm manager shows us millions of tomatoes, peppers and other vegetables being grown in 1,500 foot rows in computer controlled conditions. Spanish engineers are building the steel structure, Dutch technology minimises water use from two bore-holes and 1,000 women pick and pack 50 tons of food a day. Within 24 hours, it has been driven 200 miles to Addis Ababa and flown 1,000 miles to the shops and restaurants of Dubai, Jeddah and elsewhere in the Middle East.

Ethiopia is one of the hungriest countries in the world with more than 13 million people needing food aid, but paradoxically the government is offering at least 7.5 million acres of its most fertile land to rich countries and some of the world's most wealthy individuals to export food for their own populations.

The 2,500 acres of land which contain the Awassa greenhouses are leased for 99 years to a Saudi billionaire businessman, Ethiopian-born Sheikh Mohammed al-Amoudi, one of the 50 richest men in the world. So far, his company has bought four farms and is already growing wheat, rice, vegetables and flowers for the Saudi market. It expects eventually to employ more than 10,000 people.

But Ethiopia is only one of 20 or more African countries where land is being bought or leased for intensive agriculture on an immense scale in what may be the greatest change of ownership since the colonial era. An *Observer* investigation estimates that up to 125 million acres of land – an area more than double the size of the United Kingdom – has been acquired in the last few years.

Source: extracted from a longer article available at: www.informationclearinghouse.info/article24965.htm (accessed 7 March 2011); excellent case studies about Land Grabbing presented at Conference at IDS University of Sussex, April, 2011, available at: www.future-agricultures.org/index.php?option=com_content&view=category&layout=blog&id=1547&Itemid=978 (accessed 15 August 2011)

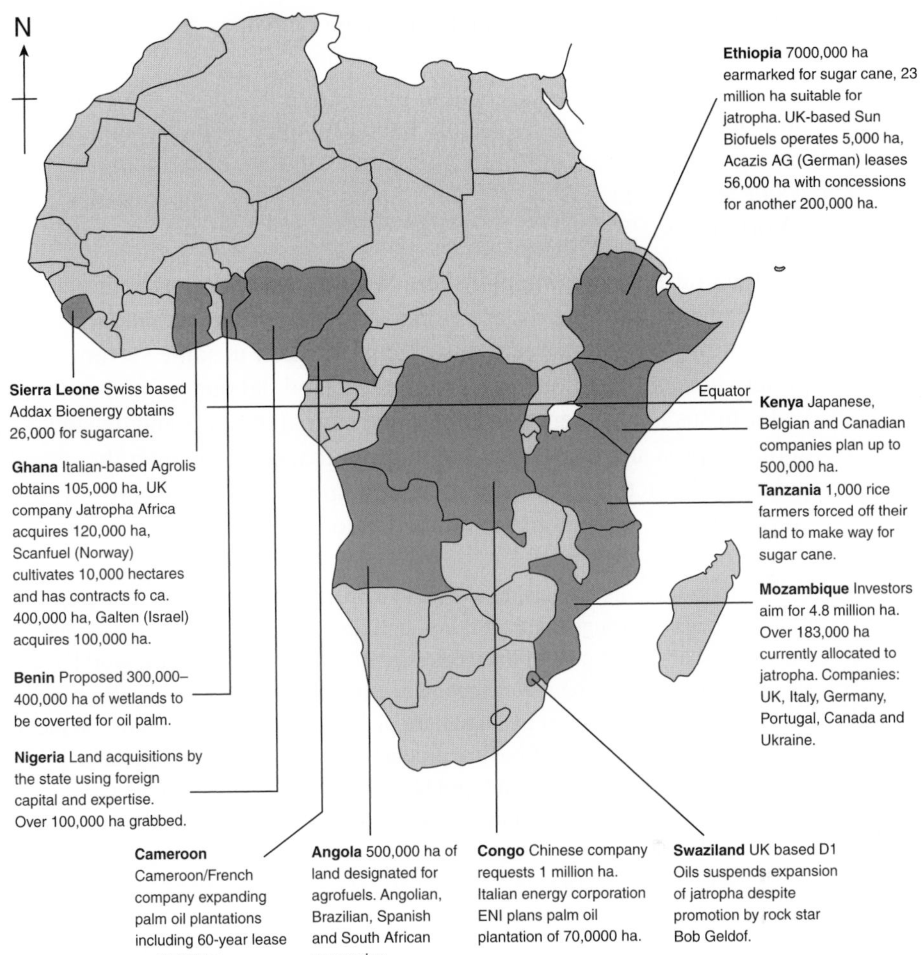

Figure 4.2 Africa: land grab.

examples of land grabbing in Africa, although as discussed in Chapter 3, this process has historical precedents in the late nineteenth century.

Borras and Franco (2010) note that the contemporary global land grab reflects the convergence of four global crises, namely, the financial, environmental, energy and food crises, that caused land, especially in the global south, to become attractive to diverse investors. The financial crisis is implicated in the land grab because global speculators and financial houses (hedge funds, pension funds, private equity groups, investment banks, holding companies, etc.), anxious to find new investment opportunities given the collapse in house and share prices, viewed cheap land in developing countries as an exciting new opportunity. '[T]hese factors have prompted a sharp increase in investor interest in acquiring significant quantities of farmland, water and forest areas in both developing and emerging countries' (World Bank, 2010).

Contract farming operations, where Western corporations contract farmers to produce goods, are increasingly important and already account for over 50 per cent of world production in some sectors. The system normally exploits local resources, land, labour, animals and water to sustain affluent consumers in external markets and debates about its pros and cons are allied to those that still rage over the policy and practice of promoting export crops. The figures are surprising and, depending on one's opinion, alarming:

> Corporations are exercising more and more direct control over farming, particularly through contract farming. In the livestock sector, for example, more than 50 per cent of the world's pork and 66 per cent of the world's poultry and egg production now takes place on industrial farms, which are generally either owned by large meat corporations or under contract to them. In Brazil, 75 per cent of poultry production is under contract; while in Vietnam, 90 per cent of dairy production is under contract. Contract production is also expanding for export commodities such as cacao, coffee, cashews and fruits and vegetables. It is even on the rise for staple foods, such as wheat and rice. In Vietnam, 40 per cent of rice production is farmed under contract with companies.
>
> (Grain, 2010)

Both the environmental and energy crises are implicated in land grabbing because of the promotion of industrial biofuels (agro fuels) which amounts to a veritable biofuels bonanza. This new biofuel adventure is truly industrial in scope: it involves the corporate sector; has high capital and energy inputs; often involves mono-culture cropping systems; has global commodity chains and involves speculative ventures. In some manifestations, it is very akin to traditional plantation production associated with tobacco, cotton, sugarcane and bananas in the tropics.

The United States and the European Union have been particularly keen to subsidize and promote industrial biofuels and justify their efforts with reference to the climate change agenda. 'In 2006 it was conservatively estimated that the EU biofuel industry was supported by financial incentives to the sum of €4.4 billion (which will rise if present trends continue) to about €13.7 billion in 2020' (Action Aid, 2010: 2). In the United States, approximately 30 per cent of maize (corn) produced is converted to ethanol, and it is estimated that '[C]orn was the largest recipient of US subsidies, taking US$51 billion of taxpayers' money between 1995 and 2005' (Lawrence, 2008: 26). In both of these cases, interventions to support industry were rationalized with reference to the green debate.

Other governments investing significantly in the sector include Brazil, China, India and Malaysia, which is the largest producer in Asia. In Africa, South Africa is the leading investor, but Nigeria, Ghana, Tanzania and Kenya are also involved. The attractiveness of sub-Saharan Africa to much of this investment has encouraged some to liken the region to a new 'green' OPEC (Organization of Petroleum Exporting Countries); envisioning a green future for all with ample production of biofuel crops from Africa serving to deliver the green agenda for Western petroleum junkies. How green, how laudable, perhaps not?

Many sceptics see other motives at its heart, namely, to reduce petroleum dependency and to generate new business opportunities. Some economists, corporations and financial arms of the BWIs view biofuel production in the global south as an attractive development opportunity. Debates surrounding biofuels and their potential contribution to modifying climate change and rural development have become very highly charged, but the most heated is the relationship between biofuels and food production. Any examination of the conflicts and tensions that typify these debates exemplify some of the core themes of this text: conflicting interests between the north and south and all manner of diverse interests within both of these categories; conflict and competition over resources, especially land (ownership and use), water and labour; tensions between different models of development; and the complicated and contested relationship between food security and food sovereignty; and then there are awkward questions allied to environmental policy. International and local environmental issues are embroiled in the promotion of the biofuel business.

Some of the best critical assessments of the process are available from international non-governmental organizations (INGOs). Grain (2009) first alerted the world to what was occurring and since then a number of excellent analyses have been published (Action Aid, 2010; Friends of the Earth, 2010; Christian Aid, 2009; and Oxfam, 2008). All these reports expose some of the damaging impacts the enthusiasm for industrial biofuels is perpetrating in poor communities and environments. Echoes from the past are loud and clear in their discussion as stories about enclosures, dispossession and exploitation abound. Food insecurity is exacerbated because people lose their entitlements to land and common property resources. There are more issues of concern.

One of the most serious associations is between the promotion of biofuels and food prices. Various proportions of wheat, sweet sorghum, maize and cassava, as well as oil seed crops, palm, soy and rapeseed, historically cultivated as food crops, are diverted for ethanol production, thus reducing the food available and increasing prices. The Food and Agricultural Organization (FAO) estimated that world cereal harvests in 2008–2009 reached record highs. However, in that year, more cereals went to feed animals and for industrial uses than to feed people. This core conflict was neatly summarized by Zoellick, president of the World Bank, in 2008:

> While many worry about filling their gas tanks, many others around the world are struggling to fill their stomachs. And it's getting more difficult every day.
>
> (Quoted in Action Aid, 2010: 3)

The problem is exacerbated when land moves out of producing food crops and is devoted instead to biofuel crops. With other resources too, competition intensifies, and several researchers have identified competition over water as a serious problem in some regions.

Action Aid harbours no doubts and simply accuses the drive for industrial biofuel as being directly responsible for the increased incidence of hunger in the developing world. They state:

- Fuel, energy and food prices are now increasing linked.
- At a global level, relatively small changes in agricultural supply or demand has a large and disproportionate effect on food prices, and so the increased demand for biofuels has driven food prices higher.
- Higher national food prices have filtered down to the local level where the majority of households are net buyers.
- This has been compounded by food growers changing over to biofuels in the anticipation of higher returns, and farmers leaving their own land uncultivated in order to work on biofuel plantations.

(Action Aid, 2010: 13)

Further damning evidence exists to link biofuel development to hunger and malnutrition, which explains why it is so vilified by some activists. As land value rises, the poorest and most vulnerable of rural inhabitants are displaced. Access to land is still essential for billions of poor across the developing world. In Africa, it is estimated that 60 per cent of the population relies on farming for survival. The profits associated with biofuel production have precipitated some unscrupulous land grab stories, where local elites and politicians deliver land without consulting the communities that have depended on it for generations.

The final global crisis that helps explain the global land grab is the food crisis of 2008–2009 when world food prices soared and the number of hungry people in the world escalated by some 100 million. This crisis was explicitly linked to the biofuels question. As food prices soared, many countries that were highly dependent on food imports suddenly appreciated their vulnerability to such commodity fluctuations. Interestingly, such dependency has intensified through the 1980s and 1990s as many countries in the global south were encouraged by the BWIs to shift resources from domestic food production to food for export markets, thereby deepening their exposure to world food prices. This policy has been called the 'Devil's Bargain' by former US president Bill Clinton, one of its earlier proponents.

Changes in world food prices, especially of staple food crops (wheat or rice) are very serious, politically, economically and socially, for any country largely dependent on the purchase of these on the open market. Since the widespread liberalization of agricultural markets in the 1980s and 1990s, many more countries fall into this highly dependent category. Experts expect such price rises to become more common as oil prices rise and climate change initiates changes in productivity.

In 2010, a similar problem emerged, and world wheat prices increased by more than 70 per cent within seven weeks (August–September 2010). Several factors help explain the price rises in 2010; poor harvests in Russia forced the government to ban all wheat exports, and floods in China, India and Pakistan meant yields of staple crops were lower than expected. While for consumers in the United Kingdom food price hikes are difficult for some low-income families, for many poor people in the developing world, such price hikes can spell disaster and hunger. Consider the wide disparities in the figures illustrated in Figure 4.3 and Table 4.5.

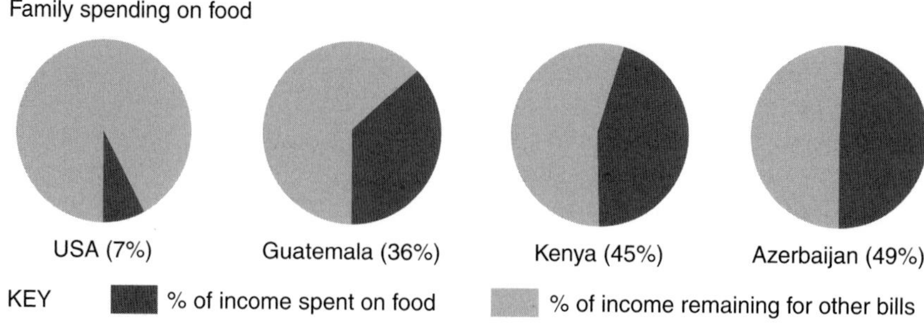

Figure 4.3 Percentage of household expenditure on food.

Table 4.5 ***Percentage of household expenditure on food, by selected countries, 2006***

Country	*Percentage of income spent on food*
Singapore	8.1
France	13.9
Bolivia	29.1
Nigeria	40.7
Azerbaijan	51.6

Source: adapted from material at: www.ers.usda.gov/briefing/cpifoodandexpenditures/data/2006table97.htm

Faced with these realities, some countries have decided to invest in land elsewhere to secure their access to adequate food stocks. 'Farming abroad' is viewed as an alternative way to secure food rather than being dependent on volatile world prices. This may be a feasible strategy for those governments with ample capital to invest, but for many more this is not an option; for them, selling that land may be attractive (Table 4.6).

If we examine the number of private companies investing in foreign land ownership, list A becomes much more extensive. Table 4.7 shows some corporate interests from the sugar business alone.

Table 4.8 shows some major corporations and their investments in farming activities across the globe. Some of these names will be familiar; the four giant trading and processing corporations are listed as major players. In addition, all the others are large corporations.

Table 4.6 ***Land deals: some major participating countries***

Source of investment funds	*Destination of investment funds*
A	***B***
Gulf States	Ethiopia
South Korea	Sudan
China	Mali
Japan	Mozambique
Libya	Philippines

Table 4.7 ***Major European sugar corporations investing in overseas production***

Company	*Countries*
Associated British Foods (United Kingdom)	China, Malawi, Mali, Mozambique, Swaziland, South Africa, Tanzania, Zambia
Tereos (France)	Mozambique, Brazil
Sudzucker (Germany)	Mauritius
JL Vilgrain (France)	Cameroon, Chad, Republic of the Congo
Tate & Lyle (United Kingdom)	Egypt, Laos, Zimbabwe
AlcoGroup (Belgium)	Brazil, Mauritius

Table 4.8 ***Some agricultural commodity trading companies investing in farms***

Corporation	*Home*	*Products*
Cargill	United States www.cargill.com	Palm oil, sugarcane, diary, cattle, poultry, pigs, aquaculture
Olam	Singapore www.olamonline.com/locations/world.asp	Dairy, almonds, palm oil
Bunge	United States www.bunge.com	Sugarcane, cereals, oil seeds, cattle
Louis Dreyfus	United States www.louisdreyfus.com/index.html	Sugarcane, cereals, oranges
Mitsui	Japan www.mitsubishicorp.com/jp/en/about/mc	Cotton, dairy, oilseeds, cereals, poultry, shrimp
Glencore	Switzerland www.glencore.com	Oilseeds, cereals
ADM	United States www.adm.com/en-US/Pages/default.aspx	Sugarcane, palm oil
Noble Group	Hong Kong www.thisisnoble.com	Oilseeds, cereals
Charoen Pokphand	Thailand www.cpthailand.com	Pigs, poultry, aquaculture, fruit and vegetables, palm oil
Wilmar	Singapore www.wilmar-international.com/about_index.htm	Palm oil, sugarcane

Source: adapted from www.grain.org/seedling/?id=693

It is interesting to visit their websites and learn about their global activities.

The arguments for such investments are usually that they 'promote development', and certainly they provide capital and technology to marginal farmers. The arrangement facilitates the introduction of new or improved crops and ensures access to new global markets for farmers. Perhaps the greatest danger is the extreme power differentials that typify such contracts and the fact that the corporation manages to burden the farmer with all the risks of the enterprise while they enjoy the bulk of the profits, not a bad bargain for the corporation.

Some view these deals as positive for all involved, arguing that such investments will improve farming practices and promote local economic growth. Many are more sceptical and understand the strategy to be neo-colonial in character, with relatively wealthy groups of individuals effectively exploiting the land, labour and resources of poorer communities. While some of the transactions occur between southern countries, South Africa buying land across the region, or Mauritius buying in Mozambique, etc., the vendors in most cases are relatively powerless and may be 'represented' by unscrupulous politicians or government officials. In some of the worst-case scenarios, local people have been intimidated, cheated and dispossessed. Understandably, many local and international NGOs are very worried about these deals and their implications for local food security. It is notable how many of the land sales occur in countries currently suffering from severe or moderate malnutrition, and it is hard to understand how these strategies are going to improve the condition of the very poor, landless or peasant farmers. However, the problem is not confined to the developing world; Australia, Ukraine, Argentina and even the United States, among others, may offer opportunities, as the following quote suggests:

> The sweep of agricultural land grabs has stripped small farmers in Africa, Latin America and Asia of control over vital tracts of fertile land. And quietly, these modern-day land marauders are coming to Canada – undermining family farms, compromising local food sovereignty, and harming the environment.
>
> (Miller, 2010)

Corporations are major participants in this global activity, but some countries are also significant players, usually in association with the corporations. In September 2010, the World Bank published a

Briefing Paper about this activity and concluded that: 'The veil of secrecy that often surrounds these land deals must be lifted so poor people don't ultimately pay the heavy price of losing their land' (World Bank, 2010). Activists have been scathing about the Bank's conclusions and accuse it of suppressing the serious implications of these policies behind 'smoke and mirrors' (Grain, 2010a; Smoke, September 2010).

Globalization and the nutritional transition: another corporate connection

The discussion so far has examined global changes in the production and distribution of food, and the final aspect to review is changes in food consumption habits and its association with corporate globalization. This discussion suggests more links between corporate control of the food system and our eating habits and, therefore, our health. Roberts (2008) in his excellent book reviews numerous changes in the food system and how they are implicated in the obesity epidemic in the United States. He describes how 'the meaning of food is being transformed: food cultures that once treated cooking and eating as central elements in maintaining social structures and tradition are slowly being usurped by a global food culture, where cost and convenience are dominant, the social meal is obsolete, and that art of cooking is fetishized in coffee-table books and on television shows' (p. xii).

My only quibble with this quote is with the suggestion that the process is slow. I think that nutritional transition is occurring across the global south at an unprecedented rate. This aside, all of Roberts observations about the role of the corporate sector in the obesity epidemic in the United States are equally valid for most countries of the global south except the very poorest, and even there, similar problems are emerging, especially in cities (see Chapter 2). Roberts describes the low-cost, high-volume food system that emerged, which helps explain the corporate emphasis on highly processed foods that deliver a good profit margin.

As our diets change, and our battle with obesity intensifies, global profits keep rising for the food manufacturers and retailers. Wilkinson (2009) identifies the 1970s as the start of significant corporate penetration of the developing world's food systems, as part and parcel

Plate 4.4 Healthy diets (Source: Wikimedia Commons).

of the liberalization of markets outlined earlier. He explains that markets in the developed world were near saturation, so corporate attention shifted to markets in the developing world 'where higher demographic growth rates and rapid urbanization were creating ideal conditions for food corporations to offset the slowdown in growth in developed country markets' (Wilkinson, 2009: 1).

At first, FDI focused on producing raw materials for export, a process that has roots in colonial economic relations, but in the 1980s a 'shift occurred and corporations started to invest in processing foods for the host markets, as transnational food corporations such as PepsiCo and Nestle invested in foreign manufacturing facilities for foods such as soft drinks, confectionary, dairy products, baked goods and snacks' (Hawkes, 2006: 6). Hawkes explains that FDI in food processing expanded, between 1980 and 2000, from US$9 billion to US$36 billion, an enormous growth rate. The picture is the same for investment in food outlets, in both supermarkets and restaurants. These facts help explain the rapid shifts in consumption patterns associated with the nutritional transition described in Chapter 3, but this is only one part.

Equally, or more important, must be the role of corporate marketing and advertising. When reviewing globalization and food retailing, it is important to understand the power and influence of food advertising, which is very effectively transforming the food we purchase. Advertising campaigns by the big food producers are furthering the nutritional transition and encouraging us to eat more meat and dairy products and consume more highly processed foods and drinks. These highly processed foods and drinks offer food manufacturers the best profits but are increasingly understood as part of the obesity problem. Roberts (2008: 38) explains that successful brands are not easily created but must be promoted with expensive advertising – 'clever packaging, celebrity endorsements, event sponsorship, coupons and relentless advertising'.

We considered the breastfeeding debate in Chapter 2. This issue illustrates how the corporate food sector intervenes early to influence our diets. Research from around the world suggests the role that advertising has on dietary change. The value-added element of food increases exponentially with the increase in the amount of food processed, and advertisements are vital to transforming our inclinations to buy and eat the stuff from their factories. Lawrence examines the history of advertising in the food processing industry in the United States and (or 'nutritional corruption industry', as she calls it). In her analysis of breakfast cereals, she describes how, from its beginnings, the food advertising industry has traded on 'our insecurity about health, manipulating our emotions' which has always been part of 'the great puff':

> Getting children hooked, making them associate breakfast cereals (and all food) with fun and entertainment, blurring the lines between advertising and programmes, exploiting new media – today it is the internet and viral marketing – was one of the main aims of competing manufacturers from the early days.
>
> (Lawrence, 2008: 19)

All manner of strategies are employed to change eating habits. 'It involves, not just advertising but a whole array of methods including sales promotions, websites, viral marketing, music and sports sponsorship, product placement in films and television, and in-school marketing' (Hawkes, 2006: 9). The emergence of the obesity pandemic is testament to the success of these campaigns (Box 4.7). New global corporations are now associated with the food industry, advertising and marketing agencies. Energy-dense, highly processed

Box 4.7

Opinion editorial by Stacey Folsom

Even with concerns about childhood obesity and related health conditions reaching new heights, fast food mascot Ronald McDonald remains the most recognizable icon to children around the world – second only to Santa Claus. But the clown is at a crossroad.

Food blogger David Koeppel recently asked whether Ronald is a 'beloved corporate mascot' or a 'sinister huckster who gets kids hooked on junk food'. The evidence indicates he is both. And because it's impossible to separate the former from the latter, it may be he should be neither. McDonald's makes its trademark clown ubiquitous and is proud of it. An animated sequence on the corporation's homepage begins with the caption, 'He's here. He's there. Man, the guy's everywhere'. He's on Saturday morning television ads. He's on Happy Meal boxes. He's featured in internet games.

And when we at Corporate Accountability International launched WheresRonald.org to gauge whether or how Ronald was hooking kids on food high in salt, fat and sugar, the clown turned up in a wide range of places in person. Ronald appeared at elementary schools to promote physical activity, at libraries to promote reading, and children's fairs to teach kids about bullying. Add in the Ronald McDonald House Charities and it's hard to see the evil in this icon.

Yet, what if we consider him in the context of a less 'beloved' character—one retired more than a decade ago: Joe Camel. What if Joe Camel had visited schools encouraging kids not to smoke? Would parents have found that acceptable, let alone sincere? I doubt it.

When the face of a corporation, and an industry, responsible for driving an epidemic of diet-related disease wants to serve as a 'health ambassador' to public schools, how is that different? As McDonald's executive R.J. Milano put it in 1998, the corporation wants, 'to own every kid transaction out there'. It has also boasted about promotions targeted at kids as young as two. Ronald remains at the center of these promotions, which are also aided by the more than 8,000 playgrounds McDonald's operates around the country. Still CEO James Skinner insists that, 'Ronald has never sold food to kids in the history of his existence'.

Right. Maybe he believes in Santa Claus too! Why make such an outlandish claim?

Well, for one it's better for sales when the clown's benevolent, philanthropic image is separated from the crass, public relations machine that ringleads his appearances. McDonald's executives know that barraging children with the clown's images from a very young age creates powerful brand identifications that can last a lifetime. Until the age of about eight, children are generally unable to understand that marketing is an attempt to sell them a product. And one 30-second commercial can influence the brand preferences of children as young as two.

McDonald's also takes full advantage of the 'pester effect'. They know that even the most responsible and strong-willed parents will cave to their children's nagging if it is persistent enough.

The company's most problematic efforts that the Where's Ronald hunt found bypass parents entirely. You'd imagine, for example, there are ways children could learn to read, lead healthy lifestyles, and get along with their classmates without being marketed fast food.

It begs a similar question of Ronald McDonald charities. Does an icon so associated with the spike in health problems related to diet have a place in caring for sick kids? Aren't there other clowns or mascots who could bring similar cheer, without the mixed messages?

Ronald McDonald's identity crisis isn't new. That's why the corporation is increasingly trying to associate him with 'healthier choices' on its kids' menu. But the fact is that a standard Happy Meal is still a hamburger or chicken 'McNuggets', fries, and a soda. The 'healthier' side offering is a pile of apples with a sugary caramel dipping sauce. Chocolate milk is an alternative beverage.

Instead of endeavoring to reinvent Ronald, it's time McDonald's did some corporate soul-searching, asking itself hard questions about the dubious practice of using a clown to promote both junk food and its charities.

Meanwhile, parents can pose these same questions to the schools, hospitals, and libraries where Ronald all too often appears.

Stacey Folsom is Corporate Accountability International's national spokesperson. Corporate Accountability International has been waging winning campaigns to challenge corporate abuse for more than 30 years.

Source: www.otherwords.org/articles/ronald_mcdonalds_identity_crisis (accessed 19 December 2011)

foods with high levels of sugar and fats are eaten by more people than ever before. Often, these foods are cheap because of the system of subsidies discussed in the preceding sections. Food manufacturers have been very effective at marketing this diet as, like smoking and alcohol, it is associated with affluence and modernity. Patel (2010) makes the point with reference to the famous McDonald clown:

> More sales for the clown mean bigger returns for Cargill and Tyson's factory farms, Archer Daniels Midland's high-fructose corn syrup processing plants, and Monsanto's pesticide production facilities. And it's

> our tax dollars that go into everything from the cheap commodities that they depend on, to the small business loans and tax credits that allow fast food franchises to breed in and around our schools. For these subsidies, and for the lax regulations around health and advertising to children, the fast food industry has spent millions in lobbying fees, and aggressively courted political favor. Ronald McDonald may have a big smile, but his shoes are steel-tipped.
>
> (Patel, 2010)

Conclusion

This chapter has examined some of the crucial factors at the global level that are responsible for altering the international food system from farm to plate. It has described and explained the comprehensive changes in how food is produced, distributed, manufactured and sold. To return to the discussion in Chapter 2, this chapter analyses the structural determinants at the global scale that explain changes in the nature and patterns of food produced and consumed. The next chapter explores how these processes, at the global level, are modified at the national level to explain the geographical variety that still exists in food consumption across the world. In other words, it examines the national-level factors that we must understand to explain the variety of malnutrition and how it varies between nation states.

Summary

- **Globalization is a complex and controversial concept.**
- **Contemporary corporate globalization can be differentiated from generic globalization.**
- **Contemporary corporate globalization is socially constructed and reflects dominant interests and ideologies in the global political economy; it is unstable and can be resisted and changed.**
- **Three broad perspectives on globalization can be identified: hyperglobalist, sceptic and transformationist. Each perspective represents different ways of understanding the process, its nature, significance and impacts.**
- **Contemporary globalization has several characteristics which differentiate it from globalization in the past.**
- **Corporate globalization has transformed the way food is produced and where it is produced. It has also transformed the type of food consumed and, frequently, where it is consumed.**

- **The global food business is very concentrated, and a great deal of power rests with a few global corporations who dominate the global food chain.**

Discussion questions

1. Review some aspects of globalization that have proved contentious.
2. Discuss the assertion that contemporary globalization is corporate in character.
3. Access the impact of recent trade liberalization to food security with specific examples.
4. Describe and explain some of the objectives of the BWIs.
5. Evaluate how contemporary globalization has influenced diets in the global south.

Further reading

On globalization in general:

Jan Aart Scholte (2005) *Globalization: A Critical Introduction*, [Basingstoke, Palgrave Macmillan]. This is an accessible and comprehensive overview of globalization.

Goldblatt, D, Held, D., Perraton, J. and McGrew, A. (1999) *Global Transformations*, [Cambridge, Polity Press]. This is an essential text for anyone trying to appreciate the complexities and debates surrounding the concept.

Richard Appelbaum and William Robinson (eds) (2005) *Critical Globalization Studies* [London, Routledge]. This is a good collection of essays from some of the most important analysts of globalization and its nature and impacts.

Hirst, P., Thompson, G. and Bromley, S. (2009) *Globalisation in Question, 3rd ed.* [London, Polity Press]. Another essential reading for those who really what to interrogate the process and if indeed there is a global economy that is significantly novel from those in the past.

On globalization and food in particular:

Roberts, P. (2008) *The End of Food: The Coming Crisis in the World Food Industry,* London, Bloomsbury. This is a really good read, excellent in depth and breadth. Roberts details the many and varied problems with the global food system and its vulnerability to numerous shocks.

Patel, R. (2007) *Stuffed and Starved*, London, Portobello Books. An excellent read which details many of the battles for the world food system. The author presents an excellent dissection of corporate control across the food chain.

Madeley, J. (2000) *Hungry for Trade*, London, Zed Books. This remains a very useful examination of trade and food security. It includes excellent material for undergraduates who are new to international political economy.

Shiva, V. (2000) *Stolen Harvest*, London, Zed Books. Another good text and an excellent brief introduction for students to some of the serious injustices perpetrated by corporate globalization.

Useful websites

www.polity.co.uk/global/default.asp This accompanies the Global Transformation text and is an excellent source for many globalization-related themes.

www.undp.org UN Development Programme, Human Development Report (various annual and regional). Statistical data about poverty and inequality can be found at this site, in addition to numerous research and policy papers.

www.worldbank.org The important World Development Report (Oxford, OUP). The website is much more user friendly now, and there is easy access to a great deal of data for most countries. In addition, useful, briefs, reports and videos and maps are available. Considerable country-specific material is available. For the themes discussed in this chapter, the recent conference is relevant at: http://live.worldbank.org/open-forum-food-crisis/response

www.who.org The World Health Organization provides useful material on global health inequalities. Special themed reports and publications, including material on nutrition and health, are available. It includes a considerable amount of material on HIV/AIDS and nutrition.

www.ifad.org This is an indispensible site. The International Fund for Agricultural Development (IFAD), a specialized agency of the United Nations, was established as an international financial institution in 1977, as one of the major outcomes of the 1974 World Food Conference. This is an excellent site for discussion papers and case studies. The website has a predominantly rural emphasis, but wide-ranging themes are considered in depth. The most recent publication of relevance, The Rural Development Report, is here at: www.ifad.org/rpr2011/index.htm

www.grain.org/front This is another very useful site. It includes a great deal of material examining the impacts of corporate globalization of small-scale producers in the global south.

www.panap.net/en/p/page/pesticides/32 To learn more about the controversies that surround the marketing of pesticides and their less favourable impacts, visit this Pesticide Action Networks site: http://landportal.info/sites/default/files/2011-04-perspectives_southern_africa_01_11.pdf (accessed 18 July 2011). This website has interesting material about land grab and other issues in the African context.

Chapter 4

Follow Up

1. A great deal of material is available on the web about globalization, but some of it is not very useful. However, there is excellent introductory material associated with the book *Global Transformations* by Held *et al.* (1999), in which one can find some theoretical material about globalization. *Global Transformations* is available at the support web page for the text at: www.polity.co.uk/global/research.asp

    - There are six indicators of globalization detailed on this page; select one to research in more detail.

2. Globalization is a very widely debated subject; a review by Martell (2007) explores some of the core debates for those who are interested in understanding the process in more depth. This discussion is available at: www.sussex.ac.uk/Users/ssfa2/thirdwaveweb.htm

    - What various perspectives does the author identify?
    - Why is an understanding of globalization relevant to the themes in this book?

3. See the presentation by Grain (2009) about what they have labelled the 'land grab' phenomena at: www.grain.org/o_files/landgrabbing-presentation-11-2009.pdf

    - Construct a table to illustrate who benefits and who may lose as a result of the process detailed in the presentation.

4. Review the variety of issues identified by the Transnational Institute (TNI) as implicated in food insecurity at: www.tni.org/category/issues/global-economic-justice/food-agriculture

5. Read a critical assessment of the new trade deals between the European Union and some countries in Latin America at: www.tni.org/sites/www.tni.org/files/download/Fritz-2010_The%20Second%20Conquest_Colombia-Peru-EU-FTA.pdf

    - Why are these analysts worried about the trade deals detailed?

5 National perspectives

Learning outcomes

At the end of this chapter, the reader should be able to:

- **Appreciate that national contrasts in malnutrition remain extreme**
- **Understand the role of historical processes to contemporary patterns of malnutrition in several national contexts**
- **Understand the potential role of national governments in reducing malnutrition (both overnutrition and undernutrition)**
- **Appreciate the roles of different models of development for malnutrition**
- **Understand the factors that help explain the nutritional transition**

Key concepts

Colonial legacies; property relations; plantation economics; core and periphery; liberalization; food sovereignty; food security; Import Substitution Industrialization; development theory

Introduction

The last chapter reviewed some of the most important global-level factors that influence what we eat and who eats what; these are known as macro-structural determinants of the food system and are considered very important in the analysis presented in this text. Global processes have been changing socio-economic characteristics and therefore our diets for centuries, but they have intensified in the recent decades. While very important, they still represent only one scale of analysis, and this chapter continues our examination of structural variables that influence the food system but focuses instead on national-level variables. Article 25 of the United Nations

Declaration of Human Rights and Article 11 of the International Committee on Economic, Social and Cultural Rights (ICESC) enshrines the right of everyone to an adequate standard of living, including food. States that have signed these agreements obliges them to 'ensure for everyone under its jurisdiction access to the minimum essential food which is sufficient, nutritionally adequate and safe, to ensure their freedom from hunger'. This chapter examines why some states, in the early twenty-first century, singularly fail to fulfil their obligations in this regard.

It is often very difficult to discern the boundaries between international and national decision-making; international actors often form alliances with nationally potent actors (consider the land grab theme in Chapter 4 where international and national interests are often implicated in the deals secured), and of course the context of national decision-making is determined by external factors. As Craig and Porter note about the contemporary modes of development, '[T]hus has capitalism always spread, as powerful external trading interests have reshaped and aligned with internal political and economic interests to de-territorialize places, in order to open them up and secure them for trade' (2006: 252). It is important to recognize the complexity of 'national' ambitions. Even in 'accountable' liberal democracies, such 'national' interests rarely give voice to the needs of the marginal or dispossessed; the prospects of them being articulated in places devoid of any democratic accountability is much less.

However, the distinction is both useful and meaningful because there are powerful national actors that play a part in designing national economic policy, including all those associated with food entitlements and diets. While recognizing that globalization has significantly altered the locus of food-related decision-making, most activists still target national governments in their efforts to improve food security and food sovereignty. National politicians are generally more accessible than bureaucrats in the WTO or in the headquarters of regional trading blocs; they may also be more accountable in democratic contexts.

Global policies are always mediated by national processes, which helps explain why globalization is experienced in diverse ways in different countries. One of the dangers associated with simplifications of the globalization process, by hyperglobalists, for example, is that they fail to appreciate the vital role that may be played by national actors and contexts. One of the reasons that global geographical diversity remains salient is because of the role of national-level processes (UNDP, 2010: 45–64). This chapter attempts to understand what national-level factors may be useful in understanding the following questions: Why does undernutrition remain a serious obstacle to development in many countries in the global south, and why does it persist in some developed countries? What countries have been successful in reducing undernutrition, and what policies help explain their success? Why is obesity now regarded as the most pressing public health problem in some developing countries, including China, when only a few decades ago they were struggling to overcome mass undernutrition? What relevance does the concept of food sovereignty have in an era of globalization, and how does it differ from traditional concerns about food security?

This chapter opens by revisiting some statistics that illustrate the great variety that exists between countries in different world regions. These statistics are evidence that, while globalization is intensifying, it is certainly not eroding national variety or leading to a homogeneous world; rather, differentiation is increasing and stark contrasts still exist between nation states, as well as within them. Statistics for traditional malnutrition, associated with not having enough to eat, are considered in addition to obesity rates. The discussion then turns to trying to understand these statistics by examining what national-level factors help explain why some people have too little in quantity (protein, calories or vitamins and minerals), and why some have too much poor-quality food (rich in sugar, fat and salt).

National contrasts

The first variable considered is the role of history. It is simply impossible to understand the diversity of malnutrition rates within or between countries without appreciating the diverse histories of incorporation into the world economy. The relevance of historically informed analysis was introduced in Chapters 2 and 3, but additional factors require further consideration. Specifically, development theory is examined in more detail to understand its influence in domestic policy and consequent socio-economic trends. This includes diverse national strategies towards globalization, and how these alter access to food and the nature of diets. We revisit the concept of 'food sovereignty' and 'food security' to understand some changes in

Table 5.1 ***A very diverse picture: percentage of undernourished population, selected countries***

Country	*1990–1992*	*2005–2007*
China	18	10
Mongolia	28	26
Cambodia	38	22
Philippines	24	15
Vietnam	31	11
Bangladesh	38	27
India	20	21
Pakistan	25	26
Sri Lanka	28	19
Tajikistan	34	30
Uzbekistan	5	11
Georgia	58	12
Nicaragua	50	19
Ecuador	23	15
Peru	27	15
Kuwait	20	5
Yemen	30	31
Chad	60	37
Democratic Republic of Congo	26	69
Ethiopia	69	41
Tanzania	28	34
Malawi	43	28

Source: adapted from 'The State of World Food Security' (FAO and WFP, 2010), available at: www.fao.org/docrep/013/i1683e/i1683e.pdf (accessed 20 October 2010)

recent years. Throughout, case studies are employed to exemplify the main issues under consideration.

The idea that globalization is creating a homogenous world is readily repudiated with reference to trends in malnutrition in the global south. The statistics in Table 5.1 refer to traditional malnutrition; that is, people who do not have access to an adequate diet. The figures reflect the fact that some countries have had some success in reducing traditional malnutrition, while others have failed miserably and are even experiencing an increase in undernutrition, and what factors help explain this diversity?

Consider the following statistics that illustrate obesity in the same countries.

Table 5.2 ***A very diverse picture: percentage of obese adults, selected countries***

Country	*Male*	*Female*
China	2	3
Mongolia	7	12
Cambodia	–	1
Philippines	3	6
Vietnam	0	0
Bangladesh	–	2
India	1	3
Pakistan	–	–
Sri Lanka	–	7
Tajikistan	–	7
Uzbekistan	5	7
Georgia	–	–
Nicaragua	33	35
Ecuador		15
Peru	11	12
Kuwait	36	48
Yemen	–	–
Chad	–	1
Democratic Republic of Congo	–	2
Ethiopia		1
Tanzania	–	–
Malawi	-	2

The most surprising cases are, perhaps, Mongolia, which suffers from a high incidence of both forms of malnutrition, and the Latin American countries, with their relatively high levels of obesity. Less than 20 years ago, this region was characterized by high levels of undernutrition. It is worth commenting that, although the percentages for obesity are low in China, this actually represents a large number of people and, as discussed in Chapter 2, obesity is now considered the most important public health crisis facing that country. Also, the percentages are rising very rapidly in China and the rest of the emerging market economies. Table 5.3 has more surprising examples, and confirms that emerging market economies and countries which have seen rapid economic growth in the last 20 years are catching up with the situation in more advanced economies. Nicaragua and

Table 5.3 ***Surprising patterns: percentage of obese adults, 2000–2010***

Country	*Males*	*Females*
Egypt	18	39
El Salvador	–	26
Guyana	14	27
Iraq	26	38
Kuwait	36	48
Malaysia	14	19
Mexico	24	35
Saudi Arabia	28	44
Nicaragua	33	35
Uruguay	18	22
United States	31	32
United Kingdom	24	24

Source: adapted from World Health Organization, 2010 at: www.who.int/gho/database/WHS2010_Part2.xls (accessed 22 October 2010)

Kuwait now exhibit higher obesity rates than the United States, a startling statistic that exemplifies how rapidly and profoundly the world's diets are changing.

Colonial legacies: implications for entitlements and undernutrition

To examine the national-level factors that help explain these statistics, it is useful to revisit the factors detailed in Table 1.3, which summarizes the core factors, historical, economic, political and social, at the national level that help explain malnutrition. It is important to realize that understanding patterns of malnutrition is about explaining patterns of poverty and privilege, because hunger and malnutrition are closely correlated to socio-economic circumstances. They cannot be explained with reference to individual characteristics. So, our first task is to appreciate how historical processes are implicated in contemporary patterns of poverty and privilege.

The following review is selected to point to a few of the most pertinent examples of how history is implicated in contemporary patterns of food provisioning. It is obviously impossible to do justice to the great variety of relevant colonial legacies, food cultures, demography, cultures, etc., all of which are of interest, and indeed of considerable relevance, in various contexts. In addition to general histories of colonialism, excellent social histories exist which explore the history and geographies of specific foods, and students should read these for their detailed insights (Salaman and Hawkes, 1985; Mintz, 1986; Kurlansky, 1999, 2003; Abbott, 2010).

One of the most significant factors that help explain the shape of the world in the twenty-first century is the history of European colonialism which emerged in the late fifteenth century. Certainly, it is important at the outset to recognize the very varied character of colonialism and therefore its diverse impacts which also vary by ethnicity, class, gender, region etc. It is important to recognize that 'the local textures of colonialism were immensely complex, but as a component of a broader, changing global process, there were immense overall repercussions' (Potter, Binns, Elliott and Smith, 2004: 73).

Many of these repercussions still explain the diversity of entitlements in some contexts and the contours of socio-economic circumstances: whether it be access to land; agricultural and industrial policies; urban bias and urban planning; multiple and often conflict-ridden ethnic circumstances. Perhaps, an inclination to presume that 'West is best' in all habits can be observed: smoking is increasing in the developing world and emerging markets; alcohol consumption is increasing rapidly; family structures and sexual habits are changing; and of course central to our concerns, Western diets are being adopted as the 'nutritional transition' gathers pace.

There are other less obvious legacies that are relevant too. Top-down health systems, with an emphasis on curative rather than preventive medicine, were established in colonies patterned on the shape of health care systems in the metropolitan cores; similar education systems were adopted where a small privileged elite, imbued with Western ideas created centralized, hierarchical and bureaucratic systems of government. A selection of examples is considered in the following to suggest some of colonialism's most significant legacies for the construction of entitlements and contemporary political and socio-economic characteristics.

When European expansion occurred in the sixteenth century, it encountered already established empires and economies that interacted with European contact in very different ways. Some areas were colonized and some, although never formally colonized, were nevertheless seriously altered by contact with European commerce and economic ambitions. Here, it is important to stress that diversity within the developing world has always existed, and responses to European and later American or Soviet hegemony during the Cold War were varied. Differential responses to globalization also helped generate a geographically variegated world. Such variations existed because the nature of capitalism, and its needs and objectives, have changed through the centuries and are always mediated by local circumstances (Bernstein, 2000; Bujra, 2000; Knox, Agnew and McCarthy, 2003).

The commercial pressures exerted in the early years of European expansion by Spain and Portugal, in Latin America, for example, were significantly different to the pressures exerted by industrial capitalism exemplified by British colonialism in the nineteenth century in South Asia and in sub-Saharan Africa. The nature of both capitalism and colonialism varied and created very diverse economic, political and institutional conditions within the European sphere of influence (Bernstein, 2000). This chapter can only summarize some broad processes that are vital to explanations of contemporary nutritional diversity; such diversity, between and within all national contexts, existed in the past and still prevails.

Bearing in mind the above caveats, some generalizations are useful. It is certainly true that in every situation 'capitalist expansion entailed a forcible transformation of pre-capitalist societies whereby their economies were internally disarticulated and integrated externally with the world economy' (Knox, Agnew and McCarthy, 2003: 250). So, despite the contrasting circumstances across and between regions exposed to European contact since the late fifteenth century, we can identify some common characteristics that, although they may do an injustice to historical specificities, nevertheless help us establish some broad patterns.

Colonial patterns: cores and peripheries

Economic and political relations established between Europe and the rest of the world between the sixteenth and twentieth centuries are often categorized as core and periphery in nature, where Europe was the core and the rest of the world was its periphery. The most important character of this relationship was its asymmetry; wants in the European market determined economic activities and political fortunes in the periphery. This power differential was remarkably stable, with only a few countries managing to achieve core status (the United States and Japan in the nineteenth and twentieth centuries, respectively). The relevance of the core–periphery nature of international relations in the post-Second World War period helps explain what countries have been classified into in the last 50 years; the terms 'First World', 'Second World' and 'Third World' all suggest unequal relations between these categories of countries.

The dominant ways of categorizing countries have their roots in the development decades after the Second World War when the tripartite classification of countries emerged. Table 5.4 shows the basis for this classification and their associated political and economic characteristics. The core, the First World, was based in the Western liberal democracies, in Western Europe and the United States, while the Second World was based in the Soviet Union. The Third World was the developing world where, throughout the 1950s, 1960s and 1970s, the First World and Second World vied for influence.

It was in the Third World, the developing world, that most of the ideological battles were fought during the Cold War years. Sometimes these materialized as real conflicts (Vietnam, Central America, Mozambique, Korea, for example). These Cold War conflicts effectively prevented development in numerous countries, as resources were diverted to military objectives, and civil wars made them ungovernable for decades. The case of Mozambique is instructive. Its civil war represents a classic 'proxy war' where internal differences were exploited and exacerbated by external powers; it became the battleground for pro- and anti-Western factions. Anti-communist and pro-communist factions sustained a civil war, with assistance from external African powers and their respective allies from just after independence until 1992. It is only since the cessation of this conflict that the country has begun to enjoy stability, economic growth and some success at reducing their alarming undernutrition and mortality rates. They have been celebrated as a success story in the political commitment to nutritional improvements.

Since the collapse of the Soviet Union, and the economic emergence of the newly industrializing countries, the traditional categories have become more differentiated and less inappropriate in some cases.

Table 5.4 ***Country classifications, post-1950***

	First World (developed)	*Second World*	*Third World (developing)*
Internal Features			
Political system	Liberal democratic All of original OECD members	Single-party Communist rule Soviet Union, China, Cuba	Mixed Rarely democratic (except India) Most of Africa, Latin America and Asia
Economic system	Market-oriented With various degrees of state welfare provision	Centrally planned Economy generally following state-led economic plans (often 5-year or 10-year plans)	Variable
Income level	High but often polarized	Mixed; generally medium	Low and often highly polarized
Economic growth rate	High	Mixed generally medium	Low
External Features			
Main trading patterns	Other First World countries: NATO	Other Second World countries: COMECOM	First World countries and occasionally Second World alliances (non-aligned movement)
Geopolitical relationships to other Worlds	Geopolitical competition with Second World; colonial power over, aid donor to, and dominant over, most of Third World	Geopolitical competition with First World; aid donor to, and influential in, parts of Third World	Aid recipient; subordinate; but actively 'non-aligned'.
Influence in main international institutions	High	Low but militarily important	Low

Source: adapted from Harris, Moore and Schmitz (2009: 11)

However, the world, in general, and the development business, in particular, are still to a large degree locked into the old language (Harris, Moore and Schmitz, 2009: 9; see this discussion for a useful history of the classification system and some tentative suggestions for improved categorizations, given the new global realities). Core and peripheral patterns in international relations are fluid, and it may be that we are entering a period where the existing realities are radically altered (as has happened throughout history).

Most recently, the BRIC (Brazil, Russia, India and China) nations, as well as some other newly industrializing nations, have been challenging the traditional status quo by their growing economic and political power.These countries have certainly become very important regional players that threaten to outperform the 'old world powers' by the middle of the twenty-first century. It is important to note that the global economic system is inherently dynamic and change is inevitable, although the nature of such change is not. The categories, core and periphery, are also very broad and have always represented a great deal of diversity, and this has increased in the last 30 years. It is debateable whether any of the terms traditionally employed are relevant in the multi-polar world that has so altered the contours of economics and politics in the last 30 years. '[T]His new world is much more complex and difficult to capture in simple distinctions' (Harris, Moore and Schmitz, 2009: 7). However, the world is still divided between rich and poor nations, and the concepts of the global north and south are still used in development discourse, although they must be considered increasingly problematic.

In most colonial regimes, a few people controlled the majority of resources, and the great majority controlled very few. Sometimes, the economically and therefore politically powerful were new arrivals or had European ancestors; other times, they were selected from the domestic elite in the colonial possession. Economic change was guided by demand in Europe, and productive resources which had, until European contact, been devoted to supplying materials for the domestic, local and regional economies, were redirected to serve external European priorities. '[T]he making of colonial economies within the international division ohf labour occurred through the production of commodities for export, above all from extractive industries and tropical agriculture. The production of these commodities took different forms' (Bernstein, 2000: 253).

The next part of this chapter describes some broad patterns of economic transformation and integration with reference to case studies from across the global south. In all cases, we find resources exploited to serve external markets; the exact nature of exploitation varied spatially and temporally but can be traced from the early sixteenth century until today. It is important to note, however, that exploitative relations which we associate with these core–periphery relations cannot be construed as simply the core exploiting the periphery. There were significant players in the peripheral regions that benefitted from these exchanges and helped to maintain what, for the majority, were extremely exploitative trading relations. Without substantial gains among some classes within the periphery, resistance to exploitation would have been more widespread and effective. The discussion now examines some specific examples of how colonialism altered entitlements and socio-economic patterns. To help contextualize some of the most important changes that occurred through the centuries, a broad summary of the chronology of European colonialism and some of its most important characteristics is shown in Box 5.1.

Colonial legacies: property relations

Social movements with important peasant support led to revolutionary regimes implementing significant land reforms in Mexico, Bolivia, Cuba and Nicaragua. Similar processes produced massive land reforms in China and Vietnam. Popularly based insurgencies in Peru and El Salvador convinced nationalist military officers wielding state power to undertake land reforms. Important land reforms by authoritarian regimes in South Korea and Taiwan had partially similar origins. Democratically elected regimes in Puerto Rico, Guatemala, Venezuela and Chile all initiated important land reforms (Barraclough, 1999: 11; available online at: www.unrisd.org/80256B3C005BCCF9/(httpAuxPages)/9B503BAF4856E96980256B66003E0622/$file/dp101.pdf (accessed 22 July 2011).

The discussion about colonialism and its relevance to contemporary food security opens with a consideration of property relations. Across the global south, land remains politically charged, and in some developed countries too, it remains politically salient; ownership or access to land is often vital for securing entitlements (Barraclough, 1991; Christodoulou, 1990). Land reform or agrarian reform has often been viewed as an important means to improve entitlements; often where land ownership is very inequitable, food security is compromised. Movements for land reform have generated political revolutions, and some of the most vociferous global resistance movements are found in regions where land ownership is unequal (Barraclough, 1991; Rossett, Patel and Courville, 2006; Desmaris, 2007). Understanding the roots of contemporary patterns of land ownership remains an important explanatory variable for patterns of

Box 5.1

Chronology of European colonialism: some major participants and characteristics

Dates	*Principle powers*	*Geography*	*Impacts and evidence*
1450–1600	*Spanish and Portuguese	*End of 1500s: Portuguese had sailed past Cape of Good Hope to reach India.	*Merchant capitalism dominated, and trade with colonial possessions was controlled and restricted. Economic activity: largely mining and tropical crop production.
		*From 1492, Spain and Portugal began to colonize the Americas.	*Sea-based empires grew, and trading bases were established in Africa, South Asia, China and Indonesia. *Demographic collapse of indigenous peoples. Conquest of pre-colonial empires was completed, and silver and gold exports established. Territory in Latin America was increasingly settled by Europeans and organized as a hierarchical system of governance based in Lisbon/Madrid.
	*Russian expansion and conquest of states to the east of the Urals.	*Russia expanded empire across Siberia in an unprecedented expansion towards the Pacific coast. Extension of their land-based empire in contrast to other European maritime empires.	*Increased trade and settlement across the eastern regions of the Russian empire.
1600–1776	*Dutch, French and British rivalry	*The Dutch achieved independence from Spain and began an aggressive campaign to establish a maritime empire; their success is associated with the seventeenth-century Dutch 'Golden Age'.	*Massive increase in international trade in spices and colonial crops, especially sugar, tobacco, cotton and tea and coffee. Plantations established in tropical possessions and associated with expansion of the slave trade.

		*In North America, competition for ascendency between the Dutch and British, and later between the French and British	*North American expansion, from eastern seaboard to western frontiers, and in Canada into the northern territories. Settlements followed to exploit the agricultural frontier and resources of northern forests, furs and trees. Conquest, removal and marginalization of indigenous peoples of North America.
	*Continued success of Russia in eastward expansion; established a port on the Pacific in 1649.	*Russia encroaching on Chinese empire in southeast and also towards the Bering Sea towards Alaska.	*Russian colonization associated with exploitation of fur and timber of Siberian forests.
1776–1820s: the first decolonization	*The Napoleonic Wars in Europe weakened and diverted resources from colonial projects. After 1810, Spanish and Portuguese colonies declared their political independence.	*13 British North American colonies declared their independence in 1776. *Early-nineteenth-century political independence in Latin America.	*Britain assumed more importance in Latin America through its economic rather than political dominance.
1820s–1870s: British hegemony	At the end of the Napoleonic Wars, British established international naval and commercial hegemony.	*Commercial dominance in Latin America.	* Infrastructural developments in Latin America, especially in Argentina, to exploit resources of region.
		*Expansion of British power in South Africa, Australia and New Zealand.	*Defeat and marginalization of indigenous peoples in New Zealand, Australia and southern Africa.
		*Most of Southeast Asia came under European control; only Thailand remained formally independent.	*Expansion of commercialization of agriculture in British Asian possessions, especially South Asia, including Sri Lanka (formerly Ceylon); tea, coffee, rubber and cotton plantations established. Indian textile industry collapsed as British-manufactured textiles imported under favourable trade rules.

Dates	*Principle powers*	*Geography*	*Impacts and evidence*
		*French expansion into North Africa and Southeast Asia, Indo-China.	*The most important legacy was commercialization of agriculture. Production of traditional crops expanded (rice and coconuts), and new crops were introduced, such as rubber and tea on plantations, for example.
1870–1919 the Age of imperialism	*Last great period of European expansion. *British hegemony challenged first by Germany, United States and Japan. *In the East, Japan began their colonial project in Korea and Taiwan	*This period is associated with the 'Scramble for Africa', as European states extended their economic and political control over inland Africa. *Britain assumed more importance in Latin America through its economic rather than political dominance.	*Formal colonialism established over all of Africa. Treaty of Berlin (1878) established formal colonies and divided the continent between European powers. Contemporary boundaries established which prevail today containing disparate ethnic groups and cultures; this legacy helps explain the problem of political legitimacy that continues to plague African regimes. *Expansion of settler communities in Africa and establishment of gold and diamond mines using exploitative labour regimes.
After 1919: the second decolonization	*Most colonies achieved political independence between the 1950s and 1960s.	*Economic dependence continues to typify exchanges between former European and former colonial states.	

Source: Based on the summary in Gwynne, Klak and Shaw (2003: 42–55)

poverty and undernutrition. As Bailey asserts in the Oxfam publication about the contemporary food crisis:

> Perhaps nothing illustrates the inequity at the heart of the food system more clearly than the case of land – the most basic resource of all. In the USA, 4 per cent of farm owners account between them for nearly half of all farm land. In Guatemala (see Box 4.3), less than 8 per cent of agricultural producers hold almost 80 per cent of land – a figure that is not atypical for Central America as a whole. In Brazil, one per cent of the population owns nearly half of all land.
>
> (Bailey, 2011: 32)

In Latin America, Portugal and Spain introduced a very inequitable system of property relations which in many of these countries is still relevant today. They transplanted the hacienda and latifundia systems. This agrarian system of production is similar to the feudal system that prevailed in the Iberian Peninsula in the centuries before and during the colonial project. These economic units covered huge areas and were associated with feudal systems of labour relations. Peasants were granted small plots (mini-fundia) to grow food for their own consumption, but were required to work on the estate to produce goods for export. Its imposition in Latin America helps explain some of the region's distinctive economic and political problems. Some experts assert that it also helps explain the relative failure of development policy in the region (Kay, 2002).

Of particular relevance for this discussion is that inequitable land ownership structures help explain why rural poverty and landlessness are prevalent across so much of Latin America. Patterns of political inequality usually mirror these property relations, with a few landed elites dominating politics and policy in the region. The prevalence of land owner interests explains why most attempts at land reform have been partial, either in their objectives or in their implementation. Examples of land reform programmes exist in Latin America, notably Mexico, but even relatively progressive governments have avoided comprehensive land reform initiatives in deference to landed interests (Barraclough, 1991; Bernstein *et al.*, 1992; Kay, 2002; for excellent examples, see www.unrisd.org/80256B3C005BCCF9/(httpAuxPages)/9B503BAF4856E96980256B66003E0622/$file/dp101.pdf (accessed 12 December 2011)).

Hacienda and latifundias are usually inefficient, and are commonly associated with extreme social polarization and consequent political unrest; such characteristics still pertain across much of Latin America, and help explain the continued calls for land reform and periodic violent rural resistance movements in the region. The roots of the 'land question' in Latin America, therefore, are deep and explain why '[L]atin American agriculture is characterised by concentrated land ownership and a production structure whereby medium and large commercial farms, while few in number, contribute the bulk of agricultural output' (CEPAL Review, 2010: 145). These grossly uneven patterns of land ownership prevail in the region and help explain the polarized nature of their societies. Most states in the region have high scores on the Gini Index, which is a measure of inequality (http://hdrstats.undp.org/en/indicators/161.html; accessed 22 July 2011). This is but one regional example of a profoundly important question; fuller discussion is required to appreciate the centrality of the land question to politics, poverty and development (Box 5.2).

Box 5.2

The importance of land in the developing world

Property relations were transformed to a greater or lesser extent throughout the colonial era; in some examples, European settlers or corporations simply expropriated land to form large plantations, as in the Americas, parts of Africa and regions of South and Southeast Asia. Elsewhere, metropolitan governments or, later, independent states with large European populations supported settler programmes and established commercial farming on land previously occupied by indigenous populations (the Americas, Australia, New Zealand, parts of Africa). Still elsewhere, the extension of international trade generated economic specialization and intensified commercialization that led to increased land values, competition and speculation, and only a few powerful players prevailed while others were marginalized.

Property relations, and associated rural labour relations, are central to analyses of poverty and entitlements across the global south. The land question helped fuel the independence movement in India, although land reforms since 1949 have been partial. In Pakistan, parts of the countryside remain feudal in character, where landlord elites exercise political power based on the support from their numerous tenants in their rural constituencies. In Africa, north and south of the Sahara, access to land was radically transformed in the nineteenth century, and politicians often have regional and ethnically specific power bases.

Land is still equated with power in many rural places across the south, from the Philippines to Honduras; everywhere, inequalities in access to land continue to have serious implications for the distribution of wealth, power and privilege. 'In most settings today, land is important socio-politically to both elite and subaltern groups' (Borras and Franco, 2009: 2). These realities help explain the continuing salience of land policies and associated discourse. However, property relations are more than simply about access to land, they are about access to other resources too, notably water and forest resources, and as pointed out earlier, everywhere, they are about politics and power. The land question has particular salience for some specific sub-groups within national communities, notably ethnic minorities, indigenous communities and females (see Chapter 6).

Issues about land reform and ownership were current after the Second World War, and were especially politically charged during the Cold War. Some argue they enjoyed a 'golden era' between the Mexican Revolution of 1910 and 1979 (Borras and Franco, 2009: 14). They were externally imposed in Japan, South Korea, and Taiwan after the Second World War, and are considered to be a crucial element of their consequent economic success.

Kay (2002), in his comparative analysis of development in Latin America and South Korea and Taiwan, concludes that the nature of agrarian reform in the Asia is paramount to explanations of their success. He argues that a crucial difference in explaining the superior economic performance of South Korea and Taiwan over Latin America is that thoroughgoing agrarian reform took place in these Asian countries before industrialization, and not the other way round as in Latin America, with the exception of Mexico. Agrarian reform preceded any significant industrialization in the Asian examples, and was a key ingredient in the subsequent industrialization process. He adds:

> Furthermore, Taiwan and South Korea's agrarian reform had a far greater redistributive impact than the Latin America agrarian reforms, with the possible exception of Cuba. It is this rural equity factor which was to have a major positive impact on Taiwan's and South Korea's industrialisation and was the missing ingredient in Latin America's industrialisation.
>
> (Kay, 2002: 1076)

The gross inequalities in Latin America helps explain rural unrest there throughout the post-colonial period. Rural unrest in Central and Latin America caused consternation in the United States during the Cold War because it was understood as being a potential catalyst for political change, namely socialist transformations in their 'backyard'. After all, both the Chinese and Russian Revolutions were, at least to some extent, fuelled by the awful injustices of rural social relations across both countries. Everywhere in the developing world, poverty represented a potential spark that might ignite communist revolutions, aided and abetted by communist allies in Russia and China. When the Cold War era passed and the threat of communist revolution retreated, interest and support for land reform policies waned. In recent years, however, a renewed interest is apparent (see Borras and Franco, 2009: 15, for details of recent institutional and NGO activity).

Most accept that '[L]and policies are crucial for poverty reduction and empowerment', and that 'there is ample evidence that the issue is back on the agenda of international development institutions as well as many nation states' (Borras and Franco, 2009: 1). Despite the fact that the world, for the first time in human history, is now predominately urban, poverty remains a predominately rural problem, and three-quarters of the world's poor live in rural places. The multidimensional character of poverty is amply illustrated with reference to the disadvantages suffered by the rural poor, '[R]ural poverty is associated with low incomes, illiteracy, social and geographical marginalisation, cultural discrimination, environmental fragility, and, political isolation and exclusion' (Borras and Franco, 2009: 4). Why then is the image of the urban poor so much more ubiquitous, from *Slumdog Millionaire* (2008) to *City of Joy* (Lapierre, 1985). Urban poverty is often considered more serious because it is more visible to visitors and development experts alike, and also perhaps because of its explosive political potential too, but poverty remains an overwhelmingly rural issue in most of the south.

The relationships between land, politics and poverty presents a dilemma identified by Borras and Franco '[I]n many agrarian societies pro-poor land policy is necessary in order to achieve democratic governance; and yet how can pro-poor land policy be implemented in settlings where land-based wealth and political power is highly concentrated in the hands of a few-private individuals, corporate power or the state?' (2009: 4). Conflicts and political activism associated with such circumstances are revisited later when we explore how the dominance of the corporate global food system is being challenged.

Source: for land issues at LDPI Research Network, visit: www.future-agricultures.org/index.php?option=com_docman&task=cat_view&gid=1552&Itemid=971&limitstart=10 (accessed on 12 December 2011)

In South Asia, after years of strategic advances and retreats, Britain emerged as the dominant colonial power in the nineteenth century. Across the continent, new property relations emerged, and new commercial pressures encouraged land owners to grow crops for market, whether sugar, tobacco (introduced from the Americas), spices, cotton, jute or indigo. Again, the processes generated winners as well as losers, and some indigenous rural power holders benefitted from the 'radical and destructive changes that had occurred' (Wolf, 1982: 249). Numerous economic transformations accompanied informally, and after 1857, formal colonialism in South Asia, notably: new landlords emerged with private property rights; new systems of taxes were implemented; and agriculture became increasingly commercialized and integrated into the world market.

In South Asia, an exploitative system of landownership was instituted which effectively exploited the peasantry and starved the agricultural economy of productive investments and created a rural economy characterized by widespread poverty and insecurity (Bernstein *et al.*, 1992). The desperate state of the South Asian countryside between the first incursions of the British East India Company and the end of formal colonialism is suggested by the frequencies of famines between 1765 and 1947. These famines, just as with contemporary examples, are the result of a conjuncture of factors, human and natural, but that large numbers of people were very poor meant that minor changes in natural conditions exacted a high death toll.

Colonial legacies: labour regimes

All colonial projects had important repercussions for the demographic composition of their colonies and the associated labour regimes.

Spanish and Portuguese colonization of Latin America proved disastrous for the indigenous populations and the pre-existing empires of the Incas, Mayans and Aztecs (Wolf, 1982: 131–157). The arrival of Europeans initiated a demographic catastrophe as millions of indigenous people died from European diseases. This initiated a problem that was to recur across all colonial possessions for the next few hundred years, specifically, the labour question (Potts, 1990). How were the colonial powers to supply the large labour force required for their new mining and agricultural projects? This problem promoted a variety of innovative and often drastic responses. In Latin America, faced with a seriously diminished indigenous population, and with conditions which proved too arduous for immigrant European labour, Portuguese and Spanish settlers looked elsewhere for labour, namely Africa.

Sub-Saharan Africa became integrated through European trading systems to the colonial project in the Americas, and one of its principal roles was to supply labour in the form of slaves. Although slavery existed long before the sixteenth century, the number of humans involved in the infamous 'Triangular Trade' connecting

Europe to sub-Saharan Africa and the 'New World' of the Americas dwarfed all previous examples. It reached its zenith in the eighteenth century, the 'golden age' of slaving, when over 6 million people were forcibly exported from Africa (Wolf, 1982: 196), and slavery continued as a formally organized labour regime until the late nineteenth century. The Triangular Trade was the cruellest manifestation of the links between the core and periphery, to produce sugar, tobacco and cotton in the Americas; black Africans were required and were delivered by European merchants. Sadly, however, various forms of extreme labour exploitation, even slavery, continue in the twenty-first century and are often associated with the global food system (Box 5.3).

Box 5.3

Unjust desserts: ADM and the chocolate industry's darkest secret

The world's largest processor of cocoa beans, ADM, has for the most part escaped charges that have tainted West Africa's multi-billion-dollar chocolate industry: the use of forced child labour to grow and process the main ingredient in chocolate. But ADM's part in industry efforts to blunt reform may implicate it in what amounts to child slavery.

Some 70 per cent of the world's child labourers are concentrated in the agricultural sector, where they are forced to toil in blazing heat, carry heavy loads and routinely exposed to deadly pesticides, machetes and other dangers, according to the US State Department. Although it is not illegal for children to work on farms, international human rights laws forbid children to work so many hours that they cannot attend school.

Field researchers estimate that hundreds of thousands of children toil in the remote cocoa fields of the Ivory Coast and Ghana, which together supply 60 per cent of the world's cocoa. Some of the children are smuggled in from Benin, Mali and other nearby countries.

In 2001, leading members of the Chocolate Manufacturers Association, along with ADM and seven other companies, signed the Harkin-Engel Protocol, an agreement that committed them to identifying and eliminating the 'worst forms of child labour' on farms in Cote d'Ivoire and other parts of West Africa. Yet, 10 years later, the reform has proven less effective than its boosters originally hoped. The protocol is non-binding, and the companies have resisted any certification regime that might allow consumers to help drive the transition to fair labour practices.

In addition, each company could implement the protocol as it saw fit, and most – including ADM – adopted only partial programs to reduce farmers' reliance on child labour. Without a binding standard, it should be no surprise that investigators from Tulane University's Payson Centre for International Development concluded in 2009 that hundreds of thousands of children are still being forced to work on remote farms in Ghana and the Ivory Coast.

In Britain, consumer pressure has been stronger. Cadbury's March 2009 announcement that all its dairy milk bars would use fair trade chocolate proves that it is possible to source cocoa entirely free of forced child labour. Although Nestlé followed with a similar announcement, its policy applies only to the UK market.

Source: extracted from Cray, C. (2010) ADM's New Frontiers: Palm Oil Deforestation and Child Labour online at: www.corpwatch.org/article.php?id=15587 (accessed at 12 December 2011)

While slavery was the answer to the labour problem in the Americas until its abolition, an allied system emerged in Asia to supply labour for agriculture and mining. Especially after the abolition of slavery, it was the indentured labour scheme. Literally millions of people, mostly from South Asia but also from China, were recruited to travel to other British colonies to work. Ferguson notes:

> [I]t should be also remembered that Indian indentured labourers supplied much of the cheap labour on which the British imperial economy depended. Between 1820 and the 1920s, close to 1.6 million Indians left India to work in a variety of Caribbean, African, Indian Ocean and Pacific colonies, ranging from the rubber plantations of Malaya to the sugar mills of Fiji. (2004: 217)

This system was riddled with corruption and coercion and was understood by some as a new system of slavery. Indian communities in East and South Africa and the Caribbean, Chinese and Indian populations in Malaysia and Tamils in Sri Lanka all date from these schemes established as part of the Empire's labour regimes. Rigg describes how the 'entire character of peninsular Malaysia was transformed over a period that spanned barely 20 years' (Rigg, 1991: 114). He estimates that over 16 million Chinese and Indians arrived in Malaya, as coolie labourers, encouraged by British colonial policy, to work in the tin mines and rubber plantations. The exact character of dislocations varied but everywhere there were economic, social, political and cultural ramifications from the rapid expansion of industrialization in Europe and its consequent appetite for raw materials.

Labour was manipulated in various ways throughout the colonial era. Various examples of forced labour schemes, slavery or indentured labour, explain the prevalence of communities of African heritage in the Americas and the Caribbean as well as Indian and Chinese communities in Africa and Southeast Asia. People were uprooted from their homes and relocated into alien economic and social landscapes and cultures where many of their descendants exist, sometimes still suffering from relative marginalization. In Latin America, the indigenous Indian communities still suffer discrimination, which is reflected in their socio-economic characteristics (see sub-national patterns).

Colonial legacies: plantation systems of agriculture

To appreciate why slavery became so lucrative, we need to understand its place in plantation agriculture, which emerged in the sixteenth and seventeenth centuries and that continues in various manifestations today. This agricultural system was based on the production of tropical crops in industrial quantities often by slave labour; the European demand for sugar in the eighteenth century explains its success in Latin America and the Caribbean.

'Plantations appear and reappear in Latin America's agrarian history, the booms and slumps of their production of sugar, sisal, rubber, cocoa and coffee following the cycles of demand for these tropical crops in the different phases of the world market' (Bernstein, Crow and Johnson, 1992: 34). Plantations often appropriated land to produce commodities for export, land that had previously been devoted to food production; this was an early and obvious way that colonial economies were transformed to serve external interests. There were obvious implications for food production and entitlements in the colonies.

The sugar plantations established in Northeast Brazil by the Portuguese (and later diffused across Brazil and the wider Latin American and Caribbean regions) are associated with some of the most inequitable systems of agricultural production in the world. Vast tracts of land were granted to a few large landowners, and the sugar was cultivated by slave labour. Elements of this system of production and its associated patterns of trade are still obvious today. Poverty is still endemic in the northeast of Brazil, and the culture and demography of Brazil reflects its participation in the slave trade.

In the Americas, North, Central or Southern, labour conditions on the plantations were appalling, long hours of arduous work, followed by poor provisioning and living quarters, meant that life for the slaves who cultivated and harvested sugar, tobacco or cotton was brutal. Sugar production was also associated with environmental destruction, and it is interesting to note that the exact same criticisms are launched against sugar production in Brazil in the twenty-first century. Habitat loss and environmental degradation are serious, and human rights abuses are severe (Martinelli and Filoso, 2008). In addition, inequitable land owning structures continue and may even be intensifying: '[T]he growth of agribusiness in Brazil, with its vast monocultures of soybean or sugarcane, is responsible for the expansion of large estates belonging to fewer and fewer landowners, according to small farmers' organisations and agricultural experts' (Source: http://ipsnews.net/news.asp?idnews=48734 (accessed 25 October 2010).

Interestingly, in the twenty-first century, sugar production, introduced into Brazil by the Portuguese in the 1500s, is enjoying a renaissance because of its use in ethanol production, domestically and increasingly globally. The concerns that surrounded sugar production in the past are proving to be of continued concern with sugarcane production in the twenty-first century. The production of sugarcane, as with some other tropical produces, appears to be inherently allied to exploitation of the environment and communities (Abbott, 2010).

British colonies were transformed to meet the needs of British capital, and in nineteenth-century South Asia, this meant the introduction of cotton, tea and jute, often grown on plantations or by peasant producers. South Asia became the source of raw materials that formed the basis of the British Industrial Revolution of the late eighteenth and early nineteenth centuries. The connections between its colonial possessions and its rise to industrial dominance were vital. Raw materials were produced in South Asia, sent to Britain for manufacture and then marketed back in South Asia. This pattern of trade became the template of subsequent trade relations between colonial powers and their possessions and is still obvious in the dominance of raw materials in the export lists of former colonial possessions. Wolf suggests the extent of economic and social transformations associated with industrial expansion in Britain and its costs abroad:

> [T]o feed the factories of Lancashire; slave plantations displaced native populations in the American South, while the growing demand for cotton burdened the slaves with ever-increasing exactions. In Egypt, peasant production yielded to cotton-growing large estates. To furnish cotton to the Bombay mills, millions of acres, formerly in food crops were given over to cotton in western India.
>
> (Wolf, 1982: 295)

The plantation system was instituted across the tropics, initially supported by colonial powers and more recently by state and corporate interests. The prevalence of plantation systems, producing raw materials for the global food system, is one of the most obvious legacies of colonialism in places as diverse as Brazil and Sri Lanka; their presence continues to explain patterns of entitlements and food provisioning into the twenty-first century. Some of the problems associated with the contemporary extension of plantation agriculture, specially associated with paper, rubber and oil palm production, are considered in Box 5.4.

Box 5.4

Plantations in the twenty-first century

Women raise their voices against tree plantations: the role of the European Union in disempowering women in the south

Vast areas of land where diverse and rich ecosystems predominate are being replaced with large-scale tree plantations in the South. These plantations – whether eucalyptus, pines, rubber, oil palm or other – have serious impacts on local communities, who see their ecosystems and livelihoods destroyed to make way for industrial tree plantations. Apart from affecting communities as a whole, they result in specific and differentiated impacts on women.

What most people in Europe are unaware of is that the European Union is a major actor in the promotion of such plantations in the south, and is therefore playing a role in disempowering women in the south. While the EU has signed a number of treaties and conventions and developed a major body of legislation aimed at achieving gender equality in the European Union, the issue of gender justice seems to lose its importance for the EU outside its borders.

In the case of Papua New Guinea, it refers to oil palm plantations that are being mainly promoted to feed the European market with palm oil (used in products such as cosmetics, soap, vegetable oil and foodstuffs), as well as for the production of agrofuels. The second case is that of Nigeria – about rubber plantations established on the lands of a local community by the France-based company Michelin for producing the rubber used in the manufacture of tyres. Finally, the Brazilian case – about eucalyptus plantations set up by three companies – the Swedish–Finnish Stora Enso, and the Brazilian Aracruz Celulose and Votorantim companies – for producing pulp for export to Europe for paper production.

There are important contrasts between these three case studies, but their similarities are more important. On of the main impacts has been the loss of food sovereignty, as land is devoted to production crops rather than food for domestic consumption. The other problems identified are that prices for plantation crops are, as with other commodities, very volatile, which means the peasant producers have insecure entitlements. In addition, land disputes are increasing as access to land becomes very competitive, for food production and for plantation agriculture. The negative social and environmental impacts of monoculture plantations have been documented in many countries all over the world, and these impacts range from human rights violations to environmental crimes. These cases are excellent case studies of these problems.

Resistance

All these projects have generated local opposition and women have been major participants in these resistance efforts. They are playing a leading role in the struggle against the expansion of tree monocultures. Unification of the action of urban women with the action of rural women will strengthen the struggle against the expansion of socially and environmentally destructive plantation agriculture.

Source: based on 'Women Raise Their Voices against the Tree Plantations' Forests and Biodiversity Program – Friends of the Earth International; World Rainforest Movement' (March 2009), summarized version of the full report available at: www.wrm.org.uy/subjects/women/fullreport.pdf (accessed 2 November 2010). Also includes extracts from Chris Lang's work 'Plantations, Poverty and Power' available at: www.wrm.org.uy/publications/Plantations_Poverty_Power.pdf (accessed 12 December 2011)

Between the sixteenth and nineteenth centuries, sugar, tea, coffee, rubber, palm oil, being among the most important crops and largely grown on plantations, became important internationally traded products, produced in the periphery and consumed in the core. Trade in these crops represent an early example of the global food system. Some might argue that their characteristics exemplify some of the worst aspects of the contemporary food system too: their production formed an economic and social link between wealthy communities and poor communities in distant places; they generated profits for a few and exploited many in their production; at their production site, they displaced and dispossessed millions; their production displaced domestic food production and replaced it with luxury foods for export markets; the system of production was often environmentally destructive. Sugar, tea and coffee, first considered luxuries for the privileged classes, rapidly became part of the birth of a new kind of economy, the 'mass consumer society' (Ferguson, 2004: 12). The popularity of tea in England from this time helps explain how it became the quintessentially British drink. Tea and tea-drinking became fashionable, which helps explain the success of companies such as Wedgwood, based in Stoke-on-Trent.

Plate 5.1a Wedgwood bone china teapot: TPOT34 – by John Cutts, showing a view of Paddington, Middlesex, c. 1815 (Courtesy: Wedgwood Museum, Stoke-on-Trent).

Plate 5.1b Wedgwood bone china teapot: TPOT21 – with stylized decoration and gilding, c. 1885 (Courtesy: Wedgwood Museum, Stoke-on-Trent).

Visit this address to listen to the story of tea: www.bbc.co.uk/programmes/b00v71qr. Also, you can listen to Neil MacGregor's (2010) story of tea in the mid-nineteenth century with reference to a Victorian tea set here: www.bbc.co.uk/podcasts/series/ahow (accessed 28 July 2011).

Colonial legacies: political connections

Colonial experience also had important political legacies that help explain the character of the colonial and post-colonial governments. The role of the landed elite in Latin America has been mentioned, but the landed class became important in other regions too. Potter (2000) describes how, in India, the British Raj depended on regionally powerful landlords or wealthy proprietors who found it convenient and profitable to assist the colonial authorities. A few British elite administrators were recruited each year and served at the pinnacle of the Indian Civil Service, staffed by well-educated English-speaking Indians, which formed a network of officials that connected the rural outposts of the empire to the metropolitan core. Indian nationalist leaders were very often drawn from these elites.

In Africa, indirect rule was frequently employed to govern colonial possessions. Overseen by European officials, local customary leaders or chiefs became the main agents of colonial control. In all colonies, the police and army were ready to maintain colonial directives and to repress dissent. Potter (2000) argues that an important legacy of colonial rule was statism, where the state largely controls the economy. Colonial provinces were controlled by directives emanating from the metropolitan power and delivered through colonial administrators; this characteristic became a defining feature of many post-colonial patterns of government. In Africa, such centralized control of the economy was eagerly adopted 'both for its resonance with socialist and developmentalist ideologies and for its obvious utility in consolidating power and personal wealth' (Diamond, Linz and Lipset, 1988: 7–8 quoted in Potter, 2000: 286). Authoritarianism, which allowed for very limited popular participation, and a political culture which relied on repression to ensure assent, unfortunately were widely adopted in former colonies with serious implications for the success of development efforts.

Colonial legacies: patterns of trade and infrastructure

Legacies of trading patterns established in colonial times are still obvious in international trade. Certainly some developing countries, former colonies of European powers, have challenged the status quo and are emerging as major producers and exporters of manufactured goods and services. However, if we examine the very poorest states, it is remarkable how many still depend on the export of primary products, mineral or agricultural, to the industrial centres. 'One of the most striking aspect of commodity flows and regionalisation of trade is the persistence of dependence of Less Developed Countries on trade with developed countries' (Knox, Agnew and McCarthy, 2003: 50). This is just an example of the dependency that evolved between developed countries and former colonies.

Many former colonies remain very dependent on the export of one or two crops or resources. Where countries have a very narrow export base, they are vulnerable to changes in the global economy; recessionary conditions mean that their export earnings decline and then suffer a fall in national revenues. Many of the very poorest states have the highest levels of export concentration: in cocoa production in Cote d'Ivoire; cotton in Burkina Faso, Chad, Mali and Benin; coffee in Ethiopia; and oil in Angola and Algeria.

Formal colonialism was accompanied by investments in infrastructure to facilitate these exchanges, the export of raw materials and the import of manufactured goods. All European powers facilitated trade by massive capital investments in their overseas possessions but the British case was exceptional in its volume and geographical extent. By the end of the nineteenth century, the British Empire had become the most extensive empire in world history, and South Asia had been joined by much of Africa and other possessions in the Far East. British capital exports accelerated and from the '1880s about 40% of British overseas investment was directed to railways, plantations, factories, government stocks and finance in the Empire' (Bernstein, 2000: 250). Ferguson asserts that Britain had effectively become the world's banker by the beginning of the twentieth century 'investing immense sums around the world' (Ferguson, 2004: 244). These investments promoted international exchange and replaced pre-existing regional and local trading patterns.

However, British capital was not restricted to its colonies. Immense sums were invested in Latin America, especially Argentina and Brazil. The role of British capital in the creation of the agricultural development of Argentina is instructive. Although not a formal British colony, the fertile agricultural land of Argentina proved lucrative for British investments because it was developed to deliver relatively cheap meat and cereals for the growing, hungry, British industrial classes. One of the most obvious manifestations of these infrastructural developments were the appearance of railway networks across the colonies; these expedited the export of raw materials to ports for export. The railway system in Argentina is the classic example, where a web of lines radiates out from the port of Buenos Aires across the rich pampas, the first part of the commodity chain that linked the haciendas of Argentina to the consumers in the United Kingdom. All forms of transport received massive investments, and transport systems and technological innovations expanded in response to these economic incentives.

The few examples selected in the preceding text begin to suggest the variety of traumas experienced by some populations as a consequence of colonial projects. Colonial authorities and their allies were following the economic and political dictates of the time and justified their actions with reference to economic, political and racial ideologies. In many cases, they were rationalized with reference to missionary objectives. To appreciate the relevance of these themes for this text, we can identify several crucial variables.

Of course, the driving force of all colonial efforts was the search for profits, and the capitalist model became its main method of extraction. However, all such efforts required ideological justification, and this was available with reference to free trade theory, racism, Darwinism, the 'civilizing mission', and intermittently religious zeal, but inherent to all projects was a concept of progress. Most colonial discourse is imbued with the concept of progress, and much of what was wrought was indeed considered progressive at the time, although this is difficult to credit today. As will become clear, however, similar processes are operative today; unfortunately, examples of injustices are not yet confined to the history books, and are often justified with reference to notions of progress. Progress is a very malleable and potent symbol which can be used to justify some very dubious interventions. The next part of this chapter examines development theory and interventions; these too offer cautionary tales about the notion of progress and how it may be employed to justify some pernicious policies. Some would argue that 'development theories' were simply another instrument through which Western interests could continue to exploit and control their former colonial peoples and places. It is important to appreciate their role in shaping the global south since decolonization.

Development theories and malnutrition

Although some colonies gained their formal independence in the early nineteenth century (most of Latin America), it was not until the twentieth century that those in Asia and Africa declared their independence. The economic disparities between the industrial world and these former colonial states, which became obvious by the middle of the twentieth century, helps explain the emergence of debates about development. Concepts of 'development' have a long and complex archaeology (Sachs, 1999) and represent 'a semantic, political and indeed moral minefield' (Adams, 1990: 6), but for our purposes I am limiting examinations of the term to the period following the end of the Second World War; from that time on 'people and whole societies could, or even should, be seen as objects of development' (Sachs, 1999: 4). Development theories have generated a considerable and controversial literature. The most comprehensive recent text which manages to chart the subject in its myriad incarnations, through time, across disciplines, and through its political battlefields, is certainly by Nederveen Pieterse (2010). Preston's text (1996) remains a very comprehensive analysis which contextualizes development discourse from its origins in social and economic theory and integrates pertinent case studies throughout. Both these texts would be challenging for undergraduates who should be encouraged to start with the very helpful introduction is by Willis (2011) in this series. Rapley (2007) delivers a very useful overview and is more accessible than Preston; but Rapley also integrates excellent case histories and useful chronology.

Shortly after the Second World War, the industrialized world was polarized into two economic systems, capitalism and Soviet communism. The Cold War that followed effectively created two major spheres of influence in the global south; one sphere was allied to the United States and the West, while the other was allied to the Soviet Union (see earlier). Clearly, economic policies in these two spheres diverged, and their development policies followed different trajectories. During the 1950s and 1960s (The first Development Decade was ushered in by the United Nations in 1961), 'development' was simply correlated with economic development and was commonly measured as gross domestic product per capita (GDP). Such a reductionist construction has the merits of being readily measured but seems absurdly partial today, although such assumptions have proved remarkably long lived. Development was equated with economic expansion, and an inherent optimism prevailed between the 1950s and 1970, the era called the 'Golden Years' of development, associated with full employment and low inflation in the West and enthusiastic 'development efforts' in the global south.

It is interesting to note that the 'science' of economics also prospered in these decades and its pronouncements about economic 'take offs' and the increased availability of statistics helped them sell their discipline as an applied natural science, ready to deliver 'sustained growth' to the underdeveloped countries; that it was to prove a very social science was then unappreciated. The road to development was understood as linear and achievable, given the correct technology and appropriate government policies. Similar optimism imbued Western and Soviet development efforts at this time; policies varied, but doubts about the development project were rare in these decades. The environmental impacts of unlimited capitalist or communist expansion or mass consumerism had not yet appeared on the global political agenda.

Surveying the world from the United States and Europe, economists and politicians viewed the 'undeveloped world regions', comprising all of the former colonies and the great majority of the world's inhabitants, as a zone ripe for rescue. In an attitude akin to the colonial mission of previous centuries, the United States, in particular, assumed the burden of driving development in its image across the globe where possible, that was, in those zones not allied to the Soviet Union. The objective was to initiate economic growth and hence reduce poverty and hardship, and in so doing, it would also usurp tradition by modernity. The Soviet satellites were viewed in a similar light by the policy-makers in Moscow, as 'underdeveloped' and ripe for modernization, Soviet style. It is important to recognize that much of the international effort then, as with previous eras of colonialism, 'was indeed altruism, and it should not be scorned' (Brookfield, 1975: 24); that this was often misguided altruism is another issue. The relative role of altruism and self-interest in development theory and practice remains contentious; certainly, self-interest is often rationalized with reference to altruistic motives.

Early development theories: modernization and Import Substitution Industrialization (ISI)

The question of development theories and their role in creating or destroying entitlements and their implications for dietary change illustrates the problem of adopting a scales approach to analysis. Development models were often generated by economists and politicians in the First World, the West or the Soviet Union, but implemented in the global south by national governments. They could, therefore, be considered in the last chapter as processes generated at the global level and reflecting hegemonic global ideologies. I have decided to discuss them in this chapter because it was at the national level that different models of development were deployed, and exactly how such policies were operationalized was coloured by national-level politics and circumstances. At the extreme level, whether a country opted for Western approaches or Soviet models reflected national histories and alliances.

Two main theories of development were common in the 1950s and 1960s; they were modernization and structuralist theories (for in-depth analyses, see Rapley, 2007; Willis, 2011; Rist, 2008) and although they differed in some respects, they shared one basic tenet, that economic growth would generate a 'trickle down' effect so that, after a time lag, all members of the developing country would enjoy the benefits of growth. These theories accepted that economic growth would be concentrated in growth poles and cores in developing countries, and then diffuse outwards across space and downwards through various classes. Unequal and uneven growth was deemed an acceptable price to pay during the initial 'stages of development' associated with the major theorist of modernization theory, Rostow (1960).

Theories current in these decades were also coloured by Keynesianism; they believed in the interventionist state. The state in developing countries would have an important place in development strategies. Capital was in short supply in newly independent states, and the state could be mobilized to accumulate capital for infrastructural and industrial expansion. These monies would be invested in projects designed to promote the dominant vision of development, often quite grandiose schemes. The other source of funds would be foreign investment.

The elites in charge of the newly independent states normally equated development with industrialization (Rapley, 1996: 12). Understandably, they viewed industrialization as necessary to gain complete independence from the former colonial powers. Appreciating that their historical economic disadvantage and dependency on the production and export of primary goods helped explain their dependence on the former colonial powers, they prioritized the industrial sector. Import Substitution Industrialization (ISI) became part and parcel of industrialization strategies in the post-colonial era. They were deployed to promote industrialization and independence from the colonial powers. The strategy became 'one of the 20th century's boldest and most widespread economic experiments' (Rapley, 1996: 25).

ISI efforts were employed in the 1920s and 1930s (Egypt, Iran, Iraq and Turkey, Mexico, Chile, Argentina and Brazil), but it was after the

Second World War that such models of development reached maturity and were employed more thoroughly by independent states. India, particularly, is often considered the paradigmatic case of this development strategy, and its neighbours followed its example (Rapley, 1996: 29). In India, there was state-led industrial development with a strictly regulated and limited private sector presence. Governments that employed this approach to development invested heavily in nationalization, industry and infrastructure. Rapley (1996: 31) notes that India, Turkey, Brazil and Mexico had large domestic markets that made ISI strategies feasible (if not necessarily effective). This was not the case for the numerous African countries which also embraced ISI policies in the 1960s and 1970s. These theories of development were commonly 'top-down' in character and allowed a central role for the state in directing development. This was true for both capitalist and socialist examples (Box 5.5).

Box 5.5

China: development strategies

Examining China's development policies since their communist revolution in 1949 is fascinating, and in many respects illustrates some of the best and worst examples of all development efforts. Circumstances in China also reflect the diversity of development challenges faced by governments everywhere: how to achieve food security; how to balance development between primary, secondary and the service sectors; how to deliver both economic growth and equity (between regions, ethnic groups and genders, and rural and urban populations); to what extent may economic development be environmentally sustainable; what the strengths and limitations of state intervention and regulation are; what degree of integration with the world economy is desirable and how much development is autonomous. These are a few, not an exhaustive list; we might include debates about human rights, democratic accountability versus authoritarianism, population policy, geopolitical strategy and military expenditure, but for our immediate purposes let us concentrate on the former, specific development challenges.

Studying China is also both useful and fascinating because its geographical scale and diversity means that every sort of issue has emerged. It is also relevant to study because approximately 20 per cent of the world's population is Chinese, and because of its potential role as a global superpower (but see Pei, 2009).

Although only briefly and partially formally colonized, informal economic and political interventions in China during the nineteenth century were extensive and quasi-colonial in nature (Gwynne, R. *et al.*, 2003: 101). European powers' competitive and occasionally aggressive efforts to exploit China, for goods and as a market, represent some of their most unsavoury policies. The Opium Wars and their social and economic consequences, for example, must rank as one of the most criminal of British imperial legacies. Ferguson remarks that '[I]t is one of the richer ironies of the Victorian value system that the same navy (British) that was deployed to abolish the slave trade was also active in expanding the narcotics trade' (Ferguson, 2004: 166).

The communist party, under Mao Zedong, which assumed control of the state in 1949, inherited a fragmented and desperately poor country that had been ravaged by Civil War. They immediately initiated state socialism, based on Mao's teachings. Their first development strategy was based on the Soviet model, and they enjoyed some financial and expertise from the Soviet Union too. This was a stage of development in spades, and it was a centralized command and control economy which promulgated 5-year plans to inform all aspects of decision-making. The revolution was fought in the name of the peasants, who supplied the majority of its army, and one of the first tasks of the government was to implement land reform policies. Feudal systems were broken up, and land was distributed to peasant farmers. They also nationalized all major sectors of the industrial economy. It is important to stress that, despite the rhetoric about 'a peasants' revolution', policy has consistently shown urban and industrial biases found in most other development strategies. During the 1950s, investment was in heavy industries, and food prices were depressed to guarantee cheap food for urban and industrial workers. Taxes have been levied on agriculture since the start of the revolution and help explain the consistently higher urban incomes.

Agricultural policies since 1949 have shifted dramatically, from the initial collectivized systems to, since 1978, more privatized market-led systems. In 1953, the first collectives were formed, and these survived until the disaster that was the Great Leap Forward in 1958, when they were combined into vast communes. In the commune period, almost all decisions were dictated from above, and there was little or no room for local decision-making. This system prevailed until the end of the Cultural Revolution in 1978. Although, with the awful exception of the famine which occurred between 1958 and 1960, when approximately 30 million died, the Chinese state had managed to improve the nutrition of most of its population, no mean achievement as the population grow very rapidly during these years. The state, although prioritizing urban industrial production, had made considerable investments in rural infrastructure including irrigation and small-scale rural enterprises. By 1978, however, agricultural productivity was falling, and the failings of the commune system were becoming obvious. At that time too, the limitations of the state-controlled industrialization policy were stark and compared very poorly with the successful economic transformations in neighbouring economies (Japan, North Korea, Taiwan, Singapore and Hong Kong). A new development strategy was adopted.

Deng Xiaoping is the name most closely associated with the transformation of China between 1978 and his death in 1997. While retaining many of the political characteristics of the revolutionary era, he began to implement dramatically new economic policies to create a complex mix of planned and free market elements. All efforts were geared to economic growth. From 1978, the direction of agricultural and industrial policy changed, and market reforms were introduced. China gradually 'opened up' to foreign trade and investment. Although government regulation has remained significant and continues to manage large elements of the economy, the liberalization of agriculture and industry helps explain the phenomenal economic growth that has occurred in China since 1978.

Of particular significance for food production was the de-collectivization of agriculture and the re emergence of the rural household as the primary unit of production; in addition, remuneration was again linked to production. This precipitated a massive increase in the volume and variety of foods available to Chinese consumers. Rising incomes in China are fuelling a very rapid 'nutritional transition', and livestock production is proving very lucrative, especially in regions close the massive urban markets. This new direction of China's economic policy was confirmed in 2001 when it joined the WTO.

So, the Chinese state has been remarkably successful, especially in recent decades, in improving the nation's food security. Traditional malnutrition has declined, and famines have been largely avoided, with that one very extreme exception in 1958–1959. Food sovereignty has altered, however, as China engages more fully in the international food system. While there is no denying the considerable success enjoyed by China, serious problems are still obvious. Growth was prioritized since the reforms of 1978, and equity has suffered. Gross differences exist within the country, and undernutrition remains significant for millions of the poorest populations, minority ethnic populations and the poorest rural communities who have been excluded from the considerable socio-economic improvements. In addition, a new problem has emerged; the environmental costs of rapid Chinese economic expansion are beginning to undermine their achievements as soil, water and air pollution intensify. In cities and in rural areas, pollution is obvious and serious, and policy interventions are still embryonic.

Sources: Minxin, Pei (2009), Becker, G. S. (2008), Thomas, S. C. (2006), Xiaoyun, L. *et al.* (2006); World Bank: see a video about a partnership project between the World Bank and Chinese government to reduce poverty in Guanxi, one of China's poor remote regions at: www.worldbank.org/en/news/2010/03/19/results-profile-china-poverty-reduction (accessed 19 July 2011)

It is crucial to appreciate the relevance of geopolitics to development in the 1950s and 1960s. The Cold War had created spheres of influence, with the West dominant in one and the Soviet Union in the other. Modernization theory and ISI models were implemented in both spheres but through quite different means. During the 1950s,

1960s and 1970s, both India and China represent socialist or communist implementations of modernization where industrialization was driven by a centralized state through a series of five-year or ten-year plans that prioritized investment in industrial expansion. The trajectories of these two examples diverged, and both adopted more market-based approaches to development by the late 1970s and 1980s (see the following text).

It is possible, in retrospect, to identify some serious flaws with the ISI and modernization models of development. In addition to poor industrial efficiency and opportunities for corruption, there were problems with their attitudes to agriculture and food production. There was an inherent urban industrial bias in such approaches, and the agricultural sector was frequently neglected; indeed, this has proved an enduring characteristic of all development policy up to and including the present. The wealth gap between rural and urban populations grew throughout the post-war decades. Agriculture was not simply neglected; it was often penalized by high taxes and artificially low food prices. If agriculture was considered, it was understood as a source of cheap labour to work in the expanding industrial sector as farming became more mechanized and peasant producers were replaced by large commercial units of production. In 2010, research concluded that 'earnings from farming in Latin American countries have been depressed by a pro-urban bias' (Anderson and Valenzuela, 2010: 144). This bias against agriculture, especially small-scale agriculture, is considered by some to be the single most important obstacle to food security in many parts of the global south, especially in Africa.

To help make industrial production more profitable, agricultural prices were artificially depressed to ensure cheap food for urban-based populations. Farming therefore became less profitable, and its failure helped fuel a massive rural exodus that helps explain the massive growth of urban populations in the global south in these decades. The challenge of delivering food security to the millions in these cities has become one of the most pressing problems of the twenty-first century. Cheap food policies had additional rationalizations associated with this urban expansion. An equally compelling reason was to avoid political unrest. Always a danger in developing countries with expanding urban poor populations, political unrest is usually more manageable among dispersed rural poor populations, although there are obvious exceptions to this generalization. This was obviously advantageous for the urban population, but in most developing countries in these decades, most of the poor lived and depended upon the rural sector.

Revised development visions: neoclassical answers

By the early 1970s, dissent was growing in development circles about the merits on modernization theory, ISI and associated industrial–urban foci. Economic growth was certainly not ‘trickling down’ in any significant way, and widespread poverty prevailed in countries, even where there had been rapid economic growth. Globally, the gap was growing between the developed world and the global south (with a few notable exceptions). Nationally, the gap between a minority of wealthy urban-based populations and the rural poor had increased in almost every context where rural populations were numerous. Even where economic growth, measured by GDP per capita, had improved, poverty had increased for the very poorest. That economic growth was synonymous with human development was exposed as a myth. A change in development thinking was obviously required.

By the late 1970s, the United States and Britain had initiated a shift to the right across the Western world which embraced free market ideologies. New approaches to development followed, and the role of markets was viewed more favourably and the role of the state more sceptically.

> Essentially, structural adjustment seeks to make the state and the market more efficient in such a way as to accelerate growth and eliminate waste. … It places the market central stage, assigns the state a secondary role in development, and puts its faith in the potential of unfettered individual initiative, creativity and ingenuity.
>
> (Rapley, 1996: 71)

This new vision must be understood with reference to the global financial crisis which ensued after the Organization of Petroleum Exporting Countries (OPEC) increased oil prices by a third overnight in 1973. Oil prices remained high for a decade, as other commodity prices collapsed. The global economic recession meant that demand fell for all other commodities, and export revenues of many developing states fell catastrophically too. The global economy was in crisis and so were development states.

Money amassed by the OPEC countries was invested in commercial banks and they, in turn, lent to developing countries eager to replace falling state revenues. Across the global south, from Mexico to Indonesia, countries borrowed money, encouraged by attractive rates of interest and eager to continue their investments in social and economic development. Other regimes borrowed money for less benevolent projects, to fund armaments build-ups or civil wars, and sums also helped expand private bank deposits of some despotic leaders. Whatever its destination, by 1982, the debt crisis had emerged on the world stage, and this was to help alter development policies in neoliberal directions, in concordance with the economic philosophies then dominant in the global financial institutions and their associated development branches.

Structural adjustment policies and liberalization

> The way neoclassical theory worked its way into the agendas of Third World countries varied from case to case. For the early implementers in the 1970s, such as Chile, Cote d'Ivoire and Sri Lanka, the new development policies were largely internally generated.
>
> (Rapley, 1996: 69)

Faced with overwhelming debt burdens (throughout the 1970s, debts rose with high interest rates so that debt repayments very often dwarfed gross domestic product), most developing countries had little option but to turn to the major global financial institutions, the World Bank (WB) and International Monetary Fund (IMF), for solutions to their problems, not always and certainly never completely of their own making. Even when their leaders were guilty of mismanagement or corruption, usually the majority of taxpayers were largely innocent of any major part of the problem which overwhelmed their countries' economies. These global financial institutions agreed on new conditional loans to the developing nations, and the conditions became known as structural adjustment policies (SAPs), which were conditional on a number of reforms.

The example of Latin American countries is typical '[R]eforms centred on macroeconomics stabilization, trade liberalization, deregulation and some privatization (or abolition) of State agencies' (Anderson and Valenzuela, 2010: 145). The specific nature of the 'medicine' varied, but they shared several elements; governments in developing countries must reduce their expenditure and increase their earnings. Decreasing their expenditure was politically problematic but in most cases was delivered by cutting social expenditure on health, education and welfare payments. The social impacts of these reforms have received considerable attention in the literature and were undoubtedly most serious for the poorest communities. The circumstances in the developing world throughout the 1980s worsened for the most part, and that decade has been called the 'lost development decade' as a consequence.

The relevance of the debt crisis for development policy lies in the fact that it laid the foundations for what would emerge in the 1980s and 1990s as the new development pathway; that is, economic growth was to be based on an enthusiastic economic liberalization and engagement with the global economy. Associated changes were to have profound implications for food security, sovereignty and entitlements in the developing world.

The new development philosophy also reconsidered the role of the state in development and deemed it to be 'part of the problem rather than the solution to the crisis of development (Potter *et al.*, 2004: 305). In some cases, the state was found to be inefficient at best and corrupt and inefficient at worst. The rapid, successful development of the Asian Tiger economies, based on export-led growth, was held as an example of the benefits of global integration, but of course their success was dependent on several other important variables, as well as exports, namely, effective state management, pre-existing rural transformation and specific geopolitical conditions. The character and capacity of post-colonial states varied, of course, but their colonial legacies meant that, in many cases, they had limited legitimacy and frequently represented narrow class interests.

Across the developing world, the WB and IMF assumed responsibility for development policy and prescribed two main solutions: that governments should retreat from development policy and that state revenues should be earned from the liberalization of economic policy and associated increases in export earnings (Box 5.6). These policies have driven economic and social change in the developing world throughout the 1990s and early years of the twenty-first century. Economic liberalization required developing countries to reduce domestic support for agriculture, reduce agricultural extension services and to reduce any protectionist policies. These have had very serious implications for food security, food sovereignty and diets; these are considered next.

Box 5.6

Aquaculture

Fishing and fish farming (aquaculture) have played crucial roles in food systems for thousands of years and continue to do so, but the relative importance of these have altered in recent decades, as fish farming has boomed. Fish farming is an ancient method of food production, and it has supported and improved local food security and diets for generations. Small-scale systems of artisanal production still play a vital part in food security strategies in many communities. However, since the 1970s, aquaculture has been transformed by new technologies which have been compared to the transformations of agricultural production associated with the 'Green Revolution' of the 1970s.

This so-called 'Blue Revolution' helps explain the phenomenal expansion of aquaculture since the 1970s, when global per capita supply was 0.7 kilograms, to 2006 when this had grown to 7.8 kg per capita. The value of world aquaculture exports grew from US$15 billion in 1980 to US$71 billion in 2004. It is now the fastest-growing animal food production sector, and it is concentrated in Asia. China dominates the sector, but Thailand, Indonesia, India, the Philippines and Vietnam have all experienced rapid expansion of aquaculture since the 1980s. Much of this expansion was encouraged by export-led development strategies promoted by the World Bank and national governments since the 1980s. Goods produced by aquaculture were low-volume, high-value goods, destined for affluent consumers in the developed world, and the United States, European Union and Japan were the main markets. However, the expansion of aquaculture benefitted domestic consumers too, as is obvious from the following data.

Table B5.6 ***Increases in fish consumption in kilograms per capita in selected countries***

Country	*1969–1971*	*2000–2002*
China	4.7	25.5
Sri Lanka	15.0	23.0
Cambodia	8.8	25.9
Indonesia	9.9	20.8
Malaysia	25.9	58.4
Thailand	23.7	31.0
Vietnam	16.4	18.2

Source: FAO (2006) State of World Aquaculture p.40 at: ftp://ftp.fao.org/docrep/fao/009/a0874e/a0874e04.pdf (accessed 25 November 2010)

Throughout the 1980s and 1990s, development strategies, based on the export of high-value, low-volume goods, from the developing world to the developed world, proliferated. These decades saw the emergence of wholly new international trade in high-value luxury foods; among the most important were horticultural and aquacultural products.

Environmental impacts

Over the last few decades, shrimp farming has been a relentless destroyer of huge expanses of tropical coastlines, particularly mangrove forests. Mangrove forest roots are bulldozed into the mud to make way for the intruding shrimp farms. The coastal equivalent of terrestrial rain forests, mangroves are home to an incredibly diverse range of life. They are the breeding grounds and nurseries for many fish, shellfish and other wildlife. Shrimp farming turns them into a barren and toxic prawn cocktail.

Once the mangroves are ripped out, the coast is rendered unstable, triggering erosion, harming coral reefs and seagrass beds, and eliminating habitat for creatures from the humble molluscs up the chain of life to the meek manatee.

While there are currently no precise figures on how great the loss of mangrove forests and other coastal wetlands is due to shrimp farms, estimates are frightening, with as high as 38 per cent of mangrove forests being lost to shrimp farming.

As the wetlands vanish, fish catches decline and ecosystems are knocked out of balance. Shrimp farms are often abandoned after only 3–5 years, leaving the once-fertile coastal ecosystem a wasteland. The proprietors then move on to destroy new territory.

The ecological damage doesn't end with the mangrove loss. To grow as much shrimp as possible and maintain overcrowded populations, large amounts of artificial feed and chemical additives, including chlorine, are added to this destructive cocktail. Malathion, parathion, paraquat and other virulent pesticides are also sprayed on the pools.

Along with the chemicals come several kinds of antibiotics, which are used to prevent shrimp disease. This resulting virulent soup is commonly dumped onto the surrounding land or into local waterways, where it harms people and other life.

Farming shrimp causes gigantic problems, even beyond the environmental harm; it can often decimate the coastal ecology that supports local communities.

Mangroves

It is common to find social resistance in many areas (in Ecuador, Honduras, Sri Lanka, Indonesia, Thailand, Philippines) where poor people live sustainably in or near mangroves by collecting shells and crabs, by fishing, by using the wood for charcoal and building materials. Mangroves are uprooted in order to provide space for commercial shrimp farming for export. The mangroves are usually public land in all countries, being in the tidal zone, but governments give concessions for shrimp farming, or the land is enclosed illegally by shrimp growers. Illegality is prevalent not only because of the public character of the mangrove land, but also because often there are specific environmental laws protecting the mangroves as valuable ecosystems. Shrimp or prawn production entails the uprooting of the mangroves, and the loss of livelihood of people living directly from, and also selling, mangrove resources (shells, fish, wood). Beyond direct human livelihood, other functions of mangroves such as for coastal defence (against sea level rise), and as carbon sinks, repositories of biodiversity (e.g. genetic resources resistant to salinity) and breeding grounds for fish, are also lost, perhaps irreversibly, together with aesthetic values. In the fight against shrimp farming, people who make a living in the mangroves have resorted, when circumstances have allowed them, to destroying the shrimp ponds, replanting rhizofora seedlings as a symbolic gesture and perhaps with some hope of reconstructing the vanished mangroves.

Source: Martinez-Alier (2000); FAO (2006); Country Fact Sheets about Aquaculture at: www.fao.org/fishery/naso-maps/fact-sheets/en (accessed 21 November 2010); FAO (2008); Khor, 'Destructive Impacts and resistance from local communities' at: www.twnside.org.sg/title/aqua-ch.htm (accessed 21 November 2010); Clay, 'Aquaculture: Greening the Blue Revolution' at: www.worldwildlife.org/what/globalmarkets/aquaculture/greeningthebluerevolution.html; White, K. *et al.* (2004); Sachs, J. (2007); FAO (2003); World Bank (2006); Szuster, B. W. (2003); also see: www.worldwildlife.org/what/globalmarkets/aquaculture/dialogues-shrimp.html (accessed 21 November 2010); www.greenpeace.org/international/campaigns/oceans/sustainable-aquaculture/shrimp-farming (accessed 21 November 2010). See a short slide show about shrimp farming in Orissa, India, at: http://news.bbc.co.uk/1/hi/8085768.stm (accessed 21 November 2010)

Neoliberalism and entitlements

Although the economic results of neoliberal reforms may be mixed, the moral critique is not. SAPs have been judged unjust, and evidence suggests that the poorest populations have suffered most from their imposition. There is evidence from across the global south that SAPs both increased food insecurity and reduced food sovereignty. Food insecurity increased in many places as the price of food increased (price support programmes were reduced). The implementation of neoliberal economic policies was associated with fiscal austerity; the state had to reduce its expenditure. In many places, this has worsened the plight of the poor who are most vulnerable to changes in welfare expenditure. Large decreases in government expenditure were correlated to increases in poverty, to an erosion of entitlements. This poverty was most obvious in the urban areas but existed in rural regions too. Food security was compromised as the poor suffered an erosion of their livelihoods.

Food sovereignty was also compromised in many places as liberalization required countries to allow imports of ‘cheap’ (subsidized) food crops. Perhaps this is the most serious accusation against structural reforms imposed between the mid-1980s and 1990s. They increased food dependency in many countries and eroded their capacity and power to make decisions about their food policies. From 1986 onwards, Haiti implemented a radical neoliberal reform package which included ‘slashing tariffs, closing state-owned industries, opening the agricultural market to US producers and cutting spending on agriculture by 30%’(Holt-Giménez and Patel, 2009: 38). The outcome has been disastrous, and food riots reflect the opposition by the poor of what they perceived to be a foreign assault on their right to affordable food. As summarized by Prakash:

> This increased reliance on markets was also concomitant to a progressive withdrawal of the state and intervention schemes from the food and agriculture sector, on the grounds that the private sector was more efficient from an economic point of view. Against these trends, public and private sectors in both developed and developing countries saw a limited need to invest in agricultural production and infrastructure, as food imports appeared an efficient way of achieving food security. Such perceptions, though, were radically changed when in 2006 prices of most internationally traded foodstuffs began to soar.
>
> (Source: Prakash, A. (2011) Safeguarding Food Security in Volatile Global Markets. Available at www.fao.org/docrep/013/i2107e/i2107e.pdf).

The example of Haiti is the worst-case scenario, but similar if not quite so serious examples are numerous. Chossudovsky (1997: 101–107) goes as far as to accuse the policies of being largely responsible for the famines in Somalia and food crises elsewhere. (See Holt-Gimenez and Patel, 2009, for examples from Ghana, the Philippines, Africa, Mexico and elsewhere.) By requiring developing countries to shift investment from domestic food production to export crops, their vulnerability to global food price movements increased. In addition, the majority of the agricultural support systems that had evolved in the south were reduced or obliterated. Numerous national agricultural support systems were dismantled, with adverse consequences for the livelihoods of millions of African farmers. It is important to note that, although such policies have proved catastrophic for some of the world's poorest people, they have had consequences for small producers in developed countries too.

In the United States and the European Union, small farm units have been in decline for over 50 years. Although part of the founding myth of the United States (family farms were understood to be at the root of the democratic system), and in spite of their political salience in parts of the European Union (France), they have been systematically excluded from the massive agricultural subsidies available in both these places. Thousands of small farms disappear every day in the United States and the European Union. As explained in the Oxfam Report (2005), 'behind the legal and technical maze that accompanies the functioning of the system is hidden a simple principle: the more you produce and the more land you own – which is to say, the richer you are, the more public assistance you receive' (quoted in Holt-Giménez and Patel, 2009: 66). The labyrinthine EU system makes it practically impossible to understand or access the process of agricultural subsidies to determine exactly how much taxpayers fund the largest farms and therefore the corporate sector (see Chapter 4). This morning (25 July, 2011), the dilemmas faced by small- or medium-sized dairy producers in Britain were news again. Effectively, small units in the present tax regime are unprofitable, and financial support systems encourage the massive extension of units. Very large dairy units are opposed by many as they are perceived to compromise animal welfare and be environmentally damaging.

Neoliberalism and overnutrition

The fact that obesity is a problem in the developed world is well known. Although its ubiquity may still surprise us, it is everywhere now (Box 5.7). The role of the 'nutritional transition' in this pandemic was introduced in Chapter 2, but there was no attempt there to examine the structural forces that were driving the dietary changes identified; now it is time to analyse these forces. Processes that have been transforming the diets of people in the developed world over the last 40 years, and which help explain the emergence of obesity, are now seen to be transforming diets in the global south. The process that took about 40 years in the West is being implanted much more rapidly in the south. Each element detailed in the following text has occurred in the West over recent decades and has been discussed elsewhere (Lang and Rayner, 2005; Popkin, 2009). The examination

Box 5.7

Australia: battling the bulge

Our image of Australians is that of a healthy population who enjoys sports and an outdoor lifestyle. A recent report describes a very different picture. Australia is today ranked as one of the fattest nations in the developed world. The prevalence of obesity in Australia has more than doubled in the past 20 years. Some alarming facts are given below.

- In Australia, more than 17 million Australians are overweight or obese.
- More than 4 million Australians are obese (BMI > 30.0 kg/m2).
- If weight gain continues at current levels, by 2020, 80 per cent of all Australian adults and a third of all children will be overweight or obese.
- Obesity has overtaken smoking as the leading cause of premature death and illness in Australia.
- Obesity has become the single biggest threat to public health in Australia.
- On the basis of present trends, we can predict that, by the time they reach the age of 20, our kids will have a shorter life expectancy than earlier generations simply because of obesity.
- Aboriginal and Torres Strait Islander Australians are 1.9 times as likely as non-indigenous Australians to be obese.

Source: See www.modi.monash.edu.au/obesity-facts-figures/obesity-in-australia (accessed 27 July 2011). Also, see a short video on Aljazeera at: http://english.aljazeera.net/video/asia-pacific/2010/05/201052343519929351.html (accessed 27 July 2011)

here is confined to understanding how the process has been replicated in the last two decades in the developing world.

Hawkes (2006) asserts that '[T]he transition, implicated in the rapid rise of obesity and diet-related chronic diseases worldwide, is rooted in the process of globalization' and that globalization is 'radically altering the nature of agri-food systems … and altering the quantity, type, cost and desirability of foods available for consumption' (Hawkes, 2006: 1). She identifies eight factors that help explain how globalization is influencing diets across the world:

1. Food trade and global sourcing
2. Foreign direct investment
3. Global food advertising and promotions
4. Retail restructuring (notably the construction of supermarkets)

5. Emergence of global agri-business and transnational food companies
6. Development of global rules and institutions that govern production, trade, distribution and marketing of food
7. Urbanization
8. Cultural change and influence

Examining the first theme, we noted earlier that SAPs, established in the 1970s and 1980s, began the liberalization of agricultural markets, and the process gathered pace in the 1990s, especially after 1994 when the Agreement on Agriculture (AoA) was adopted at the WTO. The AoA initiated a reduction in agricultural protection, although some were retained, altered but still significant in two of the world's largest agricultural producer regions, the United States and the European Union. Elsewhere, the liberalizing policies implemented after the AoA help explain some of the major changes in global agricultural activity. Since 1994, its volume has risen markedly, and the nature of the goods traded has altered.

In developing countries, 'food import bills as a share of GDP more than doubled between 1974 and 2004, and the amount of trade made up of processed agricultural products rose' (Hawkes, 2006: 2). These have two serious proximate implications for diets in the developing world. First, they have implications for food security which is reviewed in the preceding text. Second, the consumption of more processed foods is correlated with the growth in obesity, because these foods have high levels of salt, sugar and fat. Opening up their markets to global corporate food processors exposed their populations to an avalanche of cheap, fast foods.

The material in Box 5.8 illustrates some complex connections that help explain why trade in soybeans has increased so rapidly since the

Box 5.8

Soy production: China, India and Brazil

Brazil is the world's second-largest soybean producer and exporter (the United States is the largest producer and Argentina the largest exporter). Throughout the 1960s and 70s, government policies explicitly promoted the production, export and domestic consumption of soybean oil. In the 1990s, in line with the globalization agenda, the government opened up its soybean market and reduced government intervention. New policies reduced restrictions on foreign investment (to encourage the entry of more foreign capital into the soybean market), restructured farm income taxes (to encourage greater investment in soybean production), lowered import tariffs on fertilizers and pesticides (to facilitate higher soybean yields) and eliminated the soybean export tax (to promote greater exports). The government also implemented the *Real* [currency] Plan, which altered the nation's economic conditions; devaluation of the *Real* later in the decade caused the cost of Brazilian beans to fall in the world market. These policy changes spurred, as intended, acceleration of production and exports. Production costs fell, and returns to producers rose; combined with the abundant availability of low-cost land, these encouraged farmers to bring more land into production. And in light of lower production and transportation costs, vertically integrated transnational food corporations (TFCs), such as US-based Cargill (the largest soybean exporter in Brazil) and Bunge (the largest soybean processor), increased their investments in the Brazilian crushing industry. The result of these policy shifts was a 67 per cent increase in soybean oil production between 1990 and 2001, a more than doubling of exports, and one of the lowest soybean oil prices worldwide. However, somewhat ironically, the massive investment and growth in soybean oil production in the 1990s is not actually associated with increased consumption in Brazil. Although the data are difficult to interpret, per capita calorie consumption (already relatively high) appeared to decline, or at least stabilize, during the 1990s. Rather, production was set for the global market, facilitating dietary changes across the globe in countries such as China and India, who were also liberalizing their markets in line with the globalization agenda.

China implemented new tax and import regulations to encourage soybean oil imports and greater domestic production in the 1990s. Brazil, able to produce at low prices, became a major source in China of soybeans (for crushing) and soybean oil. Between 2002 and 2004, Brazil remained a crucial supplier of soy to China when greater trade openness led to a doubling of agricultural imports, of which soy formed a large proportion. Consequently, the amount of soybean oil available for consumption in China has soared. While this probably bought some benefits to under-consuming populations, consumption of vegetable oils in urban and some rural areas now exceeds recommended levels, a trend which the Chinese government has identified as a source of concern, given the rapidly rising rates of obesity and chronic diseases in the country. Recent trade policies will likely further the ready availability of soybean oil: China's accession to the World Trade Organization (WTO) has reduced import tariffs and quantitative restrictions, which is predicted to significantly increase soybean oil imports, lower prices and increase demand. Moreover, China continues to view Brazil as a good source of cheap soybeans: the Chinese government is planning to invest US$5 billion in Brazilian transportation systems to help them continue to produce soybean oil at competitive prices.

India, itself the world's fifth-largest producer of soybean oil, likewise imports Brazilian soybeans and oil. In the mid-1990s, India was a relatively small importer of vegetable oils; by 1998, the country had become the world's leading importer. This rapid change can be directly related to market liberalization. In 1994, as part of its obligations under WTO rules, India eliminated the state monopoly on imports. Facing low domestic production, imports poured in, especially of the lowest-cost oils: palm and soybean oil. Brazilian (and Argentinean) soybeans and oil were favoured, owing to their lower price and transportation costs relative to imports from the United States. Brazil also had the advantage of a high season and thus cheaper beans during the seasons of low production in India. The result was lower prices for vegetable oils, increased consumption, and increased share of consumption of imported oils. By the end of the 1990s, soybean oil accounted for 21 per cent of consumption (and palm oil at 28 per cent). This stands in stark contrast to the complete dominance of consumption of peanut, rapeseed and cottonseed oil in the 1970s, a reflection of domestic production. Today, prices of edible oils in India are more affected by soybean output in Brazil, Argentina and the United States than by domestic production.

This complex web of economic globalization illustrates how a series of policy reforms in three different countries had the effect of integrating the global soybean oil market, and, in so doing, facilitated the worldwide convergence of higher soybean oil consumption worldwide. Dietary convergence has occurred not only in the use of soybean oil in cooking, but in hydrogenated form in processed foods. Hydrogenation leads to the creation of *trans* fats, which increase the risk of coronary heart disease. Governments in Brazil and the other Mercosur countries, Canada and the United States have ruled accordingly that *trans* fats must be labelled on packaged foods. Despite these efforts to encourage consumers to eat fewer foods containing *trans* fats and high amounts of vegetable oils, dietary convergence of soybean oil consumption is likely to continue: the WTO is expected to reach an agreement in the next few years to further liberalize the vegetable oils market. Along with implications for consumption of total fat and *trans* fats, this trend introduces health concerns because it is likely to change the overall balance of fatty acids consumed in the global diet.

Importantly, though, the increasingly integrated nature of the soybean oil market is equally likely to facilitate dietary adaptation. The increased supply of soybean oil in the world market is leading to greater competition with alternative oils, thereby providing a bottom-line incentive for increased differentiation. The process is already in evidence, with TFCs adapting soybean oil to appeal to higher-value market niches, in this case, the wealthy 'health conscious consumer' wily to the detrimental health effects of *trans* fats. In September 2004, Monsanto, in partnership with Cargill, announced the development of the 'Vistive™' soybean. The bean will only require partial hydrogenation during processing, thus reducing the *trans* fat content. Cargill intends to pay producers a premium for the beans, which will be passed onto the food processors, and eventually, as a component in processed foods, onto consumers willing to pay more for a *trans* fat-free product. In October 2004, competitor DuPont, in partnership with Bunge, also introduced a soybean with similar properties, 'Nutrium™'. In the years to come, it is possible that leading companies will compete as much on high-priced oils for health as on low prices for the mass market; the former will encourage dietary adaptation while the latter its convergence. Thus, the very same processes driving the global market integration of vegetable oils may well have very different outcomes for low- and higher-income consumers.

Source: Hawkes, C. (2006) 'Uneven dietary development: linking the policies and processes of globalization with the nutrition transition, obesity and diet-related chronic diseases' *Globalization and Health*, 2:4 doi:10.1186/1744-8603-2-4 online at: www.globalizationandhealth.com/content/2/1/4

1990s, and how this is implicated in the global convergence of diets containing high levels of trans fats (chemically produced fat which are convenient for producers but not so good for consumers). These cases also illustrate the vital links between domestic and international factors to explanations of dietary change; global processes drive internal changes which similarly induce global changes. As mass consumption of soybean oil increases, prices rise initially and stimulate more production, followed by a fall in prices. These lower prices then initiate the search for new higher-cost products, and the creation of expensive 'functional foods' for a niche market. There are other important impacts when traditional oil is replaced by a new one. Shiva (2000) describes how the Indian market was conquered by soy, a process she describes as 'soy imperialism' and the consequent decline of mustard oil production, with negative implications for cultural practices, community-based food systems, biodiversity and livelihoods.

The globalization processes described in the preceding text have other potentially serious implications for food security and sovereignty because they facilitate the concentration of power throughout the global food chain. TFCs have exploited liberalization to penetrate new markets and increase their influence. These companies have expanded through vertically integrating their activities and by outsource aspects of their businesses. Vertical integration occurs as mergers and acquisitions help consolidate the power of a few large corporations across the whole food chain, from production to retailing (Box 5.9).

Box 5.9

Increased corporate control: producers squeezed and retail sector transformed

Producers squeezed in globally integrated commodity chains

The process of market consolidation has been intensifying along commodity supply chains in recent decades at the global level. Today, Transnational Corporations (TNCs) can dictate significantly the patterns of international trade through intra-firm trade under their globally integrated production and marketing strategy. TNC activities are strategically organized and integrated either horizontally or vertically. This is reflected in their dominance in commodity value chains. In agricultural commodity production and marketing, there are considerable asymmetries in market power and access to information, technology and marketing know-how between TNCs, on the one hand, and local entrepreneurs, farmers and traders in developing countries, on the other. Ironically, for small-scale producers and their governments, commodity markets have become fragmented, as TNCs have hastened the integration process of their operation globally. This parallel process of fragmentation and integration has often resulted in a hugely skewed distribution of gains from commodity trade. Under the prevailing market structures, the potential benefits of productivity improvements can be largely appropriated by the TNCs and global supermarket chains, instead of going to fragmented producers and farmers. The governance structures of primary commodity value chains have become increasingly buyer-driven with a shift in the distribution of value skewed in favour of consuming countries.

South Africa: increased corporate control

The South African food market is almost equally divided into an informal food retailing sector and a retail system based on the supermarket format. The latter accounts for about 55 per cent of the total food market. It is dominated by a small group of domestic and international firms. Pick'n Pay, Shoprite, SPAR and Woolworths together own 88.5 per cent of all supermarkets (with 33 per cent, 30 per cent, 13.5 per cent and 12 per cent, respectively). These figures indicate that consolidation and concentration are ongoing in the South African retail system, where supermarket chains are crowding out fresh produce markets.

Trans-nationalization is also on its way in the food processing sector. Unilever, Nestlé, Procter & Gamble, the Bidvest Group, Foodcorp and Pioneer Foods make up most of South African food processing sales, along with domestic firms such as Tiger Brands, National Brands, Tongaat-Hulett Group and Illovo Sugar. Market concentration is thus also high in the food processing segment, although it varies widely from one produce line to the other. On average, the market share of the top four companies is about 57 per cent. This concentration co-exists with numerous small- and medium-sized enterprises that supply the informal retail system.

Source: Nissanke and Thorbecke (2010); Dalle Mulle (2010) at: www.3dthree.org/pdf_3D/3D_ExploringtheGlobalFoodSupplyChain.pdf (accessed 27 July 2011)

Plate 5.2a, b, c, d Retailing is being transformed across the north and south (Courtesy: Rosie Duncan for Chinese photographs and Lina Krishnan for Indian photographs).

Their reach is also extended as they exploit outsourcing opportunities: 'when a company searches for inputs, production sites and outputs where costs are lower and regulatory, political and social regimes [are] favourable' (Hawkes, 2006: 2). These two processes help them reduce costs and increase profits, a process that is self-sustaining, because as they increase their power they can drive harder bargains and effectively squeeze costs out across the food chain. We may appreciate the extent of this process by reviewing the facts and figures about the concentration of power across the food chain considered in Chapters 3 and 4. Concentration has intensified at all stages of the food system, from the inputs to agriculture (seeds, machinery, chemicals) through its processing and manufacturing, through retailing and, finally, the marketing of the foods is now also in the hands of a few powerful players.

Another symptom of globalization is increased levels of foreign direct investment (FDI), as discussed in Chapter 4. As with trade in agricultural commodities, the rules and regulations governing FDI have relaxed considerably since the 1980s, allied to the liberalizing agenda. Hawkes argues that '[F]ood processing is now the most important recipient of FDI relative to other parts of the food system' (2006: 3). This investment helps explain the rise in consumption of processed foods, from retailers as well as from fast food outlets across the developing world, and particularly in the emerging market economies where obesity is growing at its most alarming rates. Hawkes (2006) examines the case of Mexico to substantiate her case, and Patel also considers the Mexican case to exemplify some of the worst aspects of 'experiments in free trade, and trade in food in particular' (Patel, 2007: 47–74); see the review of the Mexican case in Box 5.10.

Box 5.10

Mexico tackles record child obesity

> Obesity and excess weight are one of the biggest health challenges that Mexico is facing today', President Felipe Calderon said as he introduced the national public–private plan to improve eating habits, mainly among children.
>
> (*The Independent*, 1 February, 2010, available at: www.independent.co.uk/life-style/health-and-families/health-news/mexico-tackles-record-child-obesity-1885597.html?action=Popup&gallery=no)

Mexico is an excellent case study because it is typical of many emerging markets that have experienced rapid transformations in recent decades as economic liberalization has intensified. It is considered a middle-income country by the World Bank; however, extreme inequality exists among rich versus poor, urban versus rural, north versus south, and among indigenous versus other population groups (some of this is related to its colonial history). While Mexico has the highest per capita income in Latin America, 53 per cent of the country lives in poverty (living on less than US$2 per day), while 24 per cent is extremely poor (living on less than US$1 per day). This inequality exacerbates the dual burden of simultaneously higher rates of undernutrition and obesity (Cambio, 2010).

Mexico first alerted the world to the debt crisis when, in 1982, it announced that it was unable to repay its debts. In the following years, it implemented SAPs and liberalized its economy. These changes intensified, when in 1994, it signed the North American Free Trade Agreement (NAFTA) with the United States and Canada. What ensued were radical changes to its food system as it opened its borders to inflows of foreign direct investment (FDI) and subsidized corm (maize) from the United States. Both these changes were to have serious ramifications on its diet and its small farm sector.

Flows of FDI increased rapidly, especially in the food manufacturing and processing sectors. Mexico, as with so many emerging markets, has a broad-based population pyramid that reflects a very high percentage of young people. This represents a very exciting marketing opportunity to the food corporations, especially for soft drinks, snacks and fast foods which have high profit margins (unlike fruit or raw foods). Indeed, because competition in developed markets is intense and these markets are already saturated, all leading food manufactures view the emerging markets, with their high population momentum and rising incomes, as the most important areas for expansion. Across the globe, food advertising is targeting children and young people in a great variety of ways. 'Marketing explicitly involves designing strategies and implementing activities to influence consumption habits and create demand. It involves not just advertising, but a whole array of methods including sales promotions, websites, viral marketing, music and sports sponsorship, product placement in films and television, and in-school marketing' (Hawkes, 2006: 4).

Such activities are the objects of considerable concern and health workers are calling for tighter regulations, but legislation is very partial and considered woefully inadequate by most public health officials. These strategies are also very effective and have transformed the diets of whole generations in Mexico and elsewhere. General Mills, one of the world's largest food manufacturers, boasts that its 'Mexican consumers enjoy an array of General Mills products – from Betty Crocker brownies and Pillsbury cinnamon rolls, to Green Giant Niblets sweet corn' (General Mills website).

Liberalization has had other important impacts for consumers in Mexico. The retail sector and fast food landscape has been transformed, and now some of the largest food corporations from the United States enjoy a significant presence. In large- and medium-sized cities, major supermarkets, discount and convenience stores capture over 55 per cent of the market. The story of Wal-Mart in Mexico is one of year-on-year expansion since their first venture in 1991 (see the Fact Sheet at www.walmartstores.com/download/2006.pdf; accessed 12 December 2011). In rural places and smaller towns, the food manufacturers have marketed their products through the traditional roadside outlets known as '*Tiendas*' (Hawkes, 2006: 4).

Liberalization has also changed the shape of farming in Mexico. When they joined the NAFTA, the Mexican government allowed North American farmers access to their domestic market. The story is complex, but the essence is, heavily subsidized corn from the fertile heartlands of North America was sold in Mexico with devastating consequences for many small domestic producers. Patel (2007: 47–74) examines the circumstances and summarizes the cruel facts and their outcomes thus:

> [T]he price at which US farmers sold their corn was much lower than the cost to produce it. In 2002, for instance, US corn cost $1.74 per bushel to buy but $2.66 for US farmers to produce. This is because the United States had long supported its farmers and had made a range of subsidies available to them for machinery,

> fertilisers, credit and transport. With the advent of liberalisation, it was clear that US corn … would destroy the livelihoods of the poorest in the Mexican rural economy.
>
> (Patel, 2007: 49)

And so it proved; although some Mexican farmers benefitted from liberalization, such as those close to the border and with capital to invest in fruit and vegetable production for the US market, many of the poorest and those far from the border suffered a collapse of their livelihoods. It is important to mention that the Mexican political elite felt that this was a price worth paying for what they perceived as other advantages to membership of the NAFTA, and the survival of the small peasant sector was not at the top of their agenda. That Mexico is the source region for corn (which since European colonization has been taken across the globe) where more than 40 varieties are grown, and hundreds more exist, adds another ironic element to this sad story. The region's biodiversity, which has survived for thousands of years, is threatened as the peasant farmers, who traditionally cultivated a range of varieties, migrate to the cities where their knowledge of crops and expertise will be redundant.

This case study illustrates very clearly how liberalization, this time through a regional trade agreement, can simultaneously decimate rural producers and their entitlements and transform traditional diets and replace them with less nutritious options. Finally, it also suggests how such processes threaten food security and sovereignty. Food security was eroded because Mexico became much more exposed to world market prices for food, and sovereignty was eroded as control over their food choices was undermined in several respects. Corporate control over provisioning was increased, and corporate producers financed marketing campaigns targeting children and their consumption habits.

Source: *The Independent* (2010) at: www.independent.co.uk/life-style/health-and-families/health-news/mexico-tackles-record-child-obesity-1885597.html (accessed 18 November 2010); also see www.cambio-red.net/ (accessed 18 November 2010)

One of the most obvious manifestations of globalization in developing countries is the growth of advertising industries and their role in promoting Western diets. Most developed world food markets are saturated, and there is only so much food that can be consumed – well, we traditionally thought so anyway! The satisfied character of these developed markets helps explain two aspects of corporate behaviour. First, in the developed world, they are exploiting the opportunities associated with 'functional foods' and other niche markets; only by inventing 'new' foods or new consumers can corporations guarantee their profits in these markets. Diet foods, and

Plate 5.3 An important global consumer: Tina at home (Courtesy: Peter and Janet Stubbs).

the pet food market, are excellent examples; only by inventing 'new' foods and 'new' consumers (pets) can corporations extend these markets.

The pet foods story is fascinating and estimated to be worth US$73 billion in 2010. Its geography is interesting too and, just as the Western diet for humans is going global, so too is the Western diet for pets, with similar unhealthy consequences. Pets are becoming obese in developed market economies as their owners dish them up treats and gourmet meals. Across the globe, clever advertising is paying dividends:

> While Brazil is the largest dog and cat food market behind the US, India and Russia are the fastest growing. … The greatest opportunities for market growth are in Latin America and Eastern Europe. Prepared dog and cat food is on the rise in Russia, with the majority of consumers feeding table scraps in 1998 to a significant shift to packaged pet foods by 2009.
>
> (Taylor, 2010)

In 2010, pet food sales dwarfed baby food sales in the United States and Britain. Even in recessionary times apparently, some pets are well fed. Mars and Nestlé dominate the global pet food business; both have growth in recent years through a series of mergers and acquisitions, and their markets are global.

Hence, here we find another, alarming paradox in our globalized world; overweight and underweight humans are now joined by obese pets as victims of the food system. This example, about pet food, also highlights how food follows money and not need. As Sen (1981) first recognized, food goes to those who can afford to purchase it, rather than to those who need it most. It may not be an exaggeration to assert that a pet in an affluent community enjoys more purchasing power than a poor individual in the global south. It may also be the case that by using cheap cuts, such as offal, the pet food industry indirectly increases the price of food which may previously have been destined for human consumption. By opening up a lucrative 'gourmet' pet market for liver, hearts, tripe etc., the price of goods that formerly used these cheap offal foods, such as sausages, faggots, etc., may rise in tandem.

Very different market conditions exist in the developing world, which explains the second aspect of corporate behaviour. In the developing world, excellent growth potential exists, among the growing middle classes and especially among the billions of young people who make up the majority of these populations. The developing world has been inundated by sophisticated mass advertising campaigns that penetrate all aspects of work, home and play.

> Marketing explicitly involves designing strategies and implementing activities to influence consumption habits and create demand. It involves not just advertising, but a whole array of methods including sales promotions, websites, viral marketing, music and sports sponsorship, product placement in films and television, and in-school marketing.
>
> (Hawkes, 2006: 5)

Recent research, based on samples from across a number of counties, found that 'children were exposed to high volumes of television advertising for unhealthy foods, featuring child-oriented persuasive techniques' (Kelly *et al.*, 2010: 1). There is surely a relationship between the opening up of developing world markets to FDI, and associated advertising companies, and the appalling rise in childhood obesity across the developing world, a trend that is most marked in emerging market economies?

To review our analysis based on Hawkes' (2006) excellent study: we have already considered most factors except the role of urbanization and cultural change. These are allied factors, as urban populations are more exposed to advertising and international cultural influences. The world's population is now over 50 per cent urban, a novel condition for human kind. The process of rural–urban migration has been occurring for centuries, but has intensified in the recent decades. One reason that explains these trends is the transformation of agricultural conditions across the developing world, namely, the expansion of capitalist agriculture (see Chapter 4). This expansion may be rapid or gradual but everywhere its causes and consequences are similar: large numbers of the rural poor lose their entitlements; suffering economic hardship, they have little option but to migrate to the cities. Hence, urbanization rates are growing, and while some new arrivals enjoy access to a better quality of life, many remain relatively poor and survive on tenuous earnings from working in the informal sector. It is important to appreciate the links between changes in the agricultural sector and the rise of urbanization – they are linked phenomena. What are the health ramifications for those who move to the city environments? There is a need for more research, and Harvard University has recently announced new research designed to understand the nutritional transition and associated epidemiological transition in Africa (http://news.harvard.edu/gazette/story/2010/12/sick-to-death (accessed 20 July 2011).

Available research substantiates what we might expect, that urbanization facilitates the adoption of lifestyles that are conducive to obesity and cardio-metabolic risk factors. This is confirmed based on studies from Africa (Sodjinou *et al.*, 2008) and in Latin America (Cuevas, A. *et al.*, 2009). A recent book details the extent of the problem in China (French and Crabbe, 2010). They all identify elements of the obesogenic environment considered in Chapter 3. To review, essentially, obesogenic environments encourage the consumption of more food and expenditure of less energy. Urban landscapes across the world are now cluttered with the icons of fast food outlets, and in many cases, littered with street vendors selling more Westernized products than previously, more sugary drinks, crisps, fat-rich snacks, etc. Life in cities tends to be more sedentary. French and Crabbe observe that '[M]ore and more Chinese are *bai-ling*, white collar workers who, like many well-off Westerners, spend their days commuting to and from work, sitting at a desk or watching television. Sedentary lives are becoming the norm' (French and Crabbe, 2010: online 1). Unfortunately too, city planners in many developing nations are following the petroleum-dependent patterns well established in the West; car ownership is increasing, and urban spaces discourage exercise, walking or cycling.

Changing cultural norms are important too. Patterns of family life are changing as more women enter paid work and, with access to cheap pre-prepared meals, eat at home or in restaurants, spending less time preparing meals from raw ingredients. The role of advertising in explaining these changes cannot be ignored; as reviewed earlier, the entrance of the global food manufacturers into all aspects of life in the developing world plays its part in promoting all the changes identified.

After neoliberalism?

Some changes have occurred with implications for agriculture and diets. As the Doha Round of negotiations collapsed, and therefore opportunities for further globalization of the food system retreated, new avenues have become more important for integration and corporate interests. It will become increasingly important to understand how regional integration will transform both entitlements and diets. Prakash notes the shift of focus:

> The recent lack of further progress in food and agricultural trade liberalization has shifted the focus onto regional and bilateral agreements as a means of liberalizing food trade. Notable examples include the Mercado Común del Sur or Southern Common Market (MERCOSUR), the Association of Southeast Asian Nations (ASEAN), Free Trade Area in Asia, and the North American Free Trade Agreement (NAFTA). In the mid-2000s, as uncertainty about the progress of the Doha Round of WTO trade talks took hold, the number of regional trade agreements signed reached unprecedented levels. As of December 2008, 421 regional and bilateral trade agreements had been notified to the WTO and 230 agreements were in force.
>
> (Source: Prakash, A. (2011) Safeguarding Food Security in Volatile Global Markets at: www.fao.org/docrep/013/i2107e/i2107e.pdf; accessed 20 July 2011)

Development discourses have changed too. Several circumstances help explain why development theory has undergone further revisions since the late 1990s. A shift is observable from a very rigid neoliberal agenda to a softer and more 'inclusive' poverty reduction and good governance agenda. The tight embrace of free market theories has relaxed. The Asian crisis in 1997–1998 caused some of its most enthusiastic adherents to pause, and '[T]wo decades of development failure and zero net growth on whole continents had produced alarming peripheries of insecurity, disaffection and risk' (Craig and Porter, 2006: 2). Development theories were again found wanting. Even people at the highest levels of the IFIs were accepting that their prescriptions had been proved wrong and damaging to some of the poorest communities. The World Bank's Development Report (2006) accepted that their SAPs medicine has endangered the livelihoods and survival of many small holders in agriculture.

A new consensus is observable in the last decade where poverty reduction and good governance are understood to be the central tenets of development; markets alone have been judged wanting. Liberalism prevails, but is now embedded in the consensual politics of inclusivity and poverty reduction strategies. Much of this was a reaction to the crisis of government in much of the global south and the increased threat of insecurity and its association with terrorism and global criminality; maybe governments had a role after all? Just how these might impinge on the central concerns of this text is unclear.

The clearly articulated ambition to reduce poverty may be welcome but it is too early to understand just how this might be achieved or how entitlements might be supported in the new era. Now, development discourse speaks about local communities and developing countries' 'ownership' and participation of economic and poverty reduction policies. Certainly, most analysts once more accept that national governments are necessary and have a right and proper place in designing and delivering development. As will be reviewed in Chapter 9, national political initiatives have always been necessary to reduce undernutrition, and the case is the same for effective programmes to address obesity (Young, 2004). However, given the continued power of the corporate sector in agrarian and dietary change, it is difficult to see how they might be effectively policed to deliver a more just food system. Spaces are nevertheless being created that offer some room for optimisms (Chapter 9).

National responses: addressing malnutrition

It would be remiss to leave the discussion without summarizing national progress in the last few decades. Some national governments have, despite considerable obstacles, intervened to support entitlements and reduce undernutrition, whereas others have been less successful. A recent analysis by Action Against Hunger (AGH, 2011) identified some remarkably successful efforts led by national governments that have reduced undernutrition in recent decades. The five cases analysed were Brazil, Peru (summarized in Box 5.11), Malawi, Bangladesh and Mozambique (Table 5.5). The two secret ingredients in all examples were: a political determination to address hunger at the highest level of government, and a multi-sector approach.

Box 5.11

Improving child nutrition: Peru: a success story

The prevalence of underweight children worldwide fell from 31 per cent in 1990 to 26 per cent in 2008 (UNICEF, 2009). However, progress is still slow and very uneven. Half of the countries have made progress on hunger, but levels of malnutrition did not improve in 28 countries and got worse in 24. The case of Peru appears to be an encouraging exception to the rule. With the support of CARE and other organizations from civil society and the donor community, the Peruvian government has generated the political momentum to overcome obstacles and create national coordination structures and mechanisms, increase public (and private) spending on programs to tackle malnutrition and align social programs with the national nutrition strategy (known as CRECER). This included adding conditionalities on taking children to regular growth monitoring in the Conditional Cash Transfer program, JUNTOS. The international aid system has also aligned itself around CRECER. After 10 years of almost no change in child chronic malnutrition (stunting) rates (25.8 per cent in 1996, and 22.9 per cent in 2005 – with rural rates moving from 40.4 per cent to 40.1 per cent), this change in strategy has started to lead to results: malnutrition rates fell to 17.9 per cent between 2005 and 2010, with reductions mainly occurring in rural areas where malnutrition rates are highest (from 40.1 per cent in 2005 to 31.3 per cent in 2010), according to the Peruvian National Statistical Office (INEI). Over 130,000 children under five are now not chronically malnourished who would have been had rates not fallen. Indeed, there is a strong case to be made that these changes would not have occurred without the formation in early 2006 of the Child Nutrition Initiative, and its advocacy success in getting ten Presidential candidates to sign a commitment to reduce chronic malnutrition in children under five by 5 per cent in 5 years.

Source: Acosta, Andrés Mejía (2011) Analysing Success in the Fight against Malnutrition in Peru, IDS Working Paper vol 2011 no 367 at: www.ids.ac.uk/files/dmfile/Wp367.pdf (accessed 22 July 2011). Read about CARE projects in Peru at: www.careinternational.org.uk/index.php?option=com_content&view=article&id=723&Itemid=256 (accessed 22 July 2011)

Table 5.5 ***Mozambique: impressive improvements in nutrition rates****

Nutrition indicators

Indicators	*1995 (%)*	*1997 (%)*	*2000 (%)*	*2003 (%)*	*2008 (%)*
Chronic malnutrition rates	55	40	44	41–48	44
Underweight rates	27	24	26	24–20	18
Acute malnutrition rates	8			4	4
Exclusive breastfeeding for up to 6 months		30		30	37
Percentage of population below the minimum level of dietary energy consumption	199 59		2004 38		

Mortality

Children	*1990*	*1997*	*2008*	
Under 5 mortality rate	249	201	130	
Women	*1995*	*2000*	*2005*	*2008*
Maternal mortality ration	890	780	640	550

Health coverage

Immunization	*1990 (%)*		*2007 (%)*	
Measles: fully immunize 1-year-olds	59		77	
Micronutrients	*2005 (%)*	*2006 (%)*	*2007 (%)*	*2008 (%)*
Vitamin A supplements	32	33	42	83
Women	*1990 (%)*		*2007 (%)*	
Antenatal care coverage at least once	71		87	

*Data from various sources – WHO, UNICEF, Millennium Development Goal (MDG) Report Card; for more details, see original table.

Source: Undernutrition: 'What Works? A Review of Policy and Practice' Action Against Hunger (2011), at: www.actionagainsthunger.org.uk/fileadmin/contribution/0_accueil/pdf/Undernutrition%20What%20Works.pdf (accessed 26 July 2011)

In Latin America, both Mexico and Brazil were leading the fight against hunger and have implemented wide ranging, inclusive social programmes. Multiple agencies have been involved, and they enjoyed major budgetary support. The Fome Zero (Zero Hunger) programme in Brazil has been exceptionally successful. The Brazilian case may be explained with reference to its rapid economic growth, but this alone cannot explain the reduction in hunger statistics. Rapid economic expansion frequently parallels increased social polarization. More important has been the election in 2002 of Luiz Inacio Lula da Silva who was committed to hunger reduction policies.

In 2011, Lula received the World Food Prize for his personal commitment and visionary leadership while serving as president of Brazil. The statistics for Brazil are shown in Table 5.6. The award recognized his vital role in creating and implementing government policies to alleviate hunger and poverty.

Progress at reducing hunger has been disappointing in most of Africa. The case of Ghana, however, is an exception, and shows how a

Plate 5.4 President Luiz Inacio da Silva (Source: Agência Brasil, Wikimedia Commons, http://en.wikipedia.org/wiki/File:Lula_-_foto_oficial05012007_edit.jpg).

Table 5.6 ***Brazil: impressive improvements in nutrition rates***

Nutrition indicators

Indicators	*1980 (%)*	*1996 (%)*	*2006 (%)*	*Sources*
Chronic malnutrition rates	16	13.5	7	UNICEF/WHO and NCHS
Underweight rates	13	6	1.7	
Acute malnutrition rates	2	2	1.98	
Exclusive breastfeeding for up to 6 months	NA	NA	39.8	
Percentage of population below the minimum level of dietary energy consumption	***1991*** 10		***2004*** 6	

Mortality

Children	*1996*	*2003*	*2008*	
Under 5 mortality rate	42	35	22	
Women	*1995*	*2000*	*2005*	*2008*
Maternal mortality ration	98	79	64	58

Health coverage

Immunization	*1995 (%)*	*2008 (%)*
Measles: fully immunize 1-year-olds	74	99
Micronutrients		
Vitamin A supplements	No data	No data
Women	*1995 (%)*	*2008 (%)*
Antenatal care coverage at least once	86	98

*Data from various sources – WHO, UNICEF, MDG Report Card; for more details, see original table.

Source: 'Undernutrition: What Works? A Review of Policy and Practice' Action Against Hunger (2011) at: www.actionagainsthunger.org.uk/fileadmin/contribution/0_accueil/pdf/Undernutrition%20What%20Works.pdf (accessed 26 July 2011)

concerted effort at the highest political level can deliver impressive results. John Agyekum Kufuor, the president of Ghana between 2001 and 2009, was also awarded the World Food Prize in 2011 (Box 5.12). Recent initiatives in Africa include the African Union and its members endorsing the Comprehensive Africa Agricultural Development Plan (CAADP), which aims to eliminate hunger in the

Box 5.12

Ghana: a success story from Africa

A guiding principle for President John Kufuor during the entirety of his two terms as president (2001–2009) was to improve food security and reduce poverty through public- and private-sector initiatives. To that end, he implemented major economic and educational policies that increased the quality and quantity of food to Ghanaians, enhanced farmers' incomes, and improved school attendance and child nutrition through a nationwide feeding program.

Under President Kufuor's leadership, Ghana became the first sub-Saharan African country to cut in half the proportion of its people who suffer from hunger, and the proportion of people living on less than a dollar per day, on course to achieve UN Millennium Development Goal 1 before the 2015 deadline. Continuing Ghana's tradition of stability, President Kufuor prioritized national agricultural policies: Ghana saw a reduction in its poverty rate from 51.7 per cent in 1991 to 26.5 per cent in 2008, and hunger was reduced from 34 per cent in 1990 down to 9 per cent in 2004.

President Kufuor's economic reforms, including the Food and Agriculture Sector Development Policy, provided incentives and strengthened public investments in the agricultural and food sector – the backbone of Ghana's economy – which grew at a rate of 5.5 per cent between 2003 and 2008. Growth in the agricultural sector drove expansion in the national economy, with GDP quadrupling to 8.4 per cent by 2008.

Under President Kufuor, the Agricultural Extension Service was reactivated and special attention paid to educating farmers on best practices. As a result, Ghana's cocoa production doubled between 2002 and 2005, and food crops such as maize, cassava, yams and plantains increased significantly, as did livestock production.

The Ghana School Feeding Program launched by President Kufuor provided one nutritious locally produced meal a day for schoolchildren in kindergarten to junior high school (ages 4–14). By ensuring nutritious food at school, this program dramatically reduced the level of chronic hunger and malnutrition while improving attendance. By the end of 2010, approximately 1.04 million primary schoolchildren were participating and benefitting from this program.

Source: www.worldfoodprize.org/index.cfm?nodeID=33367&audienceID=1; accessed 27 July 2011

continent by supporting agricultural development. As discussed elsewhere, this is a vital priority for Africa, where the majority of the poor are still in rural areas and depend, directly or indirectly, on agriculturally based entitlements. The African Regional Nutrition Strategy 2005–2015 is another initiative that, given political commitment and investment, should help that continent improve its record in reducing undernutrition. This strategy recommends integrating nutrition as a specific issue in national policy-making; this appears as a common element of the most successful government-led efforts.

The Asian continent displays such diversity that any generalization is meaningless. China's success at reducing undernutrition has been reviewed in Box 5.5, and any change in the statistics for China has a disproportionate influence on the statistics for Asia. The same is true for India, where improvements in nutritional statistics have been more modest. A recent study of the situation in Bangladesh (AGH, 2011) judged it to have made 'impressive economic and social progress over the past decade' (p. 41). National poverty rates fell from 67 per cent in 1990 to 50 per cent in 2008, and the country is on track to meet the MDG of having extreme poverty eradicated by 2015. Bangladesh continues to suffer from frequent national disasters which, combined with its poverty, make progress more difficult. Because of its specific environmental circumstances, many fear that global warming will exacerbate an already fragile situation. In much of South Asia, conditions remain very serious, and India, Bangladesh and Pakistan account for 50 per cent of the world's underweight children (AGH, 2011: 9). Some of these statistic reflect the poor status of women in South Asia, the only region of the world where girls are more likely to be underweight than boys (see Chapter 7). In most South Asian countries, nearly 50 per cent of women of reproductive age are anaemic, and anaemia prevalence among pre-school children has not dropped below 50 per cent for over two decades (AGH, 2011: 21).

As mentioned in Chapters 1 and 2, undernutrition has not been eliminated in the developed world. Hunger continues to be found in some communities and regions in the European Union and North America. Some of these are consider in more detail in Chapter 6, but the case of hunger in the United States is worth considering here to illustrate that economic success is not necessarily correlated to social equity (Box. 5.13). Some countries in Central and Eastern Europe are struggling to eliminate undernutrition, and are battling with obesity too (Box 5.14).

Box 5.13

11 facts about hunger in the United States

1. In 2009, 50.2 million Americans (up from 35.5 million in 2006), including 17.2 million children, are food insecure, or didn't have the money or assistance to get enough food to maintain active, healthy lives.
2. In 2009, 65 per cent of adults reported that they had been hungry, but did not eat because they could not afford enough food.
3. A rise of about 6 per cent in the price of groceries in 2008 alone has led the poor to adopt a variety of survival strategies, from buying food that is beyond its expiration date to visiting food banks.
4. About 33.7 million people in America participate in the Supplemental Nutrition Assistance Program (SNAP) – a program that provides monthly benefits to poor households to purchase approved food items from authorized food stores. According to the USDA, the average benefit per person was US$124 per month, and the Federal government spent over US$53.6 billion on the program in 2009.
5. America's Second Harvest, the nation's major food bank network, annually provides food to over 23 million people. That is more than the population of the state of Texas.
6. 5.6 million households obtained emergency food from food pantries at least once during 2009.
7. The USDA recently found that about 96 billion pounds of food suitable for human consumption was thrown away by retailers, restaurants, farmers and households in the United States, over the course of 1 year. Fresh fruits and vegetables, fluid milk, grain products, and sweeteners accounted for two-thirds of these losses.
8. Hungry adults miss more work and require more health care than those who don't go hungry.
9. Kids who experience hunger are more likely to suffer from anxiety, depression, behaviour problems and other illness.
10. The total cost of hunger to American society is said to be about US$90 billion a year.
11. In contrast, it would only cost about US$10 billion to US$12 billion a year to virtually end hunger in our nation.

Sources: United States Department of Agriculture, Bread for the World, America's Second Harvest, Washington Post, Meals for Millions. Available at: www.dosomething.org/tipsandtools/11-facts-about-hunger-us (accessed 20 July 2011)

Box 5.14

Child nutrition in Central and Eastern Europe and Commonwealth of Independent States

Malnutrition in early life may damage future health and development in an irreversible way. The data available show an association between undernutrition and higher child mortality. Turkey, Armenia and Georgia, and also Kyrgyzstan, Uzbekistan, Kazakhstan, Azerbaijan, Turkmenistan and Tajikistan, are the countries that would benefit more, in terms of reduction of U5MR, from actions aimed at removing or mitigating the root causes of undernutrition. The permanence of iodine deficiency disorders (IDD) as a public health problem indicates that undernutrition will keep affecting the performance of many people for years to come.

In addition, many of these countries (namely, Armenia, Georgia, Kyrgyzstan, Uzbekistan and Kazakhstan) are already facing the double burden of undernutrition and overnutrition. The latter is probably a priority public health problem also in Serbia, Montenegro, Bosnia and Herzegovina, Ukraine, Albania and Russia. The prevention of overweight and obesity starts in infancy and includes the protection, promotion and support of exclusive breastfeeding until 6 months and continued breastfeeding until 2 years and beyond. The prevention of overweight and obesity depends also on timely and adequate complementary feeding, on a healthy family diet and on adequate levels of physical activity. These healthy behaviours depend on social factors that have to be adequately addressed in all countries.

Malnutrition is not equally distributed. In some countries, regional and ethnic differences exist, so that even where undernutrition, overnutrition, breastfeeding or micronutrient deficiencies may not be considered as public health priorities at the national level, they may be so for specific population groups or minorities. Also, there is almost always a wealth and education gradient in the rates of malnutrition within countries. This means that an equity lens should be used in both programme implementation and in monitoring.

Source: www.usaid.gov/locations/europe_eurasia/health/docs/2008_ceecis_child_nutrition_report-final.pdf (accessed 22 July 2011)

Regarding the new nutritional crisis, namely, obesity, national responses have been partial and ineffective almost everywhere. Interventions tend to address, in piecemeal fashion, proximate causes where individual responsibility is emphasized and structural causes are ignored. Most governments succumb to the power of the corporate food sector, in its various manifestations. Lang (2010) recently examined the limitations of food policy in the United Kingdom. The same accusations could be made against most Western governments. Serious efforts to challenge obesity hit powerful vested interests at every stage: agri-business, food processing corporations, car manufacturers, food advertising, to name a few. As mentioned in Chapter 3, national initiatives to reduce obesity are feeble and limited. It the structural causes are not addressed, any improvements in our health will be minimal (Young, 2004).

Most governments have supposed that corporate behaviour will respond to a 'light touch' regulatory environment and have refused to impose stringent guidelines. The position of marketing junk food to children in the United Kingdom illustrates government's reluctance to take appropriate action. Responding to the publication of the National Heart Forum's report 'An analysis of the regulatory and voluntary landscape concerning the marketing and promotion of food and drink to children', Children's Food Campaign Coordinator Christine Haigh said:

> We welcome this comprehensive review of food promotion to children, which shows the myriad ways in which children are targeted by junk food marketers and demonstrates the inadequacy of voluntary commitments to tackle this issue. Rather than pursuing the weak voluntary approach of the last government, we need decisive action from the coalition government to introduce comprehensive regulation to protect children from junk food marketing in all its forms. See the article at: www.sustainweb.org/news/new_approach_junk_food_marketing_jun11 (accessed 27 July 2011).

Conclusion

This chapter has established the very diverse patterns of development and their impacts for food and diets. It is obvious that historical legacies still help explain the patterns of power, privilege and poverty in most countries, most obviously in how they influence access to economic and political power, and land ownership. Equally vital are the development strategies employed by different countries since the 1950s, where the agricultural sector has been neglected in favour of industrial production. Everywhere, even where the political rhetoric asserts otherwise (in China and India), investment in rural infrastructure has been limited, and the rural surplus has been taxed to facilitate industrial investments. This helps explain why rural poverty is everywhere more serious than urban poverty, and why incomes and access to education and health care are both worse in rural areas.

While opportunities presented by globalization have been successfully exploited by some countries, the process has exacerbated poverty and malnutrition in other countries. National-level analysis, therefore, retains explanatory relevance. Global-level processes, examined in Chapter 4, explain part of the diversity of development, but these processes are always mediated by important national-level variables. Just how constraints and opportunities are negotiated by national governments is vital to analyses. The nature of the government, its capacities, capabilities and commitment to progressive socio-economic and political change are all relevant to explanations of the diverse paths to progress, or the lack of progress in some contexts. Many countries in the global south have seen an erosion of their food sovereignty and food security as they embraced, willingly or under duress, globalization.

Economic growth alone will not deliver equitable development, food security or a healthy diet. The relationship between economic growth and improvements in health, including nutrition and education, are complex and may result in some interesting outcomes. The United Nations Development Report (2010) illustrates this by a revealing comparison between Tunisia and China. 'In 1970 a baby girl born in Tunisia could expect to live 55 years; one born in China, 63 years. Since then China's GDP has grown at a breakneck pace of 8 per cent per annum, while Tunisia's has grown at 3 per cent. But a girl born today in Tunisia can expect to live 76 years, a year longer than a girl born in China' (UNDP, 2010: 47).

Summary

- **Despite the increased role of global-level processes, changes implemented at the national level remain important for analysis of malnutrition.**
- **National levels of nutrition are very dynamic; some countries have proved relatively successful at reducing undernutrition, and others much less so.**
- **National levels of obesity are also very dynamic and have soared in the last 20 years in some countries previously associated with extreme undernutrition.**

- **The influence of historical factors still helps explain the patterns of entitlement in many countries in the global south. Access to land is a major variable.**
- **Development theories have been important in shaping patterns of entitlement and nutrition since the 1950s.**
- **Structural adjustment policies, driven by neoliberal economic agendas in the 1980s and 1990s, have intensified corporate control over agriculture and the global food system.**
- **Corporate control over the food system has serious implications for people in the global south and north.**

Discussion questions

1. Annotate a map of the world to illustrate some marked contrasts in nutrition.
2. What relevance has colonialism to patterns of contemporary undernutrition?
3. Discuss the role of development theory to patterns of food security.
4. Review how corporate practices help create obesogenic environments.
5. Examine how national food sovereignty may be compromised by globalization, with examples.

Further reading

Knox, P., Agnew, J. and McCarty, L. (2003) *The Geography of the World Economy*, London, Hodder Arnold. This provides an in-depth introduction to the globalization of the world economy. Good read on economic transformations in the eighteenth and nineteenth centuries, and the links between developed and developing worlds.

Willis, K. (2006) *Theories and Practices of Development*, London, Routledge Development Studies Series. This is a very accessible text which introduces undergraduates to the topic.

Nederveen Pieterse (2010) *Development Theory*, 2nd Edition, London, Sage. This is one of the best; it provides a comprehensive overview of development.

Preston (1996) *Development Theory: an introduction* Oxford, Blackwell. This remains a very comprehensive analysis, but most would not consider it introductory. Rapley (1996) delivers a very useful overview and is more accessible than Preston; he too integrates excellent case histories and useful chronology.

Useful websites

www.imf.org/external/np/prsp/prsp.aspx Poverty Reduction Strategy Papers for most countries are available here. They offer interesting insights into development concerns and trends, especially new agendas about governance and poverty reduction.

www.undp.org Important source for national development data and development programmes.

www.ids.ac.uk An excellent site for debates, bulletins, discussion papers and lectures. For a recent and relevant in-depth analysis about national programmes to reduce malnutrition in Peru, see: www.ntd.co.uk/idsbookshop/details.asp?id=1232 (accessed 25 July 2011).

www.worldbank.org Numerous resources, data, publications about development. For example, for a recent publication that reviews agricultural change in Kazakhstan, see http://web.worldbank.org/WBSITE/EXTERNAL/NEWS/0,,contentMDK:22969533~menuPK:64256345~pagePK:34370~piPK:34424~theSitePK:4607,00.html.

www.survivalinternational.org/thereyougo. Here is a wry take on 'development' (accessed 28 July 2011).

Chapter 5

Follow Up

1. Visit this site for a wry look at the objectives, means and consequences of 'development': www.survivalinternational.org/thereyougo (accessed 12 January 2011).
   - Make a list of your responses to the cartoons and their messages, and then compare and contrast your responses with your peers.
   - For review of development theory, see: www.uiowa.edu/ifdebook/ebook2/PDF_Files/Part_1_3.pdf
   - Select three different tribal communities and evaluate the threats they face.
2. Visit this address and listen to the story of tea: www.bbc.co.uk/programmes/b00v71qr. Also listen to Neil MacGregor's 'History of the World in 100 Objects' (BBC, Radio 4, 2010), a story of tea in the mid-nineteenth century with reference to a Victorian Tea Set www.bbc.co.uk/iplayer/console/b00v71qr

3. Read the Report by Wen, D. 'China Copes with Globalization' at: www.ifg.org/pdf/FinalChinaReport.pdf (accessed 15 August 2011).

   - Construct an essay plan to summarize the positive and negative social consequences of the Chinese reform that started in the late 1970s.
   - Construct an essay plan to summarize the positive and negative environmental consequences of the Chinese reform that started in the late 1970s.

4. Sometimes, NGOs can significantly improve a person's livelihood by simple interventions at relatively little costs. Learn about one such intervention by NGOs, called Practical Action, at: http://practicalaction.org/transport/animal_drawn_carts (accessed 25 February 2011).

5. Innovative ways to educate farmers about new techniques include promoting them in TV soap operas. A series called 'Makutano Junction' is a very successful example from Kenya. Learn about this award-winning initiative at: www.makutanojunction.org.uk/about-the-project/makutano-junction-wins-one-world-media-award.html

6 Sub-national perspectives

Learning outcomes

At the end of this chapter, the reader should be able to:

- **Explain with reference to case studies the limitations of national-level statistics**
- **Explain why the urban–rural dichotomy is problematic**
- **Describe and explain some significant sub-national patterns of nutritional diversity with reference to case studies**
- **Understand the role of political and economic relations to the construction of sub-national entitlements**
- **Describe the diversity within the 'rural poor', category and explain the multidimensional character of their poverty**
- **Describe how some previously marginalized communities are challenging the status quo to improve their economic circumstances**

Key concepts

Core and periphery relations; national diversity; ethnicity; indigenous communities; urban–rural contrasts; dependant development; primate cities; special economic zones; the double burden of malnutrition

Introduction

The previous chapters have examined factors that influence nutritional status at the international and national scales. This chapter considers some sub-national factors that help explain patterns of nutrition. Within all countries, there are groups of people who have a higher probability of suffering from undernutrition or overnutrition.

The following analysis attempts to identify these with reference to a variety of different countries. The following categories are considered most important; regional contrasts; differences between urban and rural populations; between different classes or ethnic groups; between and within households and between age groups. All of these contrasts have proved very persistent, and they are often interrelated. As the discussion reveals, these categories tend to overlap Hence, for example, the poorest rural populations are often female, ethnically distinct and live in marginal regions; frequently, these factors interact to create double or triple jeopardy for certain people (WHO/UN Habitat, 2010: 54). The review is necessarily complicated by these considerable interconnections and correlations; regional diversity in Sudan exemplifies this issue (Box 6.1).

Regional contrasts

Regional contrasts in socio-economic circumstances are found everywhere; they may be extreme or minimal, and the processes that create such unevenness are complex (Cox, 2002: 277). Most governments attempt to redress extreme imbalances by implementing economic policies to redistribute resources, often to reduce political dissent or separatist movements. Such policies are normally designated territorially. Regional contrasts in economic development are often examined using core–periphery models (see Chapter 5). While more commonly employed at the global scale (Wallerstein, 1974; Knox, Agnew and McCarty, 2008) to understand why some world regions enjoy economic and political dominance, they may also be employed at the national scale to understand why some regions within countries prosper more than others.

The theory holds that spatial differences are structural in nature and that the advantages enjoyed by the core tend to perpetuate existing spatial inequalities. The patterns are understood to represent asymmetrical powers where political, economic and social predominance is based on the history of regional incorporation into the global economy. In most developing countries, the roots of uneven development are found in colonial systems of exploitation. Core regions emerged to service the colonial elite and their interests, while peripheral regions provided raw materials and labour to fuel colonial enterprises, normally primary activities associated with mining or agricultural production.

Box 6.1

Sudan: a country divided

These maps illustrate how regional and ethnical categories often overlap. Numerous contrasts exist between northern and southern Sudan: contrast in food security; contrasts in access to education, etc., and these are reflected in ethnic divisions.

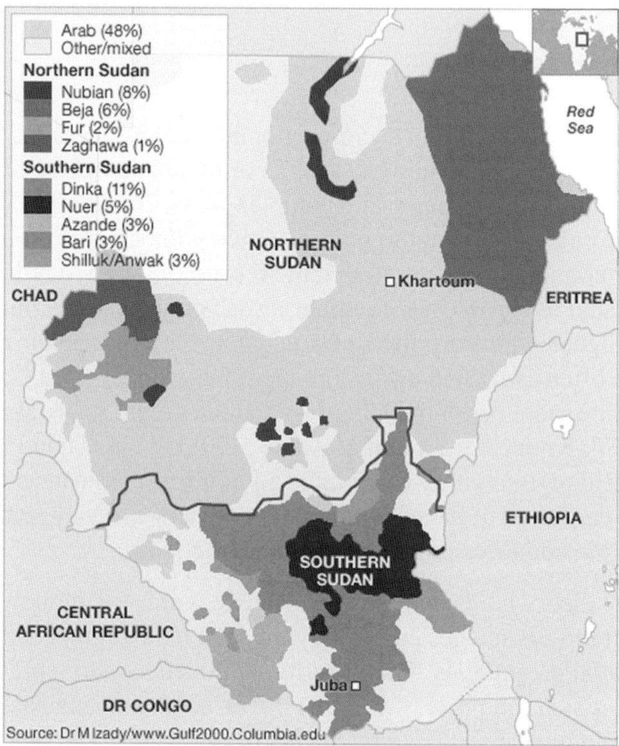

Source: Dr M Izady/www.Gulf2000.Columbia.edu

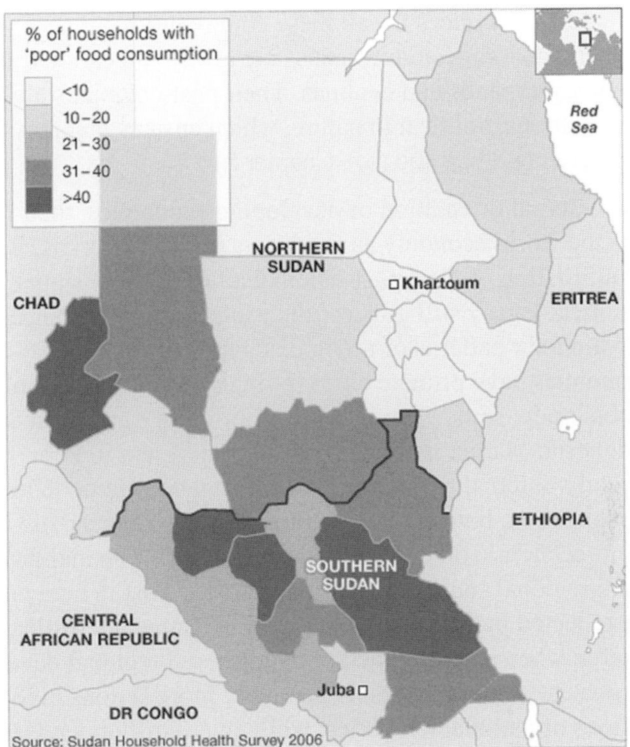

Source: www.bbc.co.uk/news/world-africa-12111730; also read more about the conflict here. More information about the situation is available at: www.insightonconflict.org/conflicts/sudan/conflict-profile (accessed 12 August 2011); this site also has information about other global conflicts.

The most extreme manifestation of such relationships is obvious in colonies where a primate city developed to administer the colonial regime and to exploit their hinterlands. Transport infrastructure, where all roads or railways led to the capital, was everywhere constructed to facilitate the export of commodities bound for imperial industrial heartlands. Examples of such 'dependent development' are found in most former colonies, but are exemplified most suitably by countries in sub-Saharan Africa which are still economically dependent on the export of raw materials. Some examples are: Cote d'Ivoire, whose export earnings are based on cocoa, coffee and recently oil; Nigeria, which earns 95 per cent of export earnings from oil; Angola, which depends on exports of oil for 85 per cent of earnings, and diamonds for 5 per cent; and the Democratic Republic of Congo, which depends on the exports from its mining sector, specifically, diamonds, copper, cobalt and gold. The problem with such 'dependent development' is that the global prices for all these commodities are very volatile, and so their economic circumstances are always precarious, as the world economy expands and declines. Their post-colonial status has other ramifications, political in nature, which ensures that they often suffer from civil conflicts too (see Chapter 8).

The external orientation of developing economies, their integration into the world economy as producers of raw materials, is an obvious manifestation of their post-colonial and periphery status. This historical legacy still governs their internal circumstances and helps explain their patterns of urban development. Many post-colonial economies have primate cities through which they are integrated into global trade; some examples are: Nairobi in Kenya; Jakarta in Indonesia; Manila in the Philippines; and Lima in Peru. More recently, internal national contrasts have been remodelled by globalization, but such restructurings tend to exacerbate regional socio-economic divergence rather than reduce spatial inequalities. Stark regional socio-economic disparities exist in many countries and are reflected in the nutritional statistics. Some interesting patterns emerge when studying dietary patterns and regional development. At present, undernutrition is generally more serious in the peripheral regions of national territories, while overnutrition tends to be more serious in urban, core regions. However, some intriguing exceptions have emerged in the recent years (see the following subsections on indigenous populations and obesity). Circumstances in China are among the most extreme.

It may be useful to review your understanding of Chinese development policy by considering the material in Box 5.5. While market reforms, instituted in China since 1978, help explain the rapid economic development that has dramatically reduced national levels of poverty, which some argue is 'the largest, sharpest, decrease in poverty recorded in human history', there is evidence that they have also intensified regional polarization. As in many countries, such contrasts have a long history, but increases in the recent years present several challenges to the central government: social cohesion is threatened, central authority is being tested and territorial integrity is also under threat in marginal regions. All these problems are associated with gross regional disparities.

Central to the Chinese development policies were the establishment of special economic zones (SEZs) and economic and technical development zones (ETDZs), beginning in the early 1980s. These territorially distinct zones represent the core industrial areas of China and are concentrated in the eastern regions and the coastal zone. Eastern coastal regions have experienced dramatic economic development, with the Pearl Delta region being the most phenomenal. On 24 January, 2011, the Chinese government announced that it would develop a new mega-city to integrate a large part of China's manufacturing heartland, stretching from Guangzhou to Shenzhen, and including Foshan, Dongguan, Zhongshan, Zhuhai, Jiangmen, Huizhou and Zhaoqing (see Figure 6.1). Together, this region accounts for nearly a tenth of the Chinese economy. This core region contrasts with the peripheral regions of the national territory in the central and western districts where economic growth has been very limited. These peripheral regions are the source of the 'greatest migration in human history', an exodus in recent decades of more than 150 million people who moved to work in the factories and construction sites of eastern China.

In the absence of regionally available nutritional statistics, contrasts in human development indicators are adequate surrogate measures of nutritional well-being. The World Health Organization (2005) reported that '[W]hile all 33 provinces of China are now in the United Nation's medium human development category, Shanghai's human development index (HDI) is almost 55% higher than Tibet's'.

Nationally, HDI are rising steadily, but in some central and western provinces, such as Tibet and Yunnan, they are deteriorating. The maternal mortality rate is 9.6 in Shanghai, 111 in Guizhou and 399 in Tibet (WHO, 2005: 11). The same paper concludes that '[A] critical health challenge in China relates to inequality in health outcomes, worsening since the 1980s' (WHO, 2005: 8). The funding of health care in China may help institutionalize health inequalities as it is extremely decentralized, and so the economic gains enjoyed in the prosperous zones, and the taxes they generate, contribute very little to the peripheral areas. In some regions, as much as 90 per cent of health funding is from local taxes; where economic growth is limited, tax revenues will be meagre. Those communities, far from the core industrial regions of China, have development indicators similar to some of the worst in the developing world. When economic disadvantage is reinforced by ethnic distinctiveness, political unrest has occasionally threatened the integrity of the national territory, in Tibet or Xinjiang, for example.

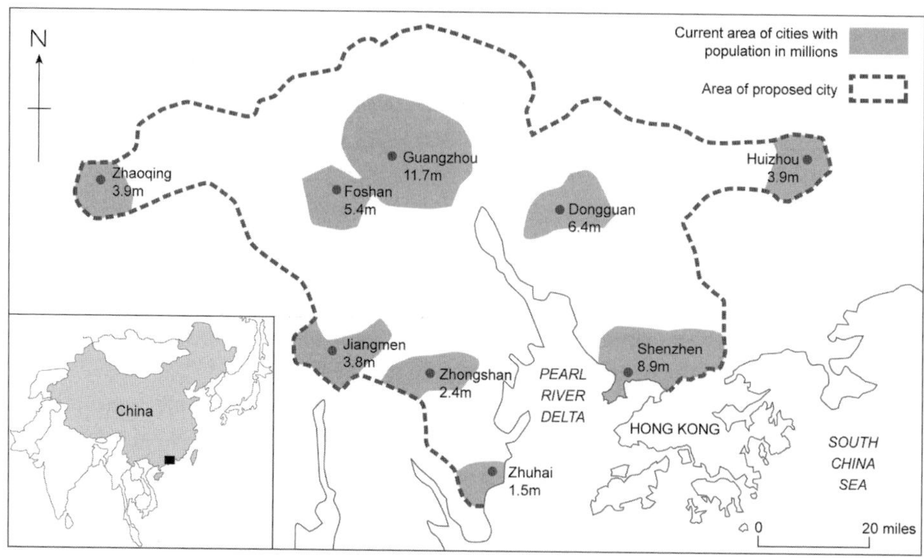

Figure 6.1 China: mega-city.

Plate 6.1a Tibet: yak grazing near a temple in South Tibet (Courtesy: Dr Louise Bonner).

Plate 6.1b Tibet: pilgrims outside Jokhang Temple: giving thanks for healthy animals and a good harvest (Courtesy: Dr Louise Bonner).

China is one of the best examples of a growing number of developing countries facing the 'the double burden of malnutrition': the persistence of undernutrition – almost 8 per cent of China's pre-school children remain underweight – along with the rapid rise of overnutrition and diseases such as diabetes, hypertension and coronary heart disease (see Chapter 2). China is often referred to as the 'dragon economy'; further examples of regional disparities are provided by Asia's other growing economy, the 'tiger economy' of India. In India, undernutrition is correlated to region, class, caste and gender.

This text rejects the idea that climate explains famine or food insecurity, and emphasizes instead the role of historically embedded socio-economic and political patterns of entitlement provision. However, granted the primacy of the social character of food security, it is important to appreciate the way that climatic variables and resource endowments may increase vulnerability to food insecurity in some regions.

Plate 6.2 Berbers: traditional nomads in North Africa (Courtesy: Dr Louise Bonner).

It is worth noting, of course, that the inhabitants of resource-poor regions, where climate is fickle, may have been removed from more fertile regions, the removal itself a symptom of their relative powerlessness. 'Seasonality continues to undermine household food security, and drought regularly triggers food crises, particularly in sub-Saharan Africa. Climate cannot therefore be totally ignored', concludes Devereux (1993: 44). Some maintain that climatic variation is intensifying due to global warming, perhaps, although the climate has always been unpredictable and people have developed a variety of coping strategies, including the cultivation of drought-resistant foods (Box 6.2). Certainly, some of the world's poorest people survive in the most inhospitable circumstances. One recent estimate asserts that:

> Almost one-third of the rural populations of developing countries – and a significantly higher proportion of the poor rural population – live in less-favoured marginal areas, many of which are either hillside or mountainous regions, or arid and semi-arid dry lands. Many of these lands are environmentally fragile, and their soils, vegetation and landscapes are easily eroded. Population growth combined with extreme poverty pushes

Box 6.2

A passion for pungent paste

Many cuisines have developed fermented protein sauces that provide flavour and nutrients year round: soy sauce and fish sauce have even become staples in kitchens internationally. In West Africa, the favoured item is soumbala, a labour-intensive preserved bean paste. In West African cuisine, soumbala has been as prevalent as soy sauce is in East Asian cuisine, for seasoning meat, vegetables, and soups. In the past, when there was nothing else to eat, soumbala was eaten with grains, and provided essential nutrients. Today, however, due to international subsidies and industrial production of substitutes, soumbala is at risk of disappearing from traditional dishes.

The African locust bean tree, indigenous to West Africa, grows from Uganda north to Chad and west all the way to Senegal. These trees produce orange and red spherical flowers, which develop long pods, each containing up to 30 protein-rich seeds. One reason for the pods' appeal across West Africa may be the tree's resilience: it can withstand droughts and grows in conditions ranging from tropical forest to rocky slopes and wooded farmlands.

Due to the labour-intensive nature of locust bean processing and the prevalence of artificially cheap cash crops, West African recipes and the soumbala recipe itself are changing. Bouillon cubes such as Maggi are widely available and are sometimes preferred over soumbala for their uniformity and, depending on the region, their lower price. The quality of soumbala is generally high, but not all producers can ensure that it is free of sand or other impurities. Unfortunately, bouillon cubes contain none of the nutritional qualities of the authentic seasoning. Likewise, the dominance of subsidized soy across the globe is beginning to affect the soumbala market. Some producers have begun substituting soy beans for the local crop, often encouraged by development agencies.

Despite the challenges to soumbala usage, the unique product is currently a source of economic innovation in West Africa. Entrepreneurs at all levels of the society are looking for ways to profit by increasing access to soumbala. Traditionally, women produced soumbala for home consumption. Today, that tradition has gained economic significance as women entrepreneurs are carving out lives for themselves as producers and distributors of soumbala. These entrepreneurs are not only supporting themselves but also are protecting the environment by encouraging the cultivation of a sustainable, native species. Additionally, they are furthering a culinary tradition that could be lost without their efforts.

Source: Campbell, W. (2009) 'A Passion for Pungent Sauce' Cultural Survival Vol 33, no 2 at: www.culturalsurvival.org/publications/cultural-survival-quarterly/uganda/passion-pungent-paste (accessed 30 January 2011)

> people into more marginal areas, and compels them to overuse the fragile resource base; the results include deforestation, soil erosion, desertification and reduced recharge of aquifers. As a result, resource degradation represents an increasing risk factor for many poor households.
>
> (IFAD, 2011: 83)

The relevance of seasonality is most marked in semi-arid areas, where access to clean water varies markedly during the year: the timing as well and the amount of rainfall is crucial in these regions. The issues raised here are relevant everywhere but are more acute and have direct ramifications for food security in semi-arid regions.

The implications are several: in these areas, the cost of water is often high and volatile, so people just managing to survive can be plunged into extreme poverty with changes in water prices. As the cost of water increases, they are required to devote a higher percentage of their effort and income to providing water for themselves and their livestock. To exacerbate matters, the cost of water often increases at the same time as food prices increase, so reducing the poor's access to both basic resources at the same time. Where water is at a premium, people tend to reduce their use of it, and sanitation and hygiene may suffer, so increasing the likelihood of disease.

Plate 6.3 Water is a precious resource: cattle enjoying a drink in a remote region in India (Courtesy: Lina Krishnan).

One of Africa's most pressing problems is associated with land management in the Sahel region, a fragile dryland ecosystem which cuts across Africa from the Red Sea to the Atlantic. In this region, complex interactions between poverty, resource use and climate are exacerbating the fuel wood crisis and desertification, which, in turn, are allied to the region's prevalence in undernutrition statistics. Politically charged debates about the causes of and cures for desertification continue. Their political character became glaringly obvious at the Rio Conference. The definition enshrined in Agenda 21 viewed land degradation in the Sahel as the result of several factors, 'including climatic and human activities'. The climatic element referred to the erratic rainfall in the region, while human activities included cutting down trees for firewood's and unsuitable farming practices. African participants in the conference objected to this partial analysis which ignored the wider context within which poverty in the Sahel must be understood, namely, the history of colonialism, misguided aid policies, falling commodity prices and pressures of debt repayments.

The food security role of livestock in semi arid areas is critical. In such regions people's entitlements are based on numerous strategies, and among the most important is the management of their animals. Livestock are relatively drought resistant compared to crops, and they are an extremely flexible asset. This flexibility is crucial to the survival strategies of rural populations. Animals, cattle, sheep, goats, camels, chickens, all provide food, and they also supply energy for agriculture in their draught power and manure. They are a form of capital which may be allowed to appreciate (grow and breed) or be realized for cash when required. Where rainfall is unreliable, another characteristic is vital – animals can be herded to new pastures and water sources when drought arrives.

Although communities may survive a 1-year drought, a pernicious interaction occurs in semi-arid areas during a prolonged drought between animal prices and grain prices. Local livestock and grain markets interact to reduce income and increase food prices simultaneously. The mechanism is deadly and simple. As a drought intensifies, grain scarcity forces up grain prices; at the same time, herders find it difficult to find water to sustain their animals and need more cash to buy food at higher prices. As more and more animals are sold, their prices fall, generating a scissors effect – high food prices and falling asset prices. The extreme outcome is famine. While attention needs to be given to the role of climate, it is, none the less,

Plate 6.4 Animal market in Sokoto, northern Nigeria (Courtesy: Dr Hamish Main, Staffordshire University).

a proximate variable, with limited explanatory power compared to the political and socio-economic context within which the communities 'at risk' are required to exist. Sadly, all these factors are exemplified in the Horn of Africa during 2010 and 2011. In this region, conflict and instability makes the situation even more protracted. For a fuller

Plate 6.5 Fulani women transporting milk to market (Courtesy: Dr Hamish Main, Staffordshire University).

discussion, see Chapter 8, but read a review of the crisis and its differential impacts by age at: www.trust.org/alertnet/news/children-under-five-suffer-the-most-in-horn-of-africa-food-crisis (accessed 28 July 2011).

Race, caste, disability and ethnic contrasts

This section summarizes some injustices suffered by people because of their race, caste, disability or ethnicity; examples are abundant, and are found in all countries. Here, a few are selected to illustrate how these factors interact with political and economic structures to explain why minority populations are frequently over-represented in malnutrition statistics. The problems suffered by these populations are frequently the result of multiple types of disadvantages. Ethnically distinct communities are often concentrated in rural areas that are marginal too, for example. Such circumstances are not coincidental but rather a symptom of their relative disadvantage, their spatial marginalization being a manifestation of their relative economic and political exclusion. Where race is an important variable in explaining poverty and disadvantage, it is often associated with spatial segregation; either formal or informal.

Members of indigenous communities frequently suffer from severe discrimination which is often racist in nature. The term 'indigenous' is problematic and contested, but here is a useful working definition:

> Indigenous communities … have a historical continuity with pre-invasion and pre-colonial societies ... and consider themselves distinct. … They are determined to preserve, develop and transmit to future generations their ancestral territories and their ethnic identity, as the basis of their continued existence as peoples in accordance with their own cultural pattern, social institutions and legal systems.
>
> (Martinéz Cobo, 1981: 5)

Resources available at the Indigenous Peoples Issues and Resources website include an interactive map which charts indigenous struggles to control their futures and access resources (http://indigenouspeoplesissues.com). The material produced by the Cultural Survival activist website is also very useful (www.culturalsurvival.org) for following indigenous struggles for dignity and cultural autonomy. Some of these communities inhabit unexploited lands which are rich in valuable resources (trees, oil, water, etc.), which explains why they are often at the front line in struggles resisting exploitative development strategies. Judging from historical experiences, local populations seldom benefit economically when these resources are exploited, although they may have to deal with the political and environmental problems that such development generates (see the section on Niger Delta in Chapter 8).

Many indigenous and tribal populations suffer from extreme social disadvantage, reflected in low levels of education, high levels of unemployment, low incomes, and a sub-standard physical environment and malnutrition. IFAD (2010) describe how although constituting only 5 per cent of the world's population, Indigenous and Tribal people account for 15 per cent of the world's poor. In Latin America, they detail how poverty rates for indigenous peoples are substantially higher than for non-indigenous: in Paraguay, poverty is almost eight times higher among indigenous peoples, in Panama almost six times higher and in Mexico three times higher. As with rural women (Chapter 7), indigenous peoples' poverty is rooted in multiple forms of disadvantage and deprivation (Box 6.3). Virtually everywhere, indigenous peoples suffer from discrimination, violation of their rights (social, political, human and economic) and exclusion (or self-exclusion) from mainstream social, economic and political processes (IFAD, 2010: 68).

Box 6.3

Indigenous populations and human rights abuses: the Batwa in East Africa and Guarani in Brazil

The Batwa in East Africa: a long history of exclusion

The Batwa, who are described in many parts as pygmy peoples, live in Eastern Uganda, eastern Democratic Republic of Congo and Rwanda. Batwa have been subject to negative stereotypes since at least 1751, when Edward Tyson concluded that pygmies were not human but rather apes or monkeys. They suffer from multiple asset depletion and wide-ranging forms of discrimination, a situation that public actions have at times made worse. Though longstanding forest dwellers, the British sought to expel them to create forest reserves in the 1930s. In 1991, the Uganda National Park authorities increased efforts to enforce this exclusion from forest areas. Although the World Bank – which was funding some of the park authorities' work – required that the government assess the impact on indigenous communities and follow defined compensation procedures, these did not take sufficient account of power differences among Batwa and other affected groups, nor consider Batwa preferences. All communities were viewed as uniform, a practice that the authorities later recognized 'did not take into account Batwa realities and left them with nothing' (Zaninka, 2003: 170). Non-Batwa locals have resisted efforts to provide more appropriate compensation to the Batwa. A Participatory Poverty Assessment (PPA) highlighted persistent discrimination, describing the Batwa as a 'group of people who are despised' and who 'have no means of production such as land, credit and training. They are regarded by other ethnic groups in Kisoro as a people with no rights'. This leads to everyday and institutionalized forms of exclusion, with the Batwa suffering discrimination in access to both public spaces and services. The same PPA reports some Batwa children saying that they did not attend school because it was so unfriendly to them. When asked what they wanted to do upon completing school, one child replied that she wished to be 'a cleaner'. Discrimination and prejudice diminish the capacity to aspire to and imagine a different future. Repudiation and discrimination can also lead the Batwa to self-exclude from the public sphere. The PPA notes that no Batwa attended PPA exercises. Non-Batwa locals explained: 'Batwa would never come to such meetings, so there is no point in mobilizing them'.

Source: Moncrieffe (2005) quoted in 'Equity and Development' World Bank Development Report, 2006 at: www-wds.worldbank.org/external/default/WDSContentServer/IW3P/IB/2005/09/20/000112742_20050920110826/additional/841401968_200508263001833.pdf (accessed 28 January 2011)

The Guaranis in Brazil: dispossession and trauma

In the last 500 years, virtually all the Guarani's land in Mato Grosso do Sul state has been taken from them. Waves of deforestation have converted the once-fertile Guarani homeland into a vast network of cattle ranches, and sugarcane plantations for Brazil's biofuels market. Many of the Guarani were herded into small reservations, which are now chronically overcrowded. In the Dourados reserve, for example, 12,000 Indians are living on little more than 3,000 hectares. The destruction of the forest has meant that hunting and fishing are no longer possible, and there is barely enough land even to plant crops. Malnutrition is a serious problem and, since 2005, at least 53 Guarani children have died of starvation.

Sugarcane plantations

Brazil has one of the most highly developed biofuels industries in the world. Sugarcane plantations were established in the 1980s, which rely heavily on indigenous labour. Workers often work for pitiful wages under terrible conditions. In 2007, police raided a sugarcane alcohol distillery and discovered 800 Indians working and living in subhuman conditions. Since many indigenous men are forced to seek work on the plantations, they are absent from their communities for long periods, and this has a major impact on Guarani health and society. Sexually transmitted diseases and alcoholism have been introduced by returning workers and internal tensions and violence have increased. Over 80 new sugarcane plantations and alcohol distilleries are planned for Mato Grosso do Sul, many of which are to be built on ancestral land claimed by the Guarani.

Trapped

The Guarani in Mato Grosso do Sul suffer from racism and discrimination, and high levels of harassment from the police. It is estimated that there are over 200 Guarani in jail with little or no access to legal advice and interpreters, trapped in a legal system they do not understand. This has resulted in innocent people being condemned. Many are serving disproportionately harsh sentences for minor offences.

The response of this deeply spiritual people to the chronic lack of land has been an epidemic of suicide unique in South America. Since 1986, more than 517 Guarani have committed suicide, the youngest just 9 years old.

Source: Survival International at: www.survivalinternational.org/tribes/guarani/despair#main (accessed on 12 January 2011)

The case of Peru is instructive to consider here. Although applauded (Chapter 5) for some recent success in reducing undernutrition, statistics for its indigenous populations are still well below the national average in this respect. Peru consists of three distinct regions: the coastal area, the Andean region and the Amazonian basin. The coastal zone is most developed and represents the core region centred in the primate city of Lima; both the Andean and Amazonian regions are peripheral economically and politically. As with other Andean countries, Peru is multi-ethnic, and approximately one-third of the population are indigenous.

The majority of these people live in the remote mountainous Andean region which formed the centre of the pre-Columbian Incan civilization. These upland communities are ethnically distinct and speak Quechua, a native language, rather than Spanish, which is spoken by the majority of Peruvians. Although these people's rights are guaranteed by the constitution, based on all socio-economic indicators, they still suffer considerable marginalization. In 2000, an estimated 62.8 per cent of indigenous families lived in conditions of poverty, while 22.2 per cent of them could be categorized as living under extreme poverty, compared with 43 per cent and 9.5 per cent of non-indigenous families, respectively. The region with the highest rates of poverty and extreme poverty are in the Central mountain

Plate 6.6 Women street traders selling crafts to tourists in the Peruvian Andes (Courtesy: Katy Stotesbury).

range (Huancavelica, Huánuco, Apurimac and Ayacucho), where populations are recorded with various symptoms of malnutrition. An EU (2007) publication also recognized that:

> Peru's poverty problem has a strong gender component. The level of women's participation in the labour world is low and their access to health and education is still inadequate, particularly in rural areas, especially those inhabited by indigenous or native people.
>
> (Source: www.eeas.europa.eu/peru/csp/07_13_en.pdf; accessed 21 January 2011)

The United States has long been associated with obesity, although, as this text details, it is not alone in suffering serious implications of obesity. It is interesting to appreciate that, in the United States, indigenous populations suffer higher than average obesity levels (Box. 6.4). Another useful case study exists for indigenous Australians (Burns and Thompson, 2008). A recent study established the health gap between members of this community and non-indigenous Australians and its link to poor nutrition. The research conclusions, that improved nutrition is central to improving the health and longevity of Australian Aboriginals, are relevant for numerous

Box 6.4

Indigenous people in the United States suffer higher rates of obesity

America is home to the most obese people in the world. According to the Center for Disease Control and Prevention, obesity in adults has increased by 60 per cent within the past 20 years and obesity in children has tripled in the past 30 years. A staggering 33 per cent of American adults are obese, and obesity-related deaths have climbed to more than 300,000 a year, second only to tobacco-related deaths. Not excluded from this statistic, native Hawaiians have alarmingly high rates of obesity, diabetes and heart disease. The number of Hawaiian children suffering from obesity is double that of children throughout the nation. In May 2001, the University of Hawaii Kinesiology and Leisure Science Department along with the Brigham Young University Exercise and Sport Science Department conducted a local study and found that more than 20 per cent of Hawaiian children were overweight.

Hawaiians aren't the only indigenous people in America that have high rates of obesity. According to Kelly Brownell, PhD, an expert on American diet and health, a study was conducted with the Pima Indians who live both in Mexico and Arizona. It was found that those Pima Indians who live in Arizona have much higher rates of obesity than their counterparts in Mexico, even though both groups of people have the same genetic and ethnic backgrounds. This is also true for many migrants in the United States who have a much higher obesity rates than their relatives back home.

Source: www.downtoearth.org/health/nutrition/obesity-america (accessed 15 July 2011)

indigenous communities across the globe in the north as well as the south. These examples illustrate the connection between diet and social status. Initially, poor nutrition among indigenous peoples is associated with undernutrition, low birth weights, and then later it manifests itself as adult obesity. Waldick (2009) summarizes the variety of issues that cause concern:

> … many Indigenous peoples are currently familiar with plants and other food sources that have never been formally documented by biologists. However, these Indigenous food systems, which have endured for generations, are being eroded. Not only does this represent a loss of precious knowledge, it is leading to a 'nutrition transition' in Indigenous homelands. As more foods from industrialized sources are consumed, chronic diseases and malnutrition are becoming increasingly prevalent in Indigenous communities. … Wherever they live in the world, Indigenous peoples are frequently the poorest of the poor and among the most marginalized groups in society.
>
> (Waldick, 2009)

In South Asia, caste remains a significant factor in determining access to a good diet and health (Box 6.5). Caste is traditionally associated with specific occupations, and is therefore allied to class. If conventional caste customs are obeyed, then whole communities are assigned a specific economic role, and their opportunities are seriously constrained. In recent decades, the importance of caste has been eroded, as legislation both criminalizes discrimination by caste and promotes educational opportunities for scheduled castes (the lowest caste groups). Discrimination has also been vigorously and successfully resisted by caste-based activism. In urban India, social interaction is less marked by caste-based strictures but traditional caste relations have proved more resilient in rural areas, where caste may still determine one's life's chances. Caste is often allied to other factors, which combine to intensify individual disadvantage.

Box 6.5

Undernutrition in India: a complex picture of place, gender, caste and wealth

Overview

Analysing the multiple causes of undernutrition in India illustrates how complex causes interact and suggests why policy interventions can be difficult. Approximately 60 million children are underweight in India. Given its impact on health, education and productivity, persistent undernutrition is a major obstacle to human development and economic growth in the country, especially among the poor and the vulnerable, where the prevalence of malnutrition is highest. The progress in reducing the proportion of undernourished children in India has been mixed but generally slower than what has been achieved in other countries with comparable socio-economic indicators. While aggregate levels of undernutrition are shockingly high, the picture is further exacerbated by the significant inequalities: regionally (gender differences tend be more extreme in northern India); across states (with different histories and social trajectories); socio-economic groups; and caste. The most vulnerable are:

- Girls
- Rural populations
- The poorest and scheduled tribes and castes

Evidence published in 2005 concludes that some of these inequalities may even be increasing. In India, child malnutrition is mostly the result of high levels of exposure to infection and inappropriate infant and young child feeding and caring practices, and has its origins almost entirely during the first 2–3 years of life. However, the commonly held assumption is that food insecurity is the primary or even sole cause of malnutrition. Consequently, the existing response to malnutrition in India has been skewed towards food-based interventions and has placed little emphasis on schemes addressing the other determinants of malnutrition.

Disaggregation of underweight statistics by socio-economic and demographic characteristics reveals which groups are most at risk of malnutrition. Underweight prevalence is higher in rural areas (50 per cent) than in urban areas (38 per cent); higher among girls (48.9 per cent) than among boys (45.5 per cent); higher among scheduled castes (53.2 per cent) and scheduled tribes (56.2 per cent) than among other castes (44.1 per cent); and, although underweight is pervasive throughout the wealth distribution, the prevalence of underweight reaches as high as 60 per cent in the lowest wealth quintile. Moreover, during the 1990s, urban–rural, inter-caste, male–female and inter-quintile inequalities in nutritional status widened.

There is also a large inter-state variation in the patterns and trends in underweight. In six states, at least one in two children are underweight, namely Maharashtra, Orissa, Bihar, Madhya Pradesh, Uttar Pradesh, and Rajasthan. The four latter states account for more than 43 per cent of all underweight children in India. Moreover, the prevalence in underweight is falling more slowly in the high-prevalence states. Finally, the demographic and socio-economic patterns at the state level do not necessarily mirror those at the national level (e.g. in some states, inequalities in underweight are narrowing and not widening, and in some states boys are more likely to be underweight than girls), and nutrition policy should take cognizance of these variations.

Economic growth alone is unlikely to be sufficient to lower the prevalence of malnutrition substantially – certainly not sufficiently to meet the nutrition MDG of halving the prevalence of underweight children between 1990 and 2015. It is only with a rapid scaling-up of health, nutrition, education and infrastructure interventions that this MDG can be met. Additional and more effective investments are especially needed in the poorest states (see examples of effective national-level interventions to reduce undernutrition in Chapter 5).

A case study of Palanpur, Northern India

The village of Palanpur, in the North Indian state of Uttar Pradesh, was the subject of intensive study by a group of development economists between the late 1950s and early 1990s. Their study confirmed the continued relevance of caste and gender to socio-economic well-being and opportunity.

Caste in Palanpur defines opportunities and determines the activities that villagers pursue, even independent of occupation, education and other standard household characteristics. The three largest castes in Palanpur are Thakurs, Muraos and Jatabs. At the top is a martial caste known as the Thakurs, which accounted for about a quarter of the population in 1993. Thakurs are disproportionately represented in jobs such as the army and police that accord well with their martial past. They are typically averse to wage employment in the village, because this would place them in a subordinate position. Alert to non-farm employment opportunities outside the village, they are well placed to take advantage of them, thanks to stronger information and social networks.

Just below the Thakurs is a cultivating caste, the Muraos, who account for a quarter of the population. Muraos are traditional cultivators who have continued to specialize in agriculture. Their fortunes have improved, and they have seen a rapid rise in wealth and economic status in the village. While they may still not enjoy the same social status as Thakurs, they have become more prosperous and now challenge their previously unquestioned political and economic dominance.

At the bottom are the scheduled castes known as Jatabs, accounting for 12 per cent of the population. Traditionally 'untouchable' leather workers who are now engaged primarily in agricultural wage labour, Jatabs have not seen any of the social mobility of the Muraos. They remain a caste apart, with little or no land, poor education and little access to non-farm employment outside the village. Despite some slight improvement over the years, Jatabs continue to endure many forms of discrimination, including that from government officials.

Gender inequalities in Palanpur are pronounced. In 1993, there were 84 females for every 100 males, strikingly lower than in most parts of the world (see Sen's 'missing women' problem, Box 7.3, Chapter 7). Child mortality rates are much higher among girls than among boys. As the researchers reported, 'We witnessed several cases of infant girls who were allowed to wither away and die in circumstances that would undoubtedly have prompted more energetic action in the case of a male child'. Young girls leave their village to join their husband's family. Marriage is 'the gift of a daughter'. In the new household, the girl is acutely vulnerable, with no income-earning opportunities, no property and no possibility of returning home permanently. Giving birth to a child improves her status – particularly if it's a boy. However, family planning practices are limited, leading to high fertility rates and short birth-spacing.

Repeated pregnancies take an enormous toll on women's general health and put their lives at risk at the time of delivery. Old age is strongly associated with widowhood, in part because of the typically large age difference between husbands and wives. To survive, widows depend overwhelmingly on adult sons.

Source: Development and Equity' World Bank Development Report (2006) at: www-wds.worldbank.org/external/default/WDSContentServer/IW3P/IB/2005/09/20/000112742_20050920110826/additional/841401968_200508263001721.pdf (accessed 28 January 2011); http://web.worldbank.org/WBSITE/EXTERNAL/COUNTRIES/SOUTHASIAEXT/0,,contentMDK:20916955~pagePK:146736~piPK:146830~theSitePK:223547,00.html (accessed 24 January 2011)

People with disabilities have received limited attention in the literature about undernutrition and overnutrition. However, data from a number of countries suggest that disabled people are much more likely to be poor and therefore more vulnerable to food insecurity. When a disabled individual is also head of a household, then the implications are important for all members of the household. Hoogeveen (2003) reports that, in Uganda, the probability of poverty for urban dwellers living in a household with a disabled head is 38 per cent higher than for those who live in a household with an able-bodied head. The Serbian Poverty Reduction Strategy reports that 70 per cent of disabled people are unemployed. In a study drawing on ten household surveys in eight countries, self-reported disability was found to be more correlated with nonattendance at school than other characteristics, including gender or rural residence. Sen (2004) emphasizes that the disabled face not only an 'earnings handicap', associated with a lower probability of employment and lower compensation for their work, but also a 'conversion handicap'. By this, he means that a physically disabled person requires more income than an able-bodied person (World Bank, 2006: 18).

There are established associations between disability and overnutrition in the United States and the European Union, and similar connections must be present in the global south. The connections between obesity and disability are complex and, as with undernutrition, obesity may be both the cause and consequence of unhealthy lifestyles (see secondary malnutrition in Chapter 2). Research suggests that, as obesity increases, so do rates of disability and then, the disability causes increased levels of obesity, and creates a complex public health issue. Disability is associated with some acute medical problems as well as considerable discomfort associated with secondary problems. Research from the United States and the European Union confirms that obesity, when allied to disability, generates secondary conditions such as pressure sores, physical inactivity, feelings of depression and fatigue, all of which interfere with performance of daily activities. For people with physical disabilities, obesity is doubly disturbing. It is not only linked to an increase in potentially disabling chronic conditions, but when paired with existing functional limitations, may also limit a person's ability to engage in physical activity and participate in social events and community activities.

Rural–urban contrasts

As with all categories we employ in social science, the urban–rural dichotomy is problematic. First of all, the definitions used to distinguish between urban and rural vary, and so comparative analysis is difficult. More important is that generalizations about the rural and urban poor conceal a great deal of diversity within each category.

In addition, many people regularly migrate between rural and urban areas, which further complicate calculations. Rural and urban places are connected by flows of people, commodities and ideas, all the time remaking the nature of each other; remittances and visits from family members who work in urban areas is just one example of such vital connections. With these caveats in mind, we employ these categories because they help inform strategic policy interventions to relieve poverty and malnutrition. It is important to stress that conditions within each category may be more extreme than the differences between them. Consider the availability of clean water, a useful surrogate for the environmental circumstances of households, shown in Table 6.1. While contrasts are obvious between rural and urban areas, differences within the urban category are also significant.

Table 6.1 ***Percentage of households with access to service***

	Piped water on premises	*Flush toilet*	*Pit toilet*
North Africa			
Rural	41.6	41.3	17.5
Urban poor	67.3	83.7	8.5
Urban non-poor	90.8	96.3	2.6
Sub-Saharan Africa			
Rural	7.8	1.1	47.6
Urban poor	26.9	13.0	65.9
Urban non-poor	47.6	27.4	67.2
Southeast Asia			
Rural	18.6	55.5	24.3
Urban poor	34.0	61.8	22.9
Urban non-poor	55.8	89.0	9.4
Latin America			
Rural	31.4	12.6	44.0
Urban poor	58.7	33.6	47.0
Urban non-poor	72.7	63.7	31.6
South Central West Asia			
Rural	28.1	4.3	55.4
Urban poor	58.0	39.8	34.1
Urban non-poor	80.2	64.0	23.2
Total			
Rural	18.5	7.5	46.6
Urban poor	41.5	28.3	51.7
Urban non-poor	61.5	48.4	46.5

Source: adapted from Montgomery, M. (2009) 'Urban Poverty and Health' in the Developing World' Population Reference Bureau at: www.prb.org/pdf09/64.2urbanization.pdf (accessed 3 February 2011)

The rural case

Students are often surprised to learn that most of the world's poor still live in rural places, and that the majority of the poorest will continue to be rural for some decades to come. 'Of the 1.4 billion people living in extreme poverty (defined as those living on less than US$1.25/day) in 2005, approximately 1 billion – around 70 per cent lived in rural areas' (IFDA 2010: 4). Upon investigation, we discover that the categories examined elsewhere, women (Chapter 7), indigenous peoples and ethnic minorities, are often disproportionately represented in poor rural communities. On every socio-economic statistic, rural populations are disadvantaged compared to their urban counterparts. The extent of the gap varies but is everywhere present. Consider the following statistics for Tanzania. Poverty rates remain highest in rural areas: 37.6 per cent of rural households live below the basic needs poverty line, compared to 24 per cent of households in other urban areas and 16.4 per cent in Dar es Salaam. Given the large proportion of Tanzanian households that rely on farming for their livelihoods and the high rate of rural poverty, the overwhelming majority (74 per cent) of poor Tanzanians are primarily dependent on agriculture or allied rural activities. Poverty in rural areas helps explain why undernutrition is still more serious in rural regions than in urban areas.

The dominance of rural poverty is often explained with reference to politics and the fact that amassed urban populations are much more effective at asserting their rights to public goods (government investment and disbursements of various kinds) than dispersed rural populations. Urban dissatisfaction and dissent is more readily galvanized and urban riots more alarming for elites living in those same cities. Political progress in Britain during the nineteenth century was often in response to urban unrest; riots help galvanize the political elite. As mentioned in Chapter 1, improvements in health and well-being in Britain in the late nineteenth and early twentieth century had more to do with political accountability than medical advances or technical change.

Urban bias in development policy was reviewed in Chapter 5. Since the 1950s to the present, development efforts have privileged urban and industrial interests, at the expense of rural or agricultural interests. One of the consequences of such policies is that food prices are depressed; while this may be useful in depressing industrial wages, it acts as a disincentive for agricultural production and rural wages.

There is some evidence that the recent food price hikes (2007–2008) and consequent food shortages have moved agricultural issues up the international and national development agendas. Certainly, the importance of pro-poor agricultural investment is recognized as a vital component of any poverty reduction strategy (OECD, 2006).

It is important to recognize the diversity of the category 'rural poor' and to appreciate the multidimensional character of their poverty. Such populations may be peasant farmers, landless labourers, artisan workers, sharecroppers, carers, wood gatherers, processors of food, street traders, etc.; often, they are also dependent on income transfers, including remittances and social transfers.

'Indeed, it is the continuous adaptation of a highly diverse portfolio of activities to secure survival that is a distinguishing feature of rural livelihood strategies in contemporary poor countries' (Ellis, 2000: 291). An example from India describes the variety of ways in which the rural poor survive and suggests why they have to be innovative:

> Madvi Madka … from India … has one thing in common with business tycoons across the globe – he is part of the construction sector that has been crippled by the global meltdown. But he is no real-estate shark. Madka is a farmer and a daily wage earner. He feeds his family of five by selling forest and agricultural produce in his remote village. … But this income is enough for only four months of the year. For the rest of the year, Madka travels to the city to work as a casual construction worker. … Over the past year, however, Madka could not find work in the cities nearby. He does not know what has led to this sudden turn of fortunes, but he is not alone.
>
> (Source: www.undp.org/poverty/projects_india.shtml; accessed 2 February 2011)

Another distinguishing characteristic is that they suffer from numerous disadvantages that preclude their economic advancement: a variety of forms of economic and political exclusion, discrimination and disempowerment; unequal access to and control over assets, especially land and credit; poor levels of education and limited collective capabilities to overcome these various constraints.

Access to common property resources (the most important being CPRs, forests, grazing lands, marine and freshwater fisheries) have been critical to the rural poor everywhere through the centuries. They may gather food and fuel from these areas for domestic consumption or to sell. Just as the 'enclosures' occurred in eighteenth-century Britain and excluded peasant farmers from access to previously 'common land', so today, across the global south, the rural poor are being excluded from resources once considered theirs by customary right. Privatization of land increases as capitalist forms of agricultural production diffuse and as all forms of resources are exploited more 'efficiently' for global markets. Such processes are also associated with rising land values which serve to further exclude the poorest rural inhabitants from access. This brief discussion illustrates how complex the roots of poverty are for many rural people and suggests the myriad interventions that are required if their circumstances are to improve. Some positive examples of state intervention are available from various countries; the one from India is instructive (Box 6.6).

Box 6.6

India: the National Rural Employment Guarantee Act and its impact on rural women

The National Rural Employment Guarantee Act is a landmark piece of legislation enacted in India in 2005. It guarantees each rural household the right to unskilled manual work for 100 days a year at the minimum wage, to be paid equally to men and women. The focus is on work in water and soil conservation, land development and forestation. With significant variation among states, the Act has stimulated the rural economy through increased incomes, demand and investments, and it has strengthened the coping mechanisms of poor rural households.

In 2007–2008, more than 30 million households were provided employment under the Act. Decreased out-migration has been recorded in areas where projects have been implemented – along with a rise in agricultural wages owing to a tightening of labour markets in some areas – although these results are preliminary. According to a 2009 study of the Act, based on a survey with women in six states, NREGA also contributed to women's access to better-paid employment. For instance, women were 44 per cent of NREGA participants in 2007–2008 in India, and considerably more in the states of Kerala, Tamil Nadu and Rajasthan. The Act itself stipulates that at least one-third of the participants should be women. Compared with the irregular, poorly paid and often hazardous labour opportunities usually available to poor rural women, NREGA offers better and more socially acceptable work, better working conditions, regularity and predictability of working hours, locations close to women's homes and better pay. The survey found that women's average wages ranged from Rs. 47–58 a day in the private labour market to an average of Rs. 85 under NREGA.

Two-thirds of respondents reported greater food security, and half reported being able to better cope with family illnesses. Some women were able to buy agricultural inputs and equipment through their wages. However, the experience of NREGA shows the persistence of social barriers to women's access to good wage opportunities. In some areas, women reported facing discrimination and being crowded out of NREGA projects by men drawn by decent wages. Elsewhere, households headed by single women have been denied registration in Act schemes.

Source: Khera and Nayak (2009) quoted in IFAD (2010: 106)

The urban case

In 2008, for the first time in human history, the majority of the world's population were classified as urban. This momentous shift reflects the fact that large numbers of rural people migrate to cities every year and swell a population already growing because of an in-built population momentum (because so many urban inhabitants are young, the population is destined to grow even if fertility rates per household decline). Migration is a major link that connects rural and urban populations. Rural poverty, detailed in the preceding text, helps

Plate 6.7 Recent migrants to Delhi.

explain the massive rural exodus recorded in recent decades. As long as development policies neglect agriculture, particularly small peasant farming sector, rural out-migration will continue at its current levels.

The massive expansion of some of the world's largest cities understandably generates considerable media interest and analyses as we speculate about how they can be sustained, politically, economically, socially and environmentally. The popularity of films such as *Slumdog Millionaire*, too, highlights the desperate living conditions that prevail in some of these new mega cities and the trials their inhabitants face in their efforts to survive (see Chapter 1 for levels of remittances). It is important to recognize the dynamism of these communities and the vital functions they provide for the city of Mumbai.

One of the largest slum settlements in Asia, with over 1 million inhabitants, Dharavi is just one of such urban settlements found in all major cities in the developing world. It is important to note that although these populations may be poor, they are not willing victims of their circumstance. People who live in the slums are economically active, political engaged and working both to provide for themselves and their extended families and to improve their living conditions. The variety of economic activities is amazing, and the economy of the slum has an annual turnover of over US$1 billion. See the

Plate 6.8 The Dharavi slum in Mumbai, India.

discussion and photos in the *National Geographic* at: http://ngm.nationalgeographic.com/2007/05/dharavi-mumbai-slum/jacobson-text (accessed 3 February 2011), and follow some of the activities of the International Slum and Shack Dwellers Alliance at: www.sdinet.org/static/upload/documents/newsletters/tgit8.pdf (accessed 3 February 2011).

The massive waves of rural–urban migrants everywhere in the global south reflects the fact that most rural poor people understand that economic opportunities are greater in urban areas than in rural communities. Migrant populations are often the youngest and most able, and their desertion of rural communities (in effect, a brain drain) compounds rural poverty, even if their remittances are also essential to the survival of those they leave behind. Socio-economic statistics for urban populations indicate that they have better access to health and education than their rural counter parts; 'this is too well documented to dispute' (Montgomery, 2009: 5).

Rural-to-urban migration is a rational choice but general statistics, such as those in Table 6.2, conceal a wide range of realities. A more careful analysis of statistics reveals a different picture; average figures may be better for urban populations, but a small number of very positive readings for wealthy households may skew results.

Table 6.2a ***Anaemia among children, urban and rural Egypt, 2005***

	Urban (%)	*Rural (%)*
Very poor	12	16.3
Poor	9	16.7
Near poor	10.6	12.7
Other non-poor	7	12.3

Table 6.2b ***Child nutrition: stunting among children, urban and rural India***

	Urban (%)	*Rural (%)*
Very poor	52.6	57.2
Poor	41.1	58
Near poor	39.3	54.1
Other non-poor	26.1	40.8

Table 6.2c ***Attendance of a physician or trained midwife at delivery, urban and rural India, 1998–2000***

	Urban (%)	*Rural (%)*
Very poor	42	14
Poor	59.8	14.2
Near poor	69.8	22.8
Other non-poor	87.2	45.8

Table 6.2d ***Composition of child mortality rates: Kenya, 2002***

All Kenya	112
Nairobi slums	151
All Nairobi	62
Other urban	84
Rural	113

Most readily available urban statistics are aggregate so 'that the different world's of city dwellers remain in the shadows' (Hidden Cities, 2010: 33). Few developing countries can supply the detailed data required to examine urban health differentials, but research is beginning. Some health statistics for the very poor urban groups are similar to those of poor rural populations, as seen in Table 6.3. When disaggregated, some very poor urban children may have equivalent health outcomes to near-poor rural children, for example.

In Hidden Cities (2010), the UN details many of the advantages enjoyed by urban dwellers before exposing some of the gross contrasts that prevail in all the world's great cities.

Contrasts between 'the haves and the have-nots' is perhaps nowhere more glaring than in the burgeoning urban conglomerations in the global south; here, slums and skyscrapers, luxury apartments and pavement dwellers live within minutes of each other. When urban data is disaggregated, the chasm that divides urban dwellers emerges. Slum dwellers are an obvious starting point. The UN report estimates that:

> … almost 900 million urban residents live in slums and squatter settlements. Housing in these settings ranges from high-rise tenements to shacks to

> plastic sheet tents on sidewalks. These settings tend to be unregulated and overcrowded. They are often located in undesirable parts of the city, such as steep hillsides, riverbanks subject to flooding or industrial areas.
>
> (Hidden Cities, 2010: 17)

Slum conditions also facilitate the rapid spread of infectious diseases. Hence, in these environments, the toxic connections between malnutrition, disease and health are evident. Here is another example of how aggregate statistics may distort our picture of urban conditions: although most urban populations are recorded as having greater access to sanitation than rural populations, in fact overall exposure to waterborne infections may be greater in urban populations because they live in such densely populated environments. Open sewers and stagnant water, often accompanied by high temperatures, attract flies, mosquitoes and rats, all of which help spread diseases. The incidence of diarrhoea is associated with socio-economic status and place of residence.

Water contamination helps explain high levels of recorded childhood diarrhoea and waterborne diseases such as malaria and worm infections. Infants and young children are particularly susceptible to these diseases and associated malnutrition problems because they

Plate 6.9a Urban gardens in Kinshasa (Courtesy: Mambu Sanduku).

have poorly developed immune systems. A recent study found that half of city dwellers in Africa, Asia and Latin America suffer from at least one disease caused by lack of sanitation and safe water (Hidden Cities, 2010: 19). Poor families in slums may use water from communal taps or spend hours carrying water from a distant source. They are occasionally forced to pay exorbitant rates for water provided by private suppliers. During the eighteenth and nineteenth centuries, improved sanitation was largely responsible for a fall in British urban death rates. Similar improvements in sanitation and water supply in the massive urban conglomerations in the global south could deliver spectacular declines in deaths from disease and malnutrition. Crowded conditions also help the spread of the two major childhood killers, namely, pneumonia and measles, as well as the still prevalent fatal disease, tuberculosis (TB).

There is evidence to suggest that poor urban inhabitants across the global south were disproportionately affected by the food price rise that started in 2006 (Box 6.7). Poor urban families spend as much as

Box 6.7

Food-price increases and the urban poor

The surge in food prices since the end of 2006 has led to increasing hunger in the world's poorest countries and made urban food security more precarious. The Food and Agriculture Organization (FAO) estimated the number of hungry people at 923 million in 2007, a more than 80 million rise since 1992. The most rapid increases in chronic hunger occurred between 2003 and 2005, and in 2007 when high food prices drove millions of people into food insecurity. Chronic hunger and malnutrition became more prevalent among the urban poor, who had to spend more to purchase not just food but other household necessities as well.

Agencies monitoring food security in Kenya, for instance, have recorded deepening urban food insecurity caused by rising prices and further compounded by conflict, floods and drought. In March 2009, an estimated 4.1 million urban poor in the country were classified as 'highly food-insecure'. High prices caused slum-dwelling households to reduce food consumption. As many as 7.6 million slum-dwellers countrywide found themselves unable to meet daily food needs, with maize prices soaring by more than 130 per cent in the capital, Nairobi, and by 85 per cent in the coastal town of Mombasa, in 2008. The cost of non-food items also increased, leading to reduced overall household consumption. Prices of cooking fuels, particularly kerosene, rose by 30–50 per cent, and the cost of water more than doubled. In neighbouring Burundi, the urban areas of Bujumbura, Gitenga and Ngozi saw a 20 per cent increase in local food prices between 2007 and 2008. The situation was exacerbated by the ongoing conflict that has hindered agricultural production in recent years. In the same period, wheat prices doubled in Senegal and quadrupled in Somalia.

Zimbabwe's highly inflationary environment has reduced purchasing power, making nearly half the population food-insecure. In Zimbabwe, as in other African countries, urban communities resort to a variety of coping strategies, including reduced frequency and content of meals, which could lead to rising malnutrition.

Stabilization or decline in food prices have occurred in several countries, but the outlook for some African countries remains bleak: they have been unable to reduce prices to pre-2006 levels, which has increased both malnutrition and household food expenditures.

Source: State of the World's Cities (UN Habitat, 2010: 105) at: www.unhabitat.org/pmss/listItemDetails.aspx?publicationID=2917 (accessed 28 July 2011)

70 per cent of their income on food, so any increase in food prices has serious implications for their nutrition and health. Most urban households are more exposed to food price fluctuations because only a minority of households produce their own food. The potential of urban agriculture to improve food security is discussed in the following text.

Evidence emerging from cities in the global south confirms that obesity is fast becoming a problem for poor urban populations, especially for those economies that are experiencing rapid economic transformations. All the global factors reviewed in Chapter 4 are transforming the urban environment, and most are obesogenic in nature. Among the most important factors is the widespread availability of cheap processed foods and new patterns of life, which mean home cooking is less common. Fresh fruit and vegetables are seldom readily available in poor urban neighbourhoods. A recent report found noticeable differences between the consumption patterns of urban and rural groups. In particular, urban residents are showing a strong trend towards consumption of non-traditional staples and animal-source foods, accompanied by a declining intake of fruits and vegetables. Finally, patterns of urban development, based on the Western model and car ownership, reduce people's opportunities to exercise at the same time as work becomes more sedentary.

In developed countries, the incidence of overweight is greatest among disadvantaged groups, and this also seems to be the case in middle-income countries. Mendez and Popkin (2004) present evidence that, in urban areas of highly urbanized countries, women of low socio-economic status (as measured by education) are more likely to be overweight than those of high socio-economic status, whereas the reverse is true in countries with low urbanization. All available analyses of obesity trends in the developing world or of transitional economies confirms that similar patterns are emerging, that the poorest are over-represented in obesity statistics.

Households

What factors help explain different household diets and, within households, how might diets vary? Households are the basic units, within which most people are born, live and die. They may consist of extended family groups of several generations, or they might be a single-head household with one child. It is worth noting Rigg's assertion that '[I]n treating the household as a single, welfare maximising decision-making unit, however, scholars have traditionally glossed over the frictions, contradictions and inequalities that are inherent in the operation of the household' (Rigg, 1991: 45). He continues to suggest that the key 'fault lines' relate to relations between men and women and between generations. Gender differences are the subject of Chapter 7 and are not considered here; other factors are worthy of attention, however. It is first necessary to understand that the household is a crucial element of the wider local food system shown in Figure 6.2.

Its characteristics and survival will be dependent on the wider local context. Local food systems are dynamic and may change drastically because of changes in the local or national economy, for better or for worse. It is important to stress how vulnerable some households are to even minor changes in their circumstances; illness, accidents, food price rises, unemployment, changes in welfare provision, flooding, family bereavement, any one, or some combination of these, can mean a family falls from being 'poor but coping' category to the 'ultra poor and not coping' category.

Households are the sites of struggle and negotiation, and usually also the site of cooperation and support. In many parts of the global south,

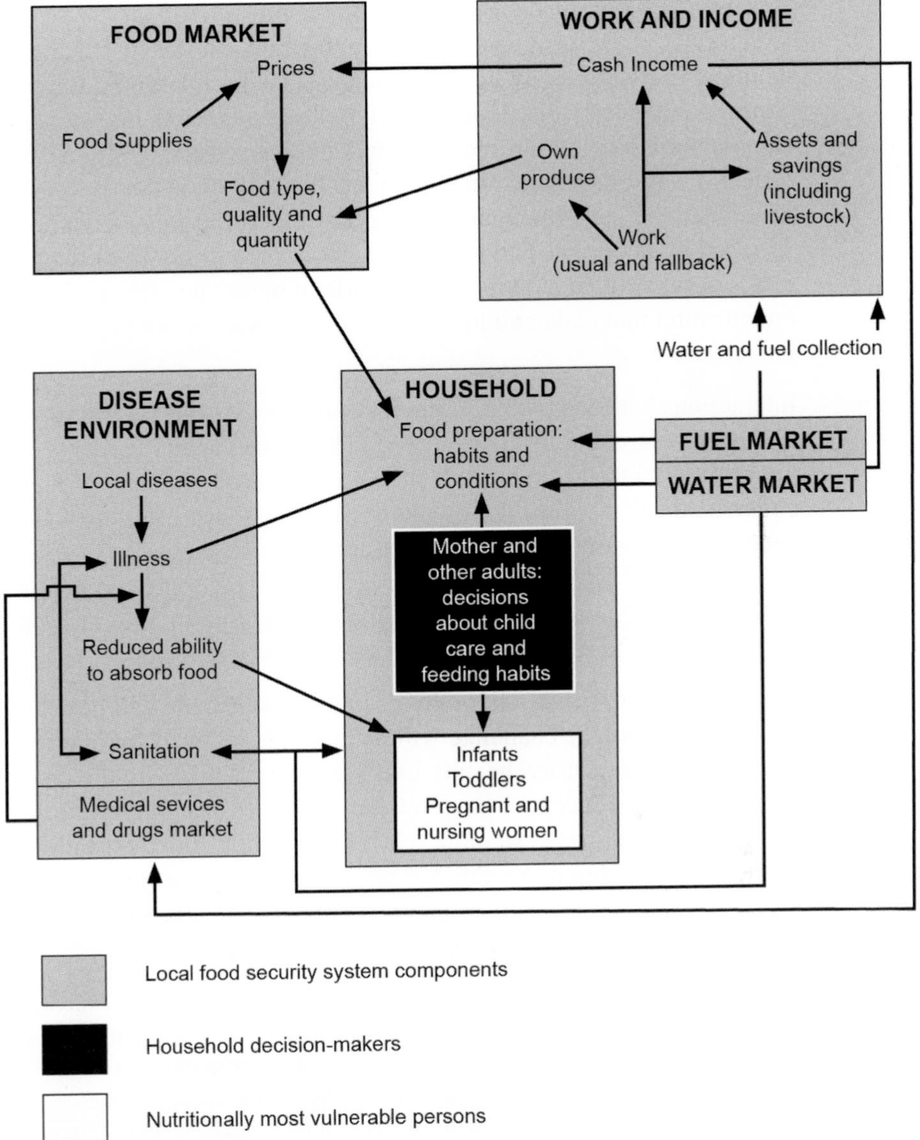

Figure 6.2 The local food security system (adapted from: Hubbard, 1995: 8).

where welfare provision is limited or non-existent, households become vital support systems for all its members throughout life, from infancy to old age. Belonging to a large family is in itself a survival strategy which helps explain higher fertility rates recorded in poor families. Smaller households are more vulnerable to sudden changes in circumstance. Figure 6.2 shows some important interactions at the local level that determine household diets; changes in one or more of these will have repercussions for household food security and diets. Household characteristics mediate the impact of external variables, and its members will experience changes in different ways, dependant on their status. In addition to distinctions by gender, the very young and the elderly may suffer more from poor nutrition in vulnerable households. There is some evidence to suggest that female children in parts of Asia suffer a higher incidence of malnutrition that male children in that region; this has not be found in the African case.

Interactions between poor households and HIV/AIDS have been reviewed before, but it is pertinent to mention this again. In many poor households, the links are insidious. If the head of household contracts the disease, all family members suffer a decline in their life chances. Where the epidemic is most serious, households become the most important support system for children who are orphaned because of the disease. Strengthening households and their capacities to cope is the best way to secure children's survival and well-being.

Many of the very poorest households in the next decade will be urban-based, and the challenge will be to design cities that promote household food security, healthy lifestyles and green environments. This is why the concept of 'Green Cities', already of interest in the developed world, has even more relevance for the rapidly expanding cities of the developing world. A major element of greening cities of the global south must be to integrate urban and peri-urban horticulture (UPH) into their very fabric. A recent study estimated that:

> … 130 million urban residents in Africa and 230 million in Latin America are engaged in agriculture, mainly in horticulture, to provide food for their families or to earn income from sales. Horticulture, in addition to contributing to the food security with high nutritional value, has short production cycles and high yields per unit of time, land and water. Fruits and vegetables are rich in micronutrients. They increase the supply of fresh, nutritious produce and improve the urban poor's economic access to food.
>
> (Growing Greener Cities, FAO, 2010)

In the past, urban agriculture was discouraged, even made illegal, in misguided efforts to 'clean up' cities. New initiatives have emerged in the last decade and enlightened policies are now beginning to support UPH. The FAO has granted assistance to more than 20 countries to support 'city farmers', including irrigated market gardens on urban peripheries, simple hydroponic micro gardens in slum areas and green rooftops in densely populated city centres. Granting poor households access to small plots of land in or near the cities provides multiple benefits. For the households, food security is increased, and they have access to more nutritious fruit and vegetables. In addition, they may supplement their income by selling surpluses in local markets. Fresh fruits and vegetables are more available for other poor urban dwellers and can be delivered without incurring food miles. Urban waste disposal is a serious challenge in all cities in the developing world, and its accumulation is an additional health hazard. UPH can turn this waste problem into a productive resource (Box 6.8).

In Cuba, a leader in UPH initiatives, the state prohibits the use of chemical fertilizers in cities and encourages organic composting instead.

Box 6.8

Urban farming against hunger

1 February, 2007, Rome – The United Nations Food and Agriculture Organization has opened a new front in its battle against hunger and malnutrition – in the world's cities where most of global population growth is set to take place over the next decades. This year will be the first time in history that the world's urban population – more than 3 billion people – exceeds the number of those living in rural areas. Currently, one-third of city dwellers, 1 billion people, live in slums. In many cities of sub-Saharan Africa, they account for three-quarters of all urban residents.

By 2030, some two-thirds of the world's people will be living in cities, according to UN projections, which also predict that the world's population will rise to 9 billion by 2050. Under its ongoing 'Food for the Cities' programme, an interdisciplinary initiative, FAO, is therefore helping a number of cities to support urban and peri-urban agriculture, so that they can increasingly contribute to the job of feeding themselves.

Allotment gardens in Africa

The Democratic Republic of the Congo, Senegal, Gabon, Mozambique, Botswana, South Africa, Namibia, Egypt and Mali are all participating in FAO-backed 'urban agriculture' initiatives in Africa. In the Democratic Republic of Congo, FAO is working alongside city authorities to help develop 800 hectares of urban land in several cities such as Kinshasa, Lubumbashi and Kisangani into allotment gardens. The aim is to produce fresh vegetables – and extra income – for 16,000 participating families, or roughly 80,000 people.

Home-grown in Latin America

As for the specific issue of slums, a novel approach is being tried by FAO in the Colombian cities of Bogota and Medellin, where a pilot project is being run to support vegetable production by internally displaced persons. With limited access to land, local experts assisted by FAO have taught hundreds of families living in 'barrios' (slums) how to produce their own vegetables right inside their homes in micro-gardens using a curious array of containers including recycled water bottles, old tyres and trays. Every month, each family's 'garden' yields some 25 kg of produce including lettuce, beans, tomatoes and onions. Any surpluses are sold off for cash to neighbours or through a cooperative set up under the project.

Source: www.fao.org/newsroom/en/news/2007/1000484/index.html (accessed 3 February 2011). See the example of Cuba at: www.youtube.com/watch?v=iGuipXzxPFY (accessed 3 February 2011)

Given the high labour inputs in horticulture, expansion of these activities could provide productive jobs for the urban poor without access to land, and the figures of those employed are impressive. Greenbelts, developed for agriculture, can help stabilize hill slopes and river banks. Their potential contribution to flood prevention is increasingly appreciated; consolidated green tracts of land reduce run-off after heavy rains and promote infiltration to recharge aquifers.

Plate 6.9b Barcelona: urban gardening is growing in popularity in the global north (Courtesy: Rosie Duncan).

However, urban and peri-urban horticulture offers even more benefits. Agriculture in cities includes animal husbandry as well as crop production; small animal production – poultry, rabbits, goats and pigs, for example – is essential to the survival strategies of many poor households (view photos and text about the benefits of small-scale rabbit production in an Egyptian village supported by the FAO at www.fao.org/Food/photo_report/Egyptra/Egyptra1_en.htm; accessed 15 December 2011). Small animals reared in this way provide relatively cheap protein with minimal environmental costs.

Urban agriculture may help fight the obesity pandemic too. In very poor communities, access to land and support for agriculture is known to promote healthier eating and lifestyle changes, which are allied to better health outcomes. Community gardens and school gardens have a proven record in promoting better nutrition and improved educational outcomes. View the videos detailing the multiple benefits of urban agriculture in a variety of developing world cities at: www.ruaf.org/book/export/html/1527.

Contrasts by age

Finally, it is important to realize that people at either end of the age spectrum tend to suffer disproportionally from malnutrition, especially undernutrition. Again, we can understand this with reference to their relative power within their communities or families, if they have families or if they still live in their communities. The very young and the old are often among the most marginalized in societies. Consider this quote about children:

> The children most at risk of missing out on the Millennium agenda, and on their rights under the Convention on the Rights of the Child, live in all countries, societies and communities. An excluded child is one who lives in an urban slum in Venezuela and takes care of her four siblings; a Cambodian girl living alone with her brothers because her mother had to go elsewhere to find a job; a Jordanian teenager working to help his family and unable to play with his friends; an orphan in Botswana who lost his mother to AIDS; a child confined to a wheelchair and unable to attend school in Uzbekistan; or a young boy working as a domestic in Nepal.
>
> (Source: State of the World's Children (2006) at: www.unicef.org/sowc/archive/ENGLISH/The%20State%20of%20the%20World%27s%20Children%202006.pdf; accessed 20 July 2011)

The incidence of poverty among children is also high because the age structures in the very poorest countries are skewed towards younger age groups. In addition, numbers are higher because the poorest families have more children as a survival strategy, quite rationally understanding children as assets who can work and improve the household's viability (Box 6.9). In such families, earning will be prioritized and educational opportunities neglected.

The term 'Children in Especially Difficult Circumstances' (CEDC) has been employed since the 1980s (Ansell, 2005: 192), and children in this category are of particular concern. Where the young or the old live in conflict situations, they will be less able to escape, and may be most vulnerable to abuse, exploitation and malnutrition. Some conflicts are notorious because of the enforced participation of children. Sometimes abandoned or orphaned, boys are required to become soldiers, and girls may be victims of sexual exploitation.

Box 6.9

Street children

An official study of street children estimated that there are 11,172 children living and working on the streets in Mexico City, the world's largest city. They wash cars and buses, run errands and carry soft drinks. Boys hate being loaders. Either they end up with spinal injuries or are run down by cars. Underlying it all is smog, heavy traffic and extreme poverty, in addition to violence, social disintegration and environmental deterioration. Drugs and delinquency are commonplace. In the streets, the children wash windshields and swallow fire. Almost all the passers-by are indifferent to the magic in their faces and hands. They wait for clients with their instruments in their hands, and in the darkness the studs on their charro outfits sparkle, their wide-brimmed hats glitter. They are guitars, violins, trumpets of Jericho, voices in search of a listener, jugglers, clowns, and magicians. The red light never stops for them, and the show goes on until three or four in the morning, later on Fridays and Saturdays, when couples feel romantic and give them a few more pesos. All those who pass by see them, but they are invisible. They do not exist. The police look at them without seeing them. Everything isolates them, denounces them.

Source: Elena Poniatowska in State of the World's Children (2006: 42), at: www.unicef.org/sowc/archive/ENGLISH/The%20State%20of%20the%20World%27s%20Children%202006.pdf (accessed 20 July 2011)

Cruelly, poor malnourished children are often more vulnerable to sexual exploitation because it offers them a livelihood. However, because of their poor nutrition, they are also more liable to contract disease, which may exacerbate their poor nutrition. Disadvantages established in early life have lifelong implications in all societies. Although apartheid no longer formally exists in South Africa, its legacies continue to shape young people's opportunities (Box 6.10).

The links between HIV/AIDS, disease and malnutrition were reviewed in Chapter 2. Although children may not be the largest number with the diseases, nevertheless millions of them suffer some of its worst ravages. Some 33.3 million (31.4–35.3 million) people were living with HIV as of 2009; 2.5 million (1.6–3.4 million) of them were children under 15 years. However, by the end of 2010, an estimated 16.6 million children had lost one or both parents to AIDS. Some 14.9 million of these children were living in sub-Saharan Africa

Box 6.10

Race in South Africa

Consider two South African children born on the same day in 2000. Nthabiseng is black, born to a poor family in a rural area in the Eastern Cape Province, about 700 kilometres from Cape Town. Her mother had no formal schooling. Pieter is white, born to a wealthy family in Cape Town. His mother completed a college education at the nearby prestigious Stellenbosch University.

On the day of their birth, Nthabiseng and Pieter could hardly be held responsible for their family circumstances: their race, their parents' income and education, their urban or rural location, or indeed their sex. Yet statistics suggest that those predetermined background variables will make a major difference for the lives they lead. Nthabiseng has a 7.2 per cent chance of dying in the first year of her life, more than twice Pieter's 3 per cent. Pieter can look forward to 68 years of life, Nthabiseng to 50. Pieter can expect to complete 12 years of formal schooling, Nthabiseng less than 1 year. Nthabiseng is likely to be considerably poorer than Pieter throughout her life. Growing up, she is less likely to have access to clean water and sanitation, or to good schools. So the opportunities these two children face to reach their full human potential are vastly different from the outset, through no fault of their own.

Source: www-wds.worldbank.org/external/default/WDSContentServer/IW3P/IB/2005/09/20/000112742_20050920110826/additional/841401968_200508263001833.pdf (accessed 28 January 2011)

(UNICEF, 2011). These children become more vulnerable to all the problems associated with poverty, and they are especially more likely to suffer from malnutrition and other forms of deprivation. 'Unite for Children, Unite Against Aids' was launched in 2005 to ensure that children and adolescents were more effectively included in HIV/AIDS initiatives (www.childinfo.org/hiv_aids.html; accessed 20 July 2011).

Often, age compounds discrimination suffered based on ethnicity. Hence, in Europe, for example, Roma children have a higher incidence of poor health and levels of education (see Chapter 5, Box 5.1). Marginalized populations are less likely to have access to public services, especially public health programmes or nutritional support programmes; this helps explain why children in these communities have poorer health and educational outcomes.

Older people may be excluded, ignored or abandoned. Shameful headlines in the United Kingdom recently exposed details about how elderly people in hospital may suffer from malnutrition. They are in hospital because they are ill, and once there, they become more ill and vulnerable to infections because they become malnourished. As detailed in Chapter 2, disease and malnutrition are very intricately connected. Where the aged are also female, they may suffer increased discrimination. In some regions, older women depend on their brothers or sons for their entitlements. In other cases, where the old have no family, they may be the most vulnerable of all except the very young.

Concern about the discrimination and poverty suffered by the poor must be balanced by recognizing their resilience and strengths. Poor communities organize to improve their conditions, and some of the most effective political campaigns have been driven by relatively poor people in the global south. Fights to protect their land rights, access to forests and ancestral lands are active across the south. People mobilize themselves to improve their livelihoods or to defend their rights when they are threatened: Chiapas in Mexico; the Chipko movement in India; Via Campesina in Brazil, are all well-known examples. Slum communities are known to be well organized and have thriving economic systems, even if they are not usually included in formal statistics. Research based on the Dharavi slum (the backdrop of the movie *Slumdog Millionaire*) concludes that the slum is an economic miracle, with an estimated turnover of approximately £500 million. All communities have hierarchies of power based on internal and external social networks. These networks offer opportunities for some but may mean oppression for others. Writing from a position of privilege, it is impossible not to admire the imaginative survival strategies and resilience of many of the world's poor. At the same time, it is important not to represent these people and communities as homogenous or necessarily noble; the dilemma of representing 'the other' complicates all portrayals.

Conclusion

This chapter reviewed distinctions at the sub-national level which influence entitlements to food and diets. All countries have contrasts within their boundaries, but in some the divisions are very stark indeed. Regional diversity may be based on inherited patterns of disadvantage and, in most countries, minorities are more likely to suffer malnutrition than the majority ethnic groups. Some household types are more prone to food insecurity than others, and some household members have a higher incidence of hunger than others. Where social and economic changes are rapid, new patterns of vulnerability emerge, sometimes exacerbating existing inequalities and sometimes creating new groups of disadvantaged people. Intense competition for land and water resources in developing countries are generating new victims of 'modernization' akin to the transformations experienced in previous centuries by colonial expropriations. In transitional economies experiencing rapid urbanization, obesity rates are alarming, especially among poor urban populations experiencing changes to patterns of diet and physical exercise. Once considered a problem only in high-income countries, overweight and obesity are increasing dramatically in low- and middle-income countries, particularly in urban settings. Numerous countries are now coping with the 'double burden' of disease associated with both the epidemiological and nutritional transitions.

Summary

- **Intra-national contrasts in nutrition may be more extreme than international contrasts.**
- **Within all countries, there are privileged regions, classes, ethnic groups and households who enjoy well-balanced diets and healthy lifestyles.**

- **Within all countries, there are regions, classes, ethnic groups and households who suffer disproportionately from malnutrition, both under and over.**
- **Understanding these intra-national patterns of malnutrition requires an appreciation of the historical and contemporary patterns of power and privilege.**
- **Patterns of malnutrition are dynamic; they change as economic and social policies change; global-, national- and local-level changes may be important at any given time.**
- **Communities organize themselves to resist negative changes to their livelihoods and to promote positive changes; their efforts may target local, national or global structures and institutions.**

Discussion questions

1. How might we explain the prevalence of undernutrition in any one region within any country?
2. What relevance do core–periphery models have for intra-national patterns of malnutrition?
3. What complications do the rural–urban category present for social scientists?
4. Why do the rural and urban poor depend on diverse entitlements?
5. In many respects, patterns of malnutrition in China illustrate all those factors that are obvious at the global level. Discuss.

Further reading

Perrons, D. (2004) *Globalization and Social Change; People and Places in a Divided World*, London, Routledge. A good introduction to uneven patterns of development associated with globalization in recent years. A good theoretical discussion presented with useful case studies.

Blaikie, P., Cannon, P., Davis, I. and Wisner, B. (2004) *'At Risk'; Natural Hazards, People's Vulnerability, and Disasters*, London, Routledge. Now a classic text, this book problematizes the concepts of risk and vulnerability.

Ellis, F. (2000) *Rural Livelihoods and Diversity in Developing Countries*, Oxford, OUP. This book explores the diversity of rural livelihoods strategies in the developing world.

Pacione, J. (2009) *Urban Geography*, London, Routledge. This is a good overview of urbanization and globalization. There are sections on developed and developing world cities, problems and prospects.

Useful websites

www.unhabitat.org The UN website for all themes connected to urbanization; for example, the recent publication about health in developing world cities at: www.hiddencities.org

www.cidse.org Christian-based global movement for equitable development; advocacy and research and resources. Number of development/justice themes including food security.

www.ifad.org Indispensible for rural development discussions and research.

www.survivalinternational.org Good website to follow issues about indigenous peoples' rights and threats to their survival.

www.amnesty.org.uk Obvious site to learn about human rights abuses and to appreciate how activists are challenging them.

Chapter 6

Follow Up

1. Read the briefing paper about Ethiopia's Gibe 111 dam at: www.internationalrivers.org/files/Gibe3Factsheet2011.pdf (accessed 15 August 2011).
   - Annotate a map of the region around the proposed dam site to illustrate the various conflicting interests that surround its construction.
   - Who will benefit from the project if completed?
   - Who and what will suffer the negative impacts if the project is completed?
   - Who are the principal financial backers of the project?
2. Meet the women and men from rural areas whose thoughts and perspectives were influential in the preparation of the International Fund for Rural Development (IFRD) Rural Poverty Report 2011 on their website and listen to some rural people and their stories: www.ifad.org/rpr2011/testimonials/index.htm
3. Learn about an international NGO which helps communities and households improve their self-sufficiency through gifts of livestock, seeds and trees and extensive training, all of which diversify their sources of food and income: www.heifer.org (accessed 26 February 2011).

4. Visit the 'State of the World's Children' (2006: 24): www.unicef.org/sowc/archive/ENGLISH/The%20State%20of%20the%20World%27s%20Children%202006.pdf

5. What national- and local-level factors help explain why some children are 'invisible' or 'excluded'?

6. Follow some of the activities of the International Slum and Shack Dwellers Alliance at: www.sdinet.org/static/upload/documents/newsletters/tgit8.pdf (accessed 3 February 2011).

7. Read the discussion about slum life in one of the world's largest slums, called Dharavi, in Mumbai, India, made famous by the movie *Slumdog Millionaire*. Visit: http://ngm.nationalgeographic.com/2007/05/dharavi-mumbai-slum/jacobson-text (accessed 3 February 2011).

7 Gender and nutrition: the female case

Learning outcomes

At the end of this chapter, the reader should be able to:

- **Describe and explain an index that measures and maps gender discrimination**
- **Understand the role of gender relations to the construction of female entitlements**
- **Detail some common misconceptions about the economic roles of females in the global south**
- **Describe some contrasts in entitlement patterns between rural and urban areas**
- **Describe and explain the 'missing women' phenomenon**
- **Describe some examples of female empowerment in the global south**

Key concepts

Gendered entitlements; formal economy; informal economy; gender equality index; empowerment; export processing zones; gender discrimination

Introduction

You should be aware now of the complexity of factors that help explain differences in nutritional status. Whatever proximate causes may be relevant in any particular case, there are usually persistent structural variables that require consideration. Using the terminology of structuralism again, we can assert that there are some communities that suffer from long-established patterns of discrimination and which are more likely to be economically and politically marginalized. Sometimes these patterns are observed through multiple generations.

Chapter 6 examined some important sub-national examples; this chapter considers arguably the most vital of all groups – women – and their role as food producers and consumers.

> In all aspects of social activity, including access to resources for production, rewards or remuneration for work, distribution of consumption, income or goods, exercise of authority and power, and participation in cultural and religious activity, gender is important in establishing people's behaviour and the outcome of any social interaction.
>
> (Pearson, 1992: 292)

One of the most important variables of relevance to diet and nutrition is gender. The crucial importance of gender as a factor in social relations began to be recognized in the 1970s, and since then a substantial theoretical literature has emerged to inform development policy and practice (see Reeves and Baden, 2000; Momsen, 2010). Considerable theoretical debates have emerged and continue to engage theorists as well as activists. At the international level, gender issues have been recognized institutionally by the United Nations and, on 2 July, 2010, the General Assembly voted to create a single UN body devoted to accelerating progress in achieving gender equality and women's empowerment – previously separate bodies were combined and now operate under the UN Women section (www.un.org/en/globalissues/women); in many national contexts, distinct bodies exist with similar remits.

Certainly, no research or analysis can ignore the part that gender plays in development and its consequences. Efforts to measure such patterns of discrimination have resulted in the construction of several development indices (Box 7.1). These, for all their limitations, help to construct maps of gender discrimination. Figure 7.1 shows a commonly employed index of gender and well-being. Access to a decent diet continues to be one manifestation of power differentials between genders. This text holds that access to a decent diet is largely determined by political, economic and social processes, and gender relations are always structured by these forces at all scales from the global to the local.

The next part of this chapter examines how differential access to resources by gender influence food production and distribution, and consequently patterns of malnutrition. The greater part of this discussion concentrates on traditional malnutrition, undernutrition, which still blights the lives of millions of women worldwide.

Box 7.1

Measuring gender discrimination

The introduction in 1995 of the Gender-related Development Index (GDI) and the Gender Empowerment Measure (GEM) coincided with growing international recognition of the importance of monitoring progress in the elimination of gender gaps in all aspects of life. While GDI and GEM have contributed immensely to the gender debate, they have conceptual and methodological limitations. In the twentieth-anniversary edition of the Human Development Report, the Gender Inequality Index (GII) has been introduced as an experimental index. It is not a perfect measure. Just as the Human Development Index (HDI) continues to evolve, the GII will also be refined over time.

The GII is a composite measure reflecting inequality in achievements between women and men in three dimensions: reproductive health, empowerment and the labour market. It varies between zero (when women and men fare equally) and one (when men or women fare poorly compared to the other in all dimensions). The health dimension is measured by two indicators: maternal mortality ratio and the adolescent fertility rate. The empowerment dimension is also measured by two indicators: the share of parliamentary seats held by each sex, and by secondary and higher education attainment levels. The labour dimension is measured by women's participation in the workforce. The GII is designed to reveal the extent to which national human development achievements are eroded by gender inequality, and to provide empirical foundations for policy analysis and advocacy efforts.

The world average score on the GII is 0.56, reflecting a percentage loss in achievement across the three dimensions due to gender inequality of 56 per cent, Regional averages range from 32 per cent in developed OECD countries, to 74 per cent in South Asia. At the country level, losses due to gender inequality range from 17 per cent in the Netherlands, to 85 per cent in Yemen. Sub-Saharan Africa, South Asia and the Arab states suffer the largest losses due to gender inequality. Regional patterns reveal that reproductive health is the largest contributor to gender inequality around the world – women in sub-Saharan Africa, with a massive 99 per cent loss, suffer the most in this dimension, followed by South Asia (98 per cent) and the Arab states, Latin America and the Caribbean (each with 96 per cent loss). The Arab states and South Asia are both also characterized by relatively weak female empowerment.

Source: United Nations Development Report, 2010, at: http://hdr.undp.org/en/mediacentre/summary/measures/faq. View a map of GDI at: http://hdr.undp.org/en/statistics/data/hd_map/gdi (accessed 6 January 2011)

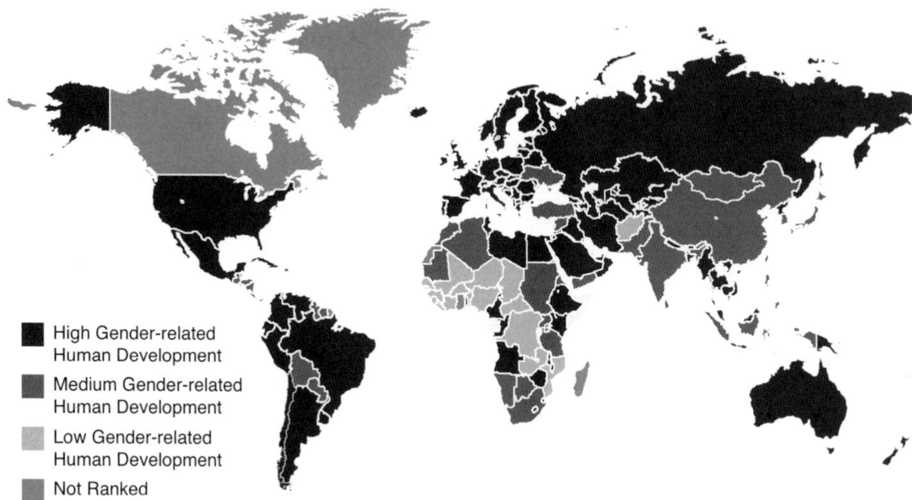

Figure 7.1 GDI rankings for the Human Development Report, 2009 (adapted from: http://hdr.undp.org/en/statistics/data/hd_map/gdi. To learn more about the index and its construction, visit: http://hdr.undp.org/en/statistics/indices/gdi_gem. This page evaluates new indices and their merits: http://hdr.undp.org/en/statistics/indices; accessed 10 August 2011).

The final section reviews the emergent literature on the relationship between gender and obesity which suggests that, as with traditional malnutrition, women are suffering disproportionately.

Gender and entitlements

What is the rationale for selecting to examine gender as a distinct category? There are three main justifications. Although the visibility of women as breadwinners has increased in recent years, their importance is still often denied or ignored. In a book published in 1997, I mentioned the invisibility of women's roles in production generally and agricultural production in particular. It seems their visibility remains partial in 2011. A report published in 2010 found that 'despite their major contribution to agricultural work and other rural economic activities, women's economic roles remain largely invisible and unrecognized in statistics and in public policy (IFAD, 2010: 60). Women are the principal food producers across the global south; they are the farmers and labourers, actually planting, harvesting and processing the food for domestic production; consider the case study in Box 7.2.

Box 7.2

Women at the centre of food security

Percentage of agricultural workforce

- *Afghanistan*: In some of the poorest and remotest areas of the mountain provinces of Bamiyan, Badakhshan and Nooristan, women are in charge of 100 per cent of the agricultural and animal-breeding activities.
- *Uganda*: It was estimated that women are in charge of 85 per cent of the sowing and of 98 per cent of the food transformation processes.
- *India*: Women constitute 82 per cent of those in charge of stocking crops and 70–80 per cent of those in charge of cattle milking.
- *Brazil*: 90 per cent of those employed in poultry are women.
- Women in the African continent spend 40 billion hours per year (combined) to fetch water.
- Women constitute 90 per cent of the rice cultivation workforce in Southeast Asia.

Source: Action Aid (March 2010) 'Her mile: Women's rights and access to land: The last stretch of road to eradicate hunger' compiled from various sources' at: www.actionaid.it/filemanager/cms_actionaid/images/DOWNLOAD/Rapporti_DONNE_pdf/HerMile_AAItaly.pdf (accessed 15 August 2011)

Multiple roles of females in food production

It is difficult for affluent consumers in the developed world, used to fast food, to appreciate just how much hard work is still required by many females in the developing world to secure an adequate diet for themselves and their children. Consider the following diagram, which details the multiple tasks required throughout the day by a female in Sierra Leone (Figure B7.2).

This also suggests how important her health is to the survival and well-being of her family. If a female food producer is ill, then catastrophe may befall her whole family, especially if she has young children or is the head of the household. This helps explain the awful problems associated with the diffusion of HIV/AIDS or other debilitating health problems in the developing world. Unable to work in food production, the family's health declines rapidly.

A study of the links between HIV/AIDS and nutrition established some alarming statistics. The disease is estimated to have killed 7 million agricultural workers since 1985 in the 25 hardest-hit countries in Africa, and it could kill 16 million more before 2020. It also found that food consumption dropped by 40 per cent in households affected by HIV/AIDS. Finally, it states that biological and social factors make females more vulnerable to HIV, especially during youth and adolescence. About 60 per cent of

4.00–6.00	*6.00–8.00*	*8.00–11.00*	*11.00–12.00*	*12.00–14.00*	*14.00–15.00*	*15.00–17.00*	*17.00–18.00*	*18.00–20.00*	*20.00–21.00*	*21.00–23.00*
Fish in local pond	Light fire	Work in rice field	Collect berries, leaves and bark	Process and prepare food	Wash clothes	Work in home garden	Fish in local pond	Process and prepare food	Wash dishes	Shell seeds
	Heat washing water	Perhaps while looking after 2 or 3 children	Collect firewood, sometimes having to walk long distances	Cook lunch	Carry water, often from some distance	Often caring for several children at the same time		Cook dinner	Bathe children	Make fish nets while chatting around a fire
	Cook breakfast			Wash dishes	Clean and smoke fish					Retire
	Wash dishes									
	Sweep compound									

Figure B7.2 A Women's Day in Sierra Leone.

Source: based on the Diagram from FAO Fact Sheet Gender and Nutrition (2001) at: http://www.fao.org/sd/2001/PE0703a_en.htm (accessed 15th August, 2011 at 12.47 GMT).

people living with HIV/AIDS in sub-Saharan Africa are women. In some countries, infection rates are three to five times higher in young women than in young men.

International appreciation of the intimate connections between development, poverty reduction and HIV/AIDS is now well established, as evidenced by President Obama's ambassador Garvelink's comments in Geneva in December 2010: As we know, reducing hunger and fighting HIV/AIDS are two of the eight objectives of the United Nations Millennium Development Goals (MDGs), with goal one to 'eradicate extreme poverty and hunger' and goal six to 'combat HIV/AIDS, malaria and other diseases'. These may seem like distinct goals, but they are closely intertwined in the lives of millions of people. To advance one goal requires advancing the other.

However, the political will and economic resources required to address these connections may be difficult to secure as the economic recession and 'austerity' programmes erode philanthropic inclinations.

Source: www.fao.org/english/newsroom/focus/2003/aids.htm (accessed 6 January 2011). View Ambassador Garvelink's speech which stresses the role of women in efforts to defeat HIV/AIDS and to deliver food security, at: http://geneva.usmission.gov/2010/12/09/amb-garvelink-food-security/ (accessed 6 January 2011)

Plate 7.1 Woman cultivating a dambo in Zimbabwe (Courtesy: Dr J. Elliott, University of Brighton).

Plate 7.2 Maize farmers in Zimbabwe (Courtesy: Dr J. Elliott, University of Brighton).

Plate 7.3 Women harvesting olives in Tunisia (Courtesy: Dr Hamish Main, Staffordshire University).

Plate 7.4 Woman winnowing guinea corn in Kano state, Nigeria (Courtesy: Dr Hamish Main, Staffordshire University).

In addition, women and children are over-represented in all statistics about malnutrition. An estimate from 2006 states that over '60% of the hungry are women and children' (FAO, 2006). This is a global problem, but it is worse in some regions than others. The specific problems faced by females in South Asia were introduced in Chapter 6, where we discussed gender as an accurate predictor for undernutrition in India (Box 6.5). Evidence suggests that similar circumstances exist in Pakistan, Bangladesh, while Afghanistan has recently been judged the worst country in the world in which to be a female.

The third reason to consider the particular circumstances of females is their crucial role in determining the nutritional well-being of children. In most communities, women are the principal carers of the young and make daily decisions about their diets. There is also disturbing evidence that female-headed households suffer more in times of food price rises. Empirical FAO research reported to the Committee on World Food Security in September 2008 that:

> ... the rise of food prices weighed more on women-headed households, since they generally spend a greater share of their income on food than those headed by a man. Furthermore, the obstacles that women usually have to face in food production in terms of access to natural resources prevented them from increasing their harvest, which would have allowed them to share in the benefits of the higher sale prices.
>
> (Quoted in Her Mile, 2010: 17)

Paying particular attention to the gender category may be justified, therefore, with reference to both production and consumption factors. Should anyone doubt the generalized pattern of discrimination, it is worth reviewing the seminal observations by Sen (1990).

On this occasion, Sen was interested in exploring statistics about male–female sex ratios, not famine or entitlements, although clearly they are allied areas of interest, as the following discussion confirms. His work exposed some of the most damming statistics that illustrate discrimination by gender, that is, the 'missing women' phenomena (Box 7.3). Discrimination is manifest in a great variety of ways, and may be obvious at various ages and in various degrees of intensity, from very serious injustices to milder types of neglect or indifference. The role of gender is obvious when analysing undernutrition or overnutrition and the concept of entitlements assumes a useful role again for our analyses.

Box 7.3

Sen's 'Missing Women': the problem continues

Amartya Sen (1990) famously claimed that millions of women were 'missing' from the population totals of Asian countries, in particular. On the basis of various assumptions, he calculated that excessive female mortality accounted for a 6–11 per cent deficiency in the total number of women, thus revealing what he called a 'terrible story of inequality and neglect'. Croll, in a paper written over 10 years later, examined the latest trends of female birth and survival in South and East Asia. She found that major problems still remained amid what amounted to continent-wide denial by governments, donors, communities and families of excessive female mortality, discrimination and disadvantage.

More recently, the United Nations Development Report (2010) also found that 'the most glaring discrimination is evident in women's low relative share in the population, a key aspect of recent demographic tends in several countries.'(p. 76). They updated Sen's earlier estimates of 'missing women' and found more than 134 million missing women in 2010 – almost a third more than in previous estimates. They understand these figures to be highly influenced by the preference for male children in China and some parts of India.

Source: Croll, E. (2001); United Nations Development Report (2010: 76): http://hdr.undp.org/en/media/HDR_2010_EN_Chapter4_reprint.pdf (accessed 6 January 2011)

It is useful to examine evidence about female entitlements at various temporal and spatial scales. The discussion opens by appreciating that historical circumstances were important in shaping some still extant patterns and then reviews contemporary transformations that continue to remake the contours of gender entitlements at the global, national, local and household levels.

Historical legacies

It is impossible to review here the multifarious ways in which colonialism has transformed gender relations. Chapter 3 summarized some of the main impacts of colonial penetration upon native people and places and their consequences for food provisioning. There was widespread commoditization of food, and traditional systems of provisioning were eroded and replaced by a market in foodstuffs. It is hard to underestimate the changes this was to make; simply put, you needed money and/or assets to access a decent diet, and the means of achieving either had implications for men and women as they do today. Secondly, the aggressive commercialization of agriculture occurred to various degrees across the colonized world, which caused equally serious implications of local and national food provisioning. Vast areas of fertile land were devoted to fuelling the Industrial Revolution and the hungry mouths of a rapidly expanding European industrial workforce. Sugar, wheat, bananas, tea, coffee and meat products were sourced from across the globe, and the roots of the contemporary global food system are found in patterns established in the seventeenth, eighteenth and nineteenth centuries.

The introduction of plantation economies in the tropics was but one of the most brutal manifestations of the changes wrought and the slave economies one of their main components. In temperate latitudes, new migrants supplanted indigenous populations in another example of what Harvey (2003) has called 'accumulation by dispossession', where wealth was amassed by some at severe costs to others. Gender was one variable among others (class, race and ethnicity) that determined one's prospects of suffering the costs of the new arrangements, or enjoying some benefits.

The process of colonization was so transformative, politically, economically, socially and culturally, that profound changes were made in all societal arrangements, including alterations in the relative power and status of men and women. Contrasts were marked between the diverse colonial regimes and their interactions with pre-existing contexts, which means that generalizations must be tentative, and impacts were certainly sometimes contradictory. However, most concur that colonial imposition altered men's and women's relative power, undermining women's legal status, control over economic resources and access to formal political roles. As the following discussion illustrates, such patterns of dispossession and marginalization are unfortunately not confined to history. Transformations allied to globalization in the present share some of the worst aspects of colonialism, and these too are gendered in character.

One legacy of colonialism that has survived into the present century is the marked differences in access to productive resources by gender across the global south; another is the centrality of women's work in food production. As men were often drawn into commercial agricultural production, the burden of domestic food production became the preserve of female family members. It is important to emphasise the central role that females have in achieving household food security in most communities. As detailed in the FAO factsheet (2001), '[I]t is usually women who grow most of the food the family eats. They also choose which foods to purchase and find ways to feed the family when supplies are low, for instance, during the dry season. In many cities, they supplement the family diet with fruits and vegetables grown in small gardens. Women process and store food, ensuring safe and sustained consumption; and they also prepare and distribute the food the family eats, collect firewood and carry water' (FAO, 2001: 1). Consider these statistics for Cambodia.

> Here 65% of agricultural labour and 75% of fisheries production are in the hands of women. In all, rural women are responsible for 80% of food production. Half the women are illiterate or have than a primary school education; 78% are engaged in subsistence agriculture, compared to 29% for men. In rural areas, only 4% of women and 10% of men are in waged employment. Households headed by women are more likely than households headed by men to work in agriculture, yet, they are also more likely to be landless or have significant smaller plots of land. Policies, programmes and budgets for poverty reduction must thus address the situation of Cambodian women.
>
> (Gender and Development Network and NGO Forum on Cambodia, 2004, quoted in OECD, 2006: 17)

If we accept that access to a decent diet is largely influenced by one's entitlement, then several factors suggest that females may suffer reduced entitlements compared to male adults. Until Boserup's (1970) classic study, the economic role of women in agriculture was ignored. In more recent analyses, the significance of their labour in all sectors is frequently minimized. Among the most important specific issues now receiving more attention are the vital role of female farmers in the south and their dominance in the subsistence crop production sector as well as the cash crop sector. In several respects, their opportunities are constrained as agricultural producers. A serious problem is that, in many countries, women are much less likely to own property and other assets than men, with negative implications for their absolute and relative status' (UNDP, 2010: 76). Even in Uganda, where women account for most agricultural production, they only own 5 per cent of the land, and their tenure is insecure (UNDP, 2010: 76). The importance of access and ownership of land is highlighted in Chapter 5; that females are particularly discriminated in this respect is important to any explanation of their poor nutritional status and their dominance in the very poor categories (Box 7.4).

Even when women enjoy land ownership, the quality of their land is often inferior. Pearson indentified the relevance of land ownership and concluded: '[S]ince … men and women have different

Box 7.4

Gender inequalities: multiple disadvantages

Women and land rights

The issues of access to and ownership of land has understandably been emphasized throughout this text. The portal www.landtenure.info provides detailed information on women's access to land and on agrarian tenure systems in Angola, Bangladesh, Benin, Bolivia, Bosnia, Brazil, Burkina Faso, Burundi, Colombia, Philippines, Guatemala, Honduras, Indonesia, Mozambique, Nepal, Niger and Peru. It was created in 2008 by the International Food Security Network, ActionAid, CERAI, aGter and COPROFAM, with the support of the Land & Water Division of the FAO. Information collected in this database is about to be merged with other data provided by the FAO on the issue. Some of the facts about women and land are:

- Less than 2 per cent of the available land worldwide is owned by women.
- In Brazil, women represent 57 per cent of the population, but just 11 per cent of the land belongs to them.
- In Nepal, women own just 10.8 per cent of the land.
- In Uganda, just 7 per cent of women own land and women's right to land is mainly considered as a mere right of use, without the possibility to make decisions (on selling, hiring, changing its use).

It is important to underline that these percentages, which provide an overview of the existing gender inequalities on land tenure, do not reveal much about the use that women make of common properties or about land management at the community level. In this respect, the lack of reliable data does not allow a complete overview of women's access to land.

Source: Action Aid (2010) Her Mile: Women's Rights and Access to Land at: www.landcoalition.org/sites/default/files/publication/762/HerMile_AAItaly.pdf (accessed 14 January 2011)

Additional disadvantages

- Men's landholdings average almost three times the size of women's landholdings (globally).
- Fertilizer is more intensively applied on men's plots, and is often sold in quantities too large for poor women to buy.
- An analysis of credit schemes in five African countries found that women received less than one-tenth of the credit that was received by men smallholders.
- In most developing countries, rural women's triple responsibilities – farm work, household chores and earning cash – often add up to a 16-hour workday, much longer than their male counterparts. However, women continue to lack access to important infrastructure services and appropriate technologies to ease their workloads.
- Women-owned businesses face many more constraints and receive far fewer services and support than those owned by men. In Uganda, women's enterprises face substantially higher barriers to entry than men's, although those that exist are generally at least as productive and efficient as men's in terms of value added per worker.
- In Guatemala, women hold only 3 per cent of snow pea production contracts but contribute more than one-third of total field labour and virtually all processing labour.

Source: World Bank, FAO and IFAD (2010) at: www.ifad.org/rpr2011/report/e/rpr2011.pdf (p. 61) (accessed 10 January 2010)

responsibilities within a household, lack of access to land can mean not just dependence on the head of household, but real difficulties in meeting her own and her children's needs for clothing and food' (Pearson, 1992: 297).

Efforts have been made to challenge traditional patterns of land ownership, but even with official and legal support, these are often undermined by local patriarchal attitudes. In situations where women are the sole cultivators of a plot of land, they frequently have limited rights to its use. It may be based on their relationship to a male family member, for example; if that relationship changes, so do their rights to the land. For the rural poor, land is vital. It helps them earn a livelihood and thereby access a decent diet; it encourages investment and helps families accumulate intergenerational assets; and for all these reasons, securing property rights is a crucial first step in reducing poverty and increasing food security. In terms of entitlements, therefore, these various issues have serious negative implications for females, from their inability to use land as collateral, to their being easily evicted from land they may have cultivated and improved for years. However, land means even more than this. Whitehead (2010: 4) emphasizes the multifaceted character of land and cites; in addition to its material basis for survival, it is important as: a site of belonging; the basis for citizenship; a site of struggle.

A study of pro-poor agricultural policies concluded that '[R]emoving gender-based barriers to growth will make a substantial contribution to realizing Africa's growth potential. Reducing gender inequalities in access and control of key resources is a concrete means of accelerating and diversifying growth, making growth sustainable and ensuring that the poor both contribute to, and benefit from, that growth. A recent publication from the Commission on Women and Development emphasizes the central role of women and their access and control of resources to eliminating food insecurity (view their recommendations at http://diplomatie.belgium.be/en/binaries/access_control_resources_tcm312-110978.pdf; accessed 16 January 2011).

However, removing such inequalities is not simple, as Whitehead's observation attests; challenging inequalities is a political process and often faces severe opposition, locally, nationally and occasionally internationally. Struggle and resistance, evidence of people's agency, is observed everywhere and is considered in more detail in the final chapter, but for an example from India, see Box 7.5.

Globalization and changing female entitlements

Some of the macro-economic changes associated with globalization are considered in Chapters 3, 4 and 5, but the discussion there is partial because it ignores the gendered character of the process. Tsikata (2010: 5), in her review of the literature on globalization, draws attention to some glaring gaps in the analyses, namely an appreciation of the gendered nature and its impacts, especially the interconnections between globalization, land tenure and gender. Her contribution is part of an excellent edited volume which attempts to redress this gap; it does so, and suggests the potential value of such perspectives. It details the momentous transformations associated with globalization, promoted through economic liberalization since

Box 7.5

Women marching forward: an example from India

Despite the demographic and economic growth of urban areas, the livelihoods of 68 per cent of the Indian population depend on land-related activities. The caste system treats Dalits (word meaning 'oppressed people') as untouchables: India counts about 100 million Dalit women, who represent 16.3 per cent of the female population and the majority of the agricultural workforce in the country. Few of them own land (estimates are 2 or 3 per cent), and many go hungry everyday and face further challenges such as sexual violence. They suffer the triple burden of caste, class and gender.

Action Aid works with Hunger Free women in India on a campaign to increase these women's access to land. From Simla in the Himalayas to Bangalore in the south, from the desert of Jaipur to Guwahati in the north-east, thousands of women gathered, talked of the wrongs they had suffered with regard to access to land and the right to food, drew up manifestos that served also as a platform for dialogue and marched to be visible and involve other women in the struggle for land. The mobilization of these women has secured some positive outcomes and is now a nation-wide movement.

Kalliammal, a female farmer from Tamil Nadu describes what legal access to land means to her. She said, 'Now I have my own plot and I harvest enough to feed my family for the whole year. I am respected by my husband and son because this land is entitled to me. And one day my daughter too will inherit this land'.

Source: Action Aid (2010) Her Mile: Women's rights and access to land: The last stretch of road to eradicate hunger' www.actionaid.it/filemanager/cms_actionaid/images/DOWNLOAD/Rapporti_DONNE_pdf/HerMile_AAItaly.pdf (accessed 13 January 2011). Action Aid India has made three short autobiographical films about the lives of three inspirational Dalit women from Andhra Pradesh, southeast India, who have joined the struggle for land and Dalit rights. Watch these at: www.actionaid.org/main.aspx?PageID=1337 (accessed 13 January 2011)

the 1970s, and how they are experienced differentially by gender. It also recognizes that impacts are very diverse and dependent on specific national and local contexts, and that there is a danger of neglecting women's 'complicated and multifaceted realities'. However, some general tendencies have emerged from the available literature.

Globalization has exulted competition and export promotion as drivers of development, but these are also responsible for driving down wages and undermining labour rights. Pressures have intensified in the agricultural sector everywhere. Given the prominence of women in agriculture, the impacts will be significant for millions of females in the developing world. A recent UN publication summarizes the situation:

> Whereas these dynamics (of globalisation) have in some ways brought benefits, in general, the largest proportion of rural women worldwide continues to face deteriorating health and work conditions, limited access to education and control over natural resources, insecure employment and low income. This situation is due to a variety of factors, including the growing competition on agricultural markets which increases the demand for flexible and cheap labor, growing pressure on and conflicts over natural resources, the diminishing support by governments for small-scale farms and the reallocation of economic resources in favor of large agroenterprises.
>
> (IAASTD, 2009: 11)

The case of China illustrates how diverse changes in the macro-economic context have differential impacts on communities by gender. Chinese industrial expansion has been very rapid since 1978 and has been associated with massive increases in employment opportunities in the urban industrial core regions (Bramall, 2009). Rural urban migration has occurred, but male adults are dominant in this exodus and often have restricted residency permits in the urban areas; hence, most male migrants become temporary labourers in cities, with agriculture as a form of insurance and retreat. 'The gender division of labour in rural households has shifted from 'the men do the tilling and the women weave' to 'the women till and the men work in industry'. This new model can be described as 'men control the outside world, women the inner'. What's new is that women's inner world is extending to include agriculture. The new gender division of labour has led to the feminization of agriculture: about 80 per cent of the rural workforce is female. In the poorer and marginal southwestern Chinese provinces of Guangxi, Yunnan and Guizhou, women make up more than 85 per cent of the agricultural labour force – and in some remote mountainous areas, about 90 per cent (Song, 1999). For an overview of gender and food security in the Asia and Pacific regions, see the report by the FAO (2005). This report establishes the substantial contribution made by rural women to the economies in the region and the persisting barriers to their advancement.

Globalization of agriculture has been accompanied by the extensive privatization of all resources, turning natural resources – such as water, minerals, forests and land – into global commodities which fuels the problem identified by Whitehead that 'competition and conflict over the access and use of land are at a historical peak'

(2010: 3). Problems of securing access have been made even more problematic for the rural poor, especially women, as competition for land and resources intensifies.

Tensions of particular relevance are the competition between land for export crops and subsistence production. In the Amazon Basin, Porro (2010) examined the consequences of globalization (graphically represented there by the trans-oceanic highway through the Pacific to integrate commodity markets in Bolivia, Brazil and Peru with those in Asia), and concludes that local livelihoods have been sacrificed in favour of global interests that favour logging, cattle ranching and agribusiness for commodity production. Such shifts result in wealth and land concentration as well as environmental degradation, and examples can be found across the global south and are usually correlated to increases in the numbers of women becoming landless labourers. Liberalization and its tendency to increase competition in land markets tend to marginalize domestic food production and privilege export production, and these policies are often explicit in development policies.

An equally important resource is water. Water is vital for health and food production everywhere, and its consumption in every type of human activity is increasing. This precious resource is under intense competitive pressure in the north and south, and its ownership is increasingly being privatized Read the excellent discussion by Yesilyurt (2011) who examines the general impacts and anomalies that surround water privatization and their implications for women in particular (Box 7.6).

As economic actors, traditional and contemporary misconceptions about 'women's roles' help explain some examples of discrimination, intentional and unintentional. Economic and social changes, often marketed as 'development policies', can inadvertently exacerbate the marginalization of females. Despite the fact that females account for the majority of agricultural labour in sub-Saharan Africa, agricultural policies in the past tended to assume that 'farmers' were male. Hence, change, improvements in technology and the extension of credit facilities, for example, were targeted at the male members of the rural community. In 2006, an analysis of agricultural growth strategies in Nigeria in recent years found:

> Another feature of the Nigerian agriculture sector is its high level of differentiation regarding the gender division of labor in production,

Box 7.6

Women and water

Although inadequate access to water and sanitation affects both women's and men's health, women are far more distressed than men by water scarcity, waterborne diseases and lack of sanitation services. Since, in most societies around the globe, women and girls are in charge of cooking, washing and cleaning, as well as family members' and home hygiene, they are considered to be responsible for collection and transportation of water for domestic use. For this, millions of women do not just open the tap – and voila – here comes clean, safe, fresh water! Instead, they have to spend a lot of time and energy in gathering water. This robs them of the chance to get a proper education, performing income-bringing work or having time for rest and recreation.

Women and girls walk long distances – 10–15 km a day – mostly barefoot, no matter whether they are pregnant or ill, young or old, no matter whether the weather is hot or cold or how dangerous the walk might be. They are exposed to dangers such as physical assault, water-related diseases, attacks from animals and physical problems due to heavy water weight.

The 2004 Consumers International report emphasizes this. Poor rural women in developing countries may spend 8 hours a day collecting water, and carry up to 20 kilos of water on their heads on each journey. One in 10 school-age girls in Africa do not attend school during menstruation or drop out at puberty because of the absence of clean and private sanitation facilities in schools. Every day, 6,000 girls and boys die from diseases linked to unsafe water, and women are the main caretakers for sick children and adults. A woman in a slum in Kenya pays at least five times more for one litre of water than a woman in the United States.

Due to gender insensitivity and restrictions of women's participation, women are barred from significant decision-making. Besides their material vulnerability, their exclusion from decision-making and management of resources makes women disproportionately distressed by the lack of adequate, clean and safe water. As the right to water is not maintained, most of women's other rights are threatened too.

Source: based on Gündüz, Zuhal Yesilyurt (2011); online at: http://monthlyreview.org/2011/01/01/water-on-womens-burdens-humans-rights-and-companies-profits (accessed 28 July 2011)

> marketing, and use of income. About 80% of the rural female population is engaged in agriculture and forestry as family labor. Most of these women have poor access to agricultural services and are living in poverty.
>
> (Fan *et al.*, 2008: 1)

The gendered division of labour in agriculture helps explain the mixed impact of new technologies. The introduction of new technologies is always associated with differential impacts: some social groups benefit disproportionately and others may suffer disproportionately. Certainly, there will be a shift in social relations and, generally speaking, the more significant the technology, the greater the differential impacts, by class or gender. The extension and promotion of industrial agriculture often erodes female entitlements and may be associated with a reduction in nutritional status.

An allied issue is that female farmers produce most of the domestically consumed food in many developing countries, in urban as well as rural areas (Box 7.7), but this sector of the economy has been very poorly supported by national development initiatives. New initiatives are surely overdue:

> There has never been a more important time to address rural poverty in developing countries. It looks likely that global food security and climate change will be among the key issues of the 21st century. As agricultural producers and custodians of a large share of the world's natural resources, poor rural people have key roles to play, contributing not only to global food security and economic growth, but also to climate change mitigation efforts. National governments have the principal responsibility for giving them the tools they need to fulfil their potential. National stakeholders and the international development community also have important supporting roles to play.
>
> (Nwanze, 2010)

This quote is from an interesting recent report about rural poverty in the south. The report suggests that there is renewed interest in supporting the rural poor and their productive capacities, an interest which may have intensified since the food price surges of 2007–2008; targeting the rural poor, many of whom are female, is crucial to any effort to reduce food insecurity and poverty across the south. Despite the world's population being, for the first time in human history, predominantly urban, the majority of the world's poor are still rural people. Rural economic activity is varied, and it is a mistake to assume that it is all agriculturally related. However, with this caveat,

Box 7.7

Investing in rural women contributes to food security

Women in agriculture play a vital role as agents of food security and rural economic growth, but often endure poor working conditions and receive limited recognition for their contributions. The International Day of Rural Women takes place each year on October 15 to honour the multiple roles that rural women play worldwide.

Rural women form the backbone of the agricultural labour force across much of the developing world. Globally, more than a third of the female workforce is engaged in agriculture, while in regions such as sub-Saharan Africa and South Asia, more than 60 per cent of all female employment is in this sector. To afford food and other basic expenses, men and women in rural areas often diversify their income by combining multiple forms of employment. Women generally work as subsistence farmers, small-scale entrepreneurs, unpaid workers on family farms or casual wage labourers – but they may take on all or a number of these activities at different times. Since, in many developing countries, women carry out a range of vital household and caring tasks, their overall working hours tend to be longer than men's. In countries such as Benin and Tanzania, for example, rural women work, respectively, 17.4 and 14 hours longer per week than their male counterparts.

Providing women with better opportunities to grow their own crops for sale, undertake paid work in an agro-industry or take on other paid activities in the rural sector is critical to increasing their bargaining power within the home, and can legitimize their control over key material resources, such as land and credit. This is important because it elevates their status within families and communities, but also because women are more likely than men to spend their income on food and other basic needs for the household.

However, many rural women continue to face obstacles that undermine their opportunities for success, such as lack of public and social infrastructure, unequal access to credit, technical equipment and other important resources, such as land and water. In Burkina Faso, Kenya, Tanzania and Zambia, for example, studies have shown that allocating land, labour, capital and fertilizer equally among men and women could increase agricultural production by between 10 and 20 per cent.

Worldwide, rural women are the key to food security.

Source: extracts from Rural Women and Development Gender, Equity and Rural Employment Division (ESW), Food and Agriculture Organization of the United Nations (FAO) at: www.un.org/womenwatch/feature/idrw/#contribution (accessed 12 January 2011). Also, see: http://diplomatie.belgium.be/en/binaries/access_control_resources_tcm312-110978.pdf for a review of the Recommendations of the Commission on Women and Development, Gender, Empowerment and Food Security Group (2010), 'Access to and Control over Resources by Women: A Challenge for Food Security. Also, FAO (2010) 'Gender and Land Rights Understanding Complexities; Adjusting Policies' at: www.fao.org/docrep/012/al059e/al059e00.pdf (accessed 12 January 2011); Tsikata, D. and Golah, P. [eds] (2010) Land Tenure, Gender and Globalisation: Research and Analysis from Africa, Asia and Latin America' (Ottawa, International Development Research) at: www.idrc.ca/openebooks/463-5 (accessed 12 January 2011) there is a vital role for small-scale agriculture in policies to reduce poverty and malnutrition in the developing world.

Agriculture plays a vital role in most countries – over 80 per cent of rural households are involved in farming to some extent. Typically, it is the poorest households that rely most on farming and agricultural labour, although, as established in Chapter 6, rural livelihoods depend on a range of very diverse activities. The role of agriculture is important everywhere but perhaps especially important in South Asia and sub-Saharan Africa where malnutrition and associated poverty are very severe. Given the centrality of female labour to agricultural production, policies must address their neglect and/or marginalization in many areas of agricultural development policy at the international and national scales. Some evidence of awareness is discernible, and the FAO, in collaboration with the International Labour Organization (ILO), is working to identify new trends in rural labour markets and to ensure that gender-sensitive measures are integrated in the policy guidance that both agencies give at the country level. The International Fund for Agricultural Development (IFAD) summarizes thus the benefits of investing in rural women:

- Women's empowerment benefits not only women themselves, but also their families and communities.
- Farm productivity increases when women have access to agricultural inputs and relevant knowledge.
- Women are dynamic organizers and participants in grass-roots organizations, and are effective in promoting and sustaining local self-help initiatives.
- Malnutrition and mortality among both boys and girls are reduced when girls obtain greater access to primary and secondary education.
- There is a strong correlation between women's literacy and lower HIV/AIDS infection rates.
- Women have a strong track record as prudent savers and borrowers in microfinance programmes, using income to benefit the entire household.

(Source: www.ifad.org/gender/index.htm; accessed 28 July 2011)

They conclude: 'Investing in women is not just about achieving the third Millennium Development Goal, which is to promote gender equality and empower women. … Investing in women by promoting gender equality is vital to achieving all the other goals as well'. Women's work in the rural sector is recognized at the highest institutional level. Ban Ki-moon, General Secretary of the United Nations, added his support: '[O]n this International Day, let us pledge to do our utmost to put the rights, needs and aspirations of rural women much higher on the global agenda' (2009). Evidence confirming the importance of women's roles is overwhelming, but obstacles are significant too. Chapter 3 reviews the vested interests attempting to usurp traditional farming and small units of production and to replace them with industrial agricultural production systems designed to fuel the corpulent global corporate appetite. However, other problems are part of the reason why women may be required to migrate from their rural communities (Box 7.8).

Box 7.8

Slum life in Kibera, Nairobi: the female experience

The vast Kibera settlement in Nairobi receives much media attention as one of the largest, most densely populated slums in Africa, but there are surprisingly few studies with reliable statistics relating to the women who live there. Kibera is a vast slum located approximately 7 kilometres southwest of Nairobi's city centre. The slum is characterized by severe poverty, poor access to clean water, overflowing open sewers, huge heaps of rubbish, overly crowded mud houses, constant threat of eviction and widespread criminality, delinquency and unemployment.

Unequal power relations between women and men generally leave Kibera women at a disadvantage in areas such as accessing land, property and other productive resources, and securing remunerated work. Indeed, research quoted in the report finds that:

> Informal settlements in Nairobi are … often home to thousands of women who were driven by in-laws out of their rural and urban homes and land upon the death of their husbands. In two separate missions to Kenya, as well as through research on women's inheritance rights in sub-Saharan Africa, COHRE found that family pressure, social stigma, physical threats and often extreme violence directed at the widow force her to seek shelter elsewhere.
>
> (UN-Habitat, 2008: 43)

The study reinforces findings from numerous others, showing that widows are particularly vulnerable to eviction because of customary and traditional practices in sub-Saharan Africa. Property grabbing from widows whose husbands have died of

AIDS is also reportedly widespread, and is one of the factors that push women to migrate to Kibera. There is also some evidence to suggest that single, unmarried mothers in informal settlements such as Kibera have the poorest choice in housing since many landlords are unwilling to rent to them.

The burden of women's domestic roles in Kibera also puts them at more of a disadvantage when basic services such as water, sanitation and electricity are lacking. As with women in many parts of the developing world, those in Kibera are usually responsible for water collection and management of waste in the home. Long queues of women waiting with rows of yellow jerrycans are a common sight in the slum. Up to 85 per cent of women in Kibera draw water from private and community-owned water kiosks.

Sources: extracts from UN-Habitat (2008)

Global processes are transforming the nature and conditions of work, formal or informal, as restructuring advances to alter economic policy and practice everywhere. Global processes have created new opportunities for female employment in the global south, and while these have certainly generated some positive changes in terms of female empowerment and status, the results are very mixed and often negative. Crowley describes its mixed results: '[W]omen make up 20–30 per cent of agricultural wage workers and often predominate in high-value industries for export – such as fresh fruits, vegetables and flowers. Although these non-traditional sectors of agriculture can present women with unprecedented income opportunities, we often find that women are hired on temporary or casual contracts to perform labour-intensive, manual tasks. They are given limited opportunities to acquire new skills and, overall, higher-level positions continue to be captured by men' (FAO, Gender Insight, 2010).

Many of these new jobs are poorly paid and insecure; they provide only fragile entitlements at best. Where these jobs are in the formal sector, female labour is cheaper and less unionized than male employment. The global garment industry is perhaps the best example to illustrate how female workers have been integrated into global production systems and 'exemplify many of the intractable issues facing today's global economy' (Dicken, 2003: 317; see Dicken, 2010, for excellent examination of global economic change in general and his analysis of the garment industry in particular) (Box 7.9).

Research across the south has identified another problem with these opportunities for females. In most places, women continue to supply

Box 7.9

The global garment industry: a quintessential case?

The global garment industry is one of the most globalized sectors, and one that employs large numbers of women under poor working conditions. They are integrated into global networks through informal and formal labour, but this discussion concentrates on their work in the formal sector. Most nations produce for the international textile and apparel market, making this one of the most global of all industries. It has often been the 'starter' industry for developing countries engaged in export-oriented industrialization: it played a leading role in East Asia's early growth. In many countries, the garment sector employs the most workers in manufacturing.

The garment sector is emblematic of the global manufacturing system, whereby production is dispersed geographically to an unprecedented number of locations across and within countries. It is typical of 'outsourcing', since much production is shifted offshore from developed and developing countries to exploit their low wages and limited labour and environmental legislation. It is also often found in specific locations called export processing zones (EPZs). The importance of EPZs in the globalization process is suggested by the figures in the following table. For example, the whole island of Mauritius is an EPZ, and it employs 65,512 workers, of which 63 per cent are female. Of all employment on the island, 42 per cent is in the manufacturing export sector, and their activities include: food, flowers, textiles, leather products, wood and paper products, optical goods, electronic watches and clocks, electrical and electronic products, jewellery, toys and carnival articles.

The ILO has defined EPZs as 'industrial zones with special incentives set up to attract foreign investors, in which imported materials undergo some degree of processing before being re-exported'. Their ubiquity is suggested in this quote from the ILO:

> EPZs have evolved from initial assembly and simple processing activities to include high tech and science parks, finance zones, logistics centres and even tourist resorts. … Madagascar, Mauritius and Hainan (China) allow factories anywhere on the respective island to apply for zone status. Port cities like Hong Kong and Singapore have enhanced their strategic trading role by providing special customs regimes for export processing and transhipment.
>
> (ILO, 2007, available at: www.ilo.org/public/english/dialogue/sector/themes/epz/epzs.htm; accessed on 16 January 2010)

Source: Much of this text was from WIEGO at: www.wiego.org/occupational_groups/garmentWorkers/index.php (accessed 16 January 2011); Tinoco, E. in Boyenge, J. P. S. (2007) 'ILO database on export processing zones' at: www.ilo.org/public/english/dialogue/sector/themes/epz/epz-db.pdf (p. 1) (accessed 16 January 2011)

all the reproductive labour in maintaining the family. Caring tasks, in all their incarnations, remain female responsibilities. Hence, while women may enjoy increases to their financial position, and consequently also their status in the family, they also suffer as they must balance the work demands in the domestic sphere as well as the formal sector. 'The pressures on all workers from the collision between these two worlds (domestic and work place) are already enormous, but for women they are particularly complex, as women do most of the housework and, at the same time, for a variety of reasons, are poorly paid and employed in precarious jobs' (ILO and UNDP, 2009: 40). Similar results are noted in a study of increased participation of Indian women in earning activities:

> Income generation programmes such as those promoted through self-help groups (SHGs) have done little to change the concentration of women's enterprises in a narrow range of low-profit activities with few assets and low productivity. This has had related impacts on women's economic participation. Any expansion in women's income earning activities is not compensated by men making a greater contribution to domestic work. Evidence on women's control over assets is scarce. … Their ability to increase their income is limited by restrictions on their interactions with men outside the household, and responsibilities for unpaid household work or childcare, which undermine their negotiating power in markets.
>
> (Thekkudan and Tandon, 2009: 14)

Processes allied to globalization that are transforming rural spaces help explain the rapid expansion of the informal sector. Increasingly marginalized in agriculture as land and labour are commandeered for commercial production, migration becomes a survival strategy for men and women. Many of these migrants find jobs in the urban informal economy. The significance of this work is illustrated by statistics gathered by Women in Informal Employment: Globalizing and Organizing (WIEGO). Informal employment, broadly defined, comprises one-half to three-quarters of non-agricultural employment in developing countries, specifically:

- 47 per cent in the Middle-East and North Africa
- 51 per cent in Latin America
- 71 per cent in Asia
- 72 per cent in sub-Saharan Africa

These statistics, collated before the economic recession of 2008–2010, have probably increased as formal workers are 'released' and rural unemployment intensifies. Indeed, a recent analysis of the relationship between globalization and the informal sector (Bacchetta, 2009) finds a negative relationship between the extent of the informal sector and the depth and negative consequences of global economic shocks, such as the 2008–2009 recession:

> … globalization has added new sources of external economic shocks. For instance, global production chains can transmit macroeconomic and trade shocks through several countries at lightning speed, as observed in the current economic crisis. … Countries with larger informal economies experience worse outcomes following adverse shocks. Indeed, estimates suggest that countries with above-average sized informal economies are more than three times as likely to incur the adverse effects of a crisis as those with lower rates of informality.
>
> (Bacchetta, 2009: 10)

As women are often over-represented in the informal sector, where earnings are minimal and conditions of employment are insecure and changeable, we can assume that they bear the brunt of such recessions. This assumption is confirmed by the findings by Action Aid (Her Mile, 2010).

In some countries, it is illegal for women to work in certain occupations, and even at different times of the day (Egypt, Jamaica and Pakistan, for example). However, in many developing countries, female earnings by work in the informal sector are crucial to household food security. The informal sector includes a fascinating diversity of activities, but among the most important in providing entitlements for poor females are street vending, sex work and home working.

Street vendors represent one of the largest elements of the informal sector. India alone is estimated to have more than 3 million people in this sector, and they are very significant in all developing economies. Broadly defined, street vendors include all those selling goods or services in public places, and while not all work without licences or legal protection, most do (WIEGO, 2011). These people play a crucial role in provisioning the poor in developing countries. Their activities require little or no capital, and their prices are low. In April 2008, the Committee for Asian Women (CAW) declared that 'the rise of food prices had a direct impact on women working as street vendors (82 per cent of the employed in the sector), causing many of them to look for other jobs, and thus adding to their daily workload' (quoted in Her Mile, 2010: 17).

Plate 7.5 Women in the informal sector (Courtesy: Dr Hamish Main, Staffordshire University).

Plate 7.6 Female vegetable trader, Jos Plateau, Nigeria (Courtesy: Dr Hamish Main, Staffordshire University).

Another significant earning opportunity is from sex work, a survival strategy in harsh economic circumstances. Prostitution remains a survival strategy for millions of girls and women, and there is evidence that the global recession of 2007–2008 has increased its incidence. A study carried out in March 2009 in five countries pointed out that:

> … the food crisis led to an increase of prostitution among adolescents and young women in Kenya and Zambia. The interviewees in both countries, in rural and urban areas, declared they had to look for new income-generating activities, beyond housework, care and livelihood. In all the communities surveyed by the research, it was found that when food is scarce, it is always men or male children who have the priority and never women, even when pregnant or breastfeeding.
>
> (quoted in Her Mile, 2010: 17)

This is another business where the workers are often exploited by their managers, and it presents very specific dangers of its own in a world where HIV/AIDS is ever present (Box 7.10). Sex work also increases when food security is at risk; this may be in conflict situations (Chapter 8) but not necessarily so.

Weiser (2007) asserts that HIV and food insecurity are the two leading causes of illness and death in sub-Saharan Africa; indeed, there is an ominous connection. Being food insecure increases the likelihood that women will get involved in sex work and therefore increase their exposure to HIV/AIDS. However, when malnourished, women may be more vulnerable to infection, so this is another example of the deadly connection between malnutrition and disease reviewed in Chapter 2. Weiser reports disturbing research findings from Botswana and Swaziland which illustrates the relationship between food insecurity and risky sexual behaviour. Among women, the lack of food was associated with a 70 per cent increase in the likelihood of unprotected sex. Food insecurity was associated with an 84 per cent increase in the likelihood of a woman engaging in transactional sex and a doubling in the odds of a woman reporting forced sex. The work found that the association between food insecurity and sexual risk taking were much weaker for men. There is every reason to assume that such statistics are similar across sub-Saharan Africa, and that they illustrate the awful links between poverty, survival strategies and disease.

WIEGO recently estimated that there are over 100 million people (note, this category of worker is, by definition, very difficult to count,

Box 7.10

Secrets and lies: tackling HIV among sex workers in India

Protecting the rights of women who work as prostitutes is one of the best ways to prevent the spread of HIV. And in Orissa, India, that approach is working – even amid deep-rooted taboos.

Damanjodi is a small mining town nestled in the verdant hills of Orissa, in India. The skyline is dominated by a sprawling network of mines, refineries and factory buildings. They are owned and run by Nalco, Asia's largest aluminium producer. Men come from all over the country to seek employment, creating a large migrant workforce. In their wake, women come too, seeking work of a different kind. There are more than 500 women engaged in prostitution in Damanjodi and its satellite towns. Their poverty drives them to sex work out of desperation and in the terrible knowledge that the risk of contracting HIV goes with the territory.

Ambica Das, an energetic woman with a commanding presence, is hosting a women's meeting. They do not – and cannot – discuss openly the main reason why Das organizes these meetings. Das is employed as a peer educator by a targeted intervention programme funded by the Orissa state Aids Control Society and run by Ekta, a local non-governmental organization (NGO).

Das's job is to distribute condoms and educate the women about safe sex practices, but in reality she does much more – from arranging safe places where they can find some refuge from harassment, to suggesting alternative ways of earning a living. In a world without social workers and where the police can cause as much trouble as the customers, she is often the only friend the women can turn to.

The risk of HIV infection is ever-present. There are more than 18,000 people living with HIV in Orissa. Some 88 per cent of new infections are a result of sexual contact, and a further 7 per cent are a result of mother-to-child transmission. Despite the hardships that Das encounters on a daily basis, the good news is that targeted interventions such as these are working. In Orissa, the HIV infection rate for sex workers remains below 1 per cent. The contrast with other areas is marked. In neighbouring Andhra Pradesh, nearly 10 per cent of sex workers are HIV positive.

Source: extracted from Perry (2010)

so statistics are poor) working as home workers, and that they have the least security and lowest earnings. The vast majority of these workers are female and may be self-employed or contract workers. Again, globalization of production in recent years explains the phenomenal increase in this type of work, as competition drives down costs.

Women working at home offer numerous advantages for manufacturers, namely, by passing off responsibility for rent, electricity, equipment and other production costs onto the workers. These workers are engaged in a great number of sectors 'including personal services such as shoe repair and childcare, clerical services such as data processing and invoicing, handicraft production, and manufacturing – especially of textiles, garments, electronics, and other consumer goods' (WIEGO, Factsheet). The extract from Markkanen (2009) gives a glimpse into the working conditions endured by many home workers at the bottom end of the global production chain (Box 7.11). Such conditions are reminiscent of those exposed by Mayhew's writings in nineteenth-century century Britain (republished, 2010).

Box 7.11

Shoes, glues and homework: dangerous work in the global footwear industry

My first visit to a home-based shoe workshop in Thailand, in 2001, was an epiphany. The visit was the first chance I had to observe how the global market turns homes into manufacturing sites. It appeared that I had arrived in a residential neighbourhood with apartment units in three- and four-storey buildings. But a shoe workshop was tucked away in one of the units. At the front door, stacks of shoeboxes were ready to be transported to a subcontractor. At the street-level floor were two women sitting cross-legged on the floor cleaning and polishing dozens of pairs of shoes. On the second floor, workers toiled, surrounded by raw materials, boxes, tools, cooking equipment, glue cans, sewing machines, finished and unfinished shoes and electrical wiring. Some of them ate a meal, drank a cup of coffee or smoked cigarettes amid the clutter. As pieceworkers, they could not step away to eat. Some mothers had children with them. The clutter represented a serious fire hazard, and toxic fumes emanated from open glue bowls, glue containers and cleaning and polishing chemicals. Little by little, reality unfolded.

Home-based footwear manufacturing is fraught with serious safety, health and environmental hazards. The International Labour Organization's (ILO) project on the Elimination of Child Labour (IPEC) in the footwear sector wanted me to explore why the adhesives used in shoe manufacturing are hazardous solvent-based ones, as opposed to available safer water-based ones. Reading the labels on the adhesive containers, I found that most of them explained how to apply the product to get a solid bond. None of them provided information on how to use the solvent safely and about the ingredients, and their health effects. Soon, I witnessed comparable conditions in Indonesia and the Philippines. After observing over 50 shoe workshops, I concluded that working conditions in the informal shoe production economy are extraordinarily dangerous.

The places where people in developing countries live have become home factories where a variety of goods are produced in very hazardous conditions.

All family members – wives, husbands and children – may participate in shoe production and be exposed to solvents, dust and a range of other hazards. Because home-based shoemaking belongs to the informal economy, its workforce remains largely beyond the reach of labour protection laws, social security benefits and systematic labour union organizing efforts.

Source: extract from Markkanen, P. (2009). Available at: http://baywood.com/intro/328-4.pdf (accessed 19 January 2011)

Efforts to control conditions in this sector are notoriously difficult. One of the reasons why they have grown so phenomenally in recent decades is that they evade legislative control. Workers are difficult to organize in this sector too, as they are dispersed and often in competition. The Self Employed Women's Association (SEWA) in India is an admirable exception that has grown remarkably since its inception in 1972 and has proved very successful at promoting and supporting home-based workers across India (Box 7.12). They claim that 'of the female labour force in India, more than 94 per cent are in the unorganised (informal) sector. However, their work is not counted and hence remains invisible' (SEWA, 2011; http://wiego.org/wiego/self-employed-women%E2%80%99s-association-sewa).

The multifaceted character of women's labour in the informal sector is established, but its invisibility remains a problem. The work these women do is vital to their households' entitlements and thus for food security across the developing world. It continues to suffer from very limited recognition, understanding and support; this situation inspired the formation of WIEGO, introduced earlier. This is a global research-policy network that seeks to improve the status of the

Box 7.12

Self-Employed Women's Association (SEWA): a success story

SEWA was born in 1972 and developed from the established Textile Labour Association because women found that the union was not serving their interests. The first struggle that SEWA undertook was in obtaining official recognition as a trade union. The Labour Department refused to register SEWA, because they felt that since there was no recognized employer, the workers would have no one to struggle against. However, their efforts prevailed, and SEWA was registered as a trade union in April 1972.

SEWA grew continuously from 1972, increasing its membership and including more and more different occupations within its fold. The beginning of the Women's Decade in 1975 gave it a boost. In 1977, SEWA's general secretary, Ela Bhatt, was awarded the prestigious Ramon Magsaysay Award, and this brought it international recognition. In recent years, SEWA has expanded its work. In particular, the growth of many new co-operatives, a more militant trade union and many supportive services has given SEWA a new shape and direction.

Supportive services such as savings and credit, health care, child care, insurance, legal aid, capacity building and communication services are important needs of poor women. If women are to achieve their goals of full employment and self-reliance, these services are essential. Recognizing the need for supportive services, SEWA has helped women take a number of initiatives in organizing these services for themselves and their SEWA sisters. Many important lessons have been learnt in the process of organizing supportive services for and by poor women. They provide these services in a decentralized and affordable manner, at the doorsteps of workers. Further, supportive services can be and are themselves a source of self-employment. For example, midwives charge for their services and crèche workers collect fees for taking care of young children.

Source: www.sewa.org/About_Us.asp (accessed 15 August 2011)

working poor, especially women, in the informal economy (Box 7.13).

The preceding discussion reviews some significant processes in the past and present that have constrained women's entitlements and, therefore, their food security. Evidence exists of their resistance to

Box 7.13

WIEGO: an International Non-Governmental Organization (INGO)

Women in Informal Employment: Globalizing and Organizing (WIEGO) is a global research-policy network that seeks to improve the status of the working poor, especially women, in the informal economy. It does so by highlighting the size, composition, characteristics and contribution of the informal economy through improved statistics and research; by helping to strengthen member-based organizations' informal workers; and by promoting policy dialogues and processes that include representatives of informal worker organizations. The common motivation for those who join the network is the relative lack of recognition, understanding and support for the working poor in the informal economy, especially women, by policy-makers, economic planners and the international development community.

Most activities of WIEGO fall under one of its five major programmes: see programme areas here: www.wiego.org/program_areas

- Global Trade
- Organization & Representation
- Social Protection
- Statistics
- Urban Policies

The component activities under these programmes involve some mix of research studies, data analysis, case study documentation and policy dialogues. Conscious efforts are made to involve member-based organizations of informal workers in the identification, prioritization and design of all activities; and to disseminate the findings, data and case studies generated – and related lessons learned – as widely as possible. WIEGO's Organization and Representation Programme also seeks to strengthen – and build networks of – member-based organizations of informal workers. And, its Statistics Programme seeks to promote improved statistics and statistical analysis of the informal economy.

The WIEGO network is comprised of 150 active Members and several hundred Associates (who have participated in one WIEGO meeting or activity) from over 100 countries around the world. The Members and Associates of the WIEGO network are drawn from three broad constituencies: member-based organizations of informal workers; research, statistical and academic institutions; and development agencies of various types (non-governmental and inter-governmental). Between them, the 20 member-based organizations in the WIEGO network have organized about 1 million informal workers.

Source: www.wiego.org/about (accessed 15 January 2011)

marginalization and their determined efforts to overcome discrimination. It is important to recognize these efforts. Millions of them are local or national in nature; others are evidence of a more beneficial outcome of globalization, that is, in the emergence of global networks and alliances united to change some of the worst forms of gender bias. First, however, it is important to explore some recent work that examines overnutrition and gender relations.

Most research on gender and nutrition to date has been devoted to exploring the themes discussed in the preceding text, namely, how we may understand the dominance of women and children in so many poverty statistics, including those on food insecurity. However, in the last decades of the twentieth century, new problems were emerging in the developed world that hinted at the fraught relations between gender and food. Several issues surround the politics of gender and food. The increased incidence of 'body image' issues and associated anorexia and bulimia among affluent young women exposed some poorly understood but clearly disturbing connections between women and food. This problem was appearing at the same time as obesity was becoming a serious public health worry – surely too much of a coincidence, something strange is influencing our attitudes too, and relationships with food, and they are gendered. This is not a theme to explore further here; central to this text, however, is the emergence of obesity in the global south and north and its gendered character.

Gender and obesity: a preliminary review

The problem of obesity, its causes and consequences were considered in Chapter 3, the following discussion does not revisit these; rather, here we begin to explore its gendered character. Research is limited but that which exists certainly confirms that obesity is a gender issue; in many countries in the north and south, we can identify a male–female obesity gap of significant proportions. '[I] n all but a handful of countries (primarily in Western Europe), obesity is more prevalent among women than men', and 'in 138 of 194 countries, for which the WHO reports obesity statistics, women were more than 50% more likely to be obese' (Case and Menendez, 2009: 271). Brooks and Maklakov (2010: 2) observe '[O]nly a very few countries have higher levels of male than female obesity, and where there are large disparities between men and women in obesity, far more women are obese than men'. In some places, the contrasts are extreme; Case and

Menendez (2009) report that, among black South Africans, the rate of obesity for women is five times higher than that for men (271).

Understanding the diverse picture of gender and obesity across the developing world is problematic because it is dynamic and statistics are limited. It is dynamic because countries' progress along the 'nutritional transition' will vary, because of their varied interactions with global processes and divergent national policies. We can certainly agree with Brooks and Maklakov (2010) that the gender 'differences between obesity rates are particularly poorly understood' and that research that attempts to explain why countries differ in the size of the male–female obesity gap are non-existent. A paper (Monteiro *et al.*, 2004) reviews the literature and describes how the incidence of obesity, for populations generally and by gender, changes as countries experience improvements in socio-economic circumstances. Before 1989, obesity in the developing world was associated with the economic elite, the more prosperous populations. However, they observe that, between 1989 and 2003, the situation changed and by then the burden of obesity in each developing country had shifted towards groups with lower socio-economic status. The emergence of obesity and its association with global processes of liberalization have been mapped (Hawkes, 2006; Rayner *et al.*, 2007). Hence, persuasive theories exist to explain why obesity has occurred since the 1980s but what is intriguing, and as yet little understood, is why the impacts are so dramatically gendered.

If we accept the thesis proposed in Chapter 3 about the structural determinants of obesity, and if we accept that obesity is a problem, then our attention should perhaps turn to analysis of female poverty and how this is allied to obesity. In the developed world, obesity is highly correlated with poverty and disadvantage and, while the association might vary in intensity and timing, surely we should expect a similar correlation in the developing world?

The work by Brooks and Maklakov (2010: 1) suggests the potential power of this perspective. They find that, within developing countries, female bias in obesity 'is strongly associated with socio-economic disadvantage and low socio-economic and educational status'. They also implicate reproduction as a risk factor for obesity in women that 'interacts with socio-economic and educational factors'. Most analyses understand high fertility as a symptom of poverty – a very rational survival strategy for poor families who rely on children to supplement household income and to guarantee support in old age.

If we employ high fertility as a surrogate for poverty, then the connection between high fertility, poverty and obesity is established.

The central paradox then emerges. Just as economic and political marginalization is implicated in patterns of undernutrition, so too they appear to be correlated to emerging patterns of obesity. When severe undernutrition is reduced, as development increases and access to food becomes more secure, a different logic emerges, as in more developed countries.

Relatively disadvantaged populations are more exposed to poor diets and less likely to exercise. Cheap, fast or processed foods, high in salts and saturated fats, are more available in poor urban neighbourhoods in the developed and developing world. In every respect, we can map the creation of obesogenic environments in developing countries (see Chapters 3 and 5).

Education must also play a role; again, educational achievement is correlated to falls in fertility, and we may expect, if patterns confirm to those experienced in the developed world, that as female education improves, obesity rates may fall. Educated women are more likely to be aware of the health implications of obesity and more effectively respond to government health messages. It is safe to assume that they will be more aware of the links between lifestyle and health and more pro-active in monitoring and changing the diets and exercise regimes of their children, too, thereby improving their children's longevity. Education is allied to female empowerment and self-esteem. May these also be implicated in the male–female obesity gap? No research exists employing these variables in the developing world, although some work exists from the developed world that suggests some linkages and finds the correlation stronger for female children. My experience of female graduates in India leads me to be optimistic about change there; some of the most articulate and politically motivated graduates I have encountered were the female students in India.

In parts of North Africa and the Middle-East, where female obesity rates are very high, there might be links to lifestyle, specifically, social taboos that surround girls in public spaces and their relative lack of access to exercise. In some of these cultures, women may be more sedentary. Elsewhere, in developed and developing countries, there is evidence to suggest that teenage girls are less inclined to play sports or exercise than their male counterparts. Hence, social and

Plate 7.7 Education and empowerment: the author with Indian students pursuing MA in Sustainable Development with Staffordshire University's distance learning programme.

cultural elements are relevant in all contexts, in that they play a part in governing eating and exercise behaviour. A recent study of Cuba found some disturbing trends; being a country where adult female obesity is already almost double male levels, the problem starts early. The study measured physical activity in children:

> [O]ur findings in across Cuba show cause for alarm: only pre-school boys carry out the required 60 minutes of moderate physical activity, and then only boys in rural areas (68 minutes) and small towns (60 minutes). Boys in urban settings spend only 37 minutes, the same as city girls. However, girls in rural settlings (at 31 minutes) are the most sedentary of all, followed by urban girls, than those in small towns (at 38 minutes). … In other words, the influence of habitat and gender on Cuban lifestyles conducive to unhealthy childhood weight gain is evident.
>
> (Hermandez-Triana, 2010: 1)

The article continues to detail all those aspects of the modern urban environment that conspire to make them unhealthy places to live, from readily available cheap poor food to lack of green spaces to play. Quite understandably, Dr Hermandez-Triana is alarmed by these statistics. Cuba has been celebrated over the decades for providing exceptionally good health with limited financial resources. Their emphasis has been on preventative medicine, but all these improvements are at risk from the new obesity trends. He concludes by stating that 'the stakes are high, as an abundance of overweight children foreshadows a growing population of chronically ill adults', and the majority of them will be female.

A further factor may be of relevance. If we accept the role of global forces to the emergence of the obesity pandemic, then attention should be given to a country's exposure to globalization – not simply to the nature and type of food readily available, but to corporate marketing campaigns. Understanding how these are gendered must surely be a factor in explaining the consistent dominance of women in the obesity statistics. Perhaps these help explain the variations in ideal body shapes, revealed by Case and Menendez (2009).

Conclusion

This chapter tries to reveal some of the most important factors that help explain why females have the unhappy distinction of being over-represented in the statistics for undernourished and overnourished across the developing world. The thesis developed here is that their relative economic marginalization may be the most important explanatory variable for their prominence in both types of statistics. It is important to finish with two important caveats. Such general overviews as this chapter presents tend to simplify circumstances and homogenize the category 'female'; it is vital to realize that females across the developing world live in varied and dynamic contexts; there are dangers in such broad generalizations. As emphasized by Mackenzie (2010) globalization opens as well as closes spaces for female actions and activism. Their activism is everywhere apparent and often remarkably successful; they are not willing victims and often find very creative strategies for resisting the roles occasionally imposed upon them.

Summary

- **Understanding the patterns of malnutrition and disadvantages requires a gendered perspective.**

- **Gendered processes operate at global, national, local and household levels; they may support or erode female entitlements and opportunities.**
- **Several indices have been generated to measure gender inequality; they all illustrate a great deal of national diversity.**
- **Females suffer from structural disadvantages that help explain their lack of entitlements to food.**
- **Female workers in the formal and informal economy are vital to household entitlements, but are often 'invisible' to policy-makers and politicians.**

Discussion questions

1. Discuss the value of an historical perspective to the analyses of gender entitlements.
2. Has neoliberal globalization promoted or eroded gender entitlements?
3. Compare and contrast the objectives and strategies of WIEGO and SEWA.
4. How might we understand the 'central paradox' that economic and political marginalization are implicated in patterns of both undernutrition and obesity?
5. Why has female work remained relatively invisible for so long?

Further reading

Momsen, J. (2010) *Gender and Development,* London, Routledge. This new edition is the best available introduction to the topic, containing interesting material on numerous central themes, including environment and resources.

Pearson, R. and Sweetman, C. (2011) *Gender and Economic Crisis,* Oxford, Oxfam. A timely volume that offers a gendered analysis of the 2007–2008 crisis and its effects on the lives of women and men in developing countries; excellent case studies.

Visvanathan, N., Duggan, L., Wiegersma, N. and Nisonoff, L. [eds] (2011) *The Women Gender and Development Reader*, 2nd edition, London, Zed Books. This provides a comprehensive reader which includes theoretical debates and case studies from across the global south on diverse themes.

Tsikata, D. and Golah, P. [eds] (2010) 'Land Tenure, Gender and Globalisation: Research and Analysis from Africa, Asia and Latin America', Ottawa, International Development Research, at: www.idrc.ca/openebooks/463-5/ (accessed 12 January 2011). Excellent theoretical overview and case studies.

Websites

www.un.org/womenwatch First stop for a great deal of material and resources. Recent example: 'Gender dimensions of agricultural and rural employment: Differentiated pathways out of poverty', at: www.fao.org/docrep/013/i1638e/i1638e.pdf (accessed 28 July 2011)

www.oxfam.org.uk/oxfam_in_action/issues/gender.html As with other development-related themes, Oxfam produces some excellent material on gender issues in development.

http://wiego.org Very helpful with lots of interesting material about women and the informal economy.

www.gadnetwork.org.uk A United Kingdom-based NGO devoted to promoting gender equality.

www.ifad.org/gender/index.htm Excellent material with good case studies from across the global south.

Chapter 7

Follow Up

1. Explore the variety of issues raised about gender and development at the following web page: http://uk.oneworld.net/guides/gender?gclid=CKm2n-3X06oCFRQLfAod1w9vyw#Women's_Voices (accessed 15 August 2011)
2. Read about some of the complexities about gender and economics at: www.guardian.co.uk/global-development/poverty-matters/2011/may/18/difficult-issue-gender-economics (accessed 16 August 2011)
3. Evaluate the role of micro finance to female empowerment and the fight against hunger at: www.thehungerproject.co.uk/whatwedo/keyinitiatives/microfinanceprogrammeinafrica
4. Review the discussion in 'Her Mile' at: www.landcoalition.org/sites/default/files/publication/762/HerMile_AAItaly.pdf (accessed 16 August 2011)
   - Summarize their proposals in Chapter 3.
5. Visit the WIEGO site at: www.wiego.org/about and select the research by Man-Kwun Chan (2011), detailed at: http://wiego.org/sites/wiego.org/files/

resources/files/WIEGO_Commodities_VCA_Review_REPORT_V2_FINAL.pdf (accessed 15 January 2011)

- What is VCA?
- What does an analysis of their operations reveal? Summarize some main conclusions and recommendations of the research detailed in this particular study.

6. At the same website, review the case of home-based workers in South Asia by Sudarshan and Sinha (2011): http://wiego.org/sites/wiego.org/files/publications/files/Sudarshan-Home-Based-Work.pdf (accessed 16 August 2011)
   - What particular problems are faced by home-based workers?
   - Describe some important types of work that occurs in the home in South Asia.
   - What are the main recommendations of this research?

8 Conflict and hunger

> Conflict is a significant source of vulnerability and risk for poor people. In 2006, for example, 25 out of 39 food emergencies were linked to conflict. Livelihoods are eroded in conflict, as are institutions, including those regulating land and other natural resources, local government and markets.
>
> (IFAD, 2010: 98)

Learning outcomes

At the end of this chapter, the reader should be able to:

- **Explain with reference to case studies the relevance of historical analyses to explanations of contemporary conflict**
- **Explain what you understand by the concepts 'protracted crisis' and 'complex emergency', and their relevance for food security**
- **Describe and explain some of the most important ways that conflict reduces food security with reference to case studies**
- **Appreciate how food may be used as a weapon of war**
- **Describe who may benefit from conflict situations**
- **Evaluate the complex relationship between food aid and food security**

Key concepts

Cold War; protracted crisis; complex emergency; international arms trade; refugee; internally displaced persons; humanitarian assistance; food aid

Introduction

This chapter is one of the hardest to write, since analysing the links between hunger and conflict is depressing. Here, poverty is exacerbated by militarization and human carnage. Conflict situations are always very complex, deeply embedded in historical, economic and political circumstances. They often appear intractable too. Conflict is caused by a conjuncture of factors from the international to the local. It can be serious in places where resources are scarce, and hence largely ignored by the international community (Northern Uganda and neighbouring regions). Indeed, it may be consequent upon such scarcity, or it may be found in areas rich in resources (see Box 8.3 on Niger Delta), which may also be part of the explanation. Sometimes, conflicts have agrarian roots, and they always have agrarian implications (see the *Journal of Agrarian Change* online for themed issue on this subject, at: http://onlinelibrary.wiley.com/doi/10.1111/joac.2011.11.issue-3/issuetoc accessed 28 July 2011). Wherever conflict is experienced, it has serious implications for food security of all inhabitants, but its impacts are frequently experienced differentially by gender, age, ethnicity and type of household and class.

While there has been a decline in traditional interstate war since the end of the Cold War, and evidence suggests that civil wars and conflicts between non-state actors have also declined, conflict remains a significant factor to the analyses of hunger (Messer and Cohen, 2006). These conflicts exert an awful toll in human suffering. Analyses of entitlements, a core element of this text, makes the examination of conflict unavoidable; conflict erodes, destroys, despoils and transforms the entitlement patterns of all those caught in its scope. Conflict has relevance for explanations of undernutrition only; it has no role to play in explanations of obesity. Although we may note that some people always have access to adequate food even in conflict situations, guns, in these circumstances, are a form of entitlement. In its widest interpretation, analyses of the links between hunger and conflict would include appreciating the connections between food aid and hunger, and the role of international assistance of all kinds, this literature is extensive and specialist and is only summarized here. My concern is to illustrate some very obvious connections between conflict and hunger with reference to contemporary examples.

History and contemporary incidence of conflict

Convincing analyses of conflict situations require an appreciation of the broad historical and structural circumstances within which they occur, as well as recent global changes that have intensified the economic, environmental and social processes of marginalization. Consider the following examples and limitations of analyses that adopt a narrow spatial or temporal perspective: Afghanistan; Angola; Sri Lanka; Somalia; Iraq; Tajikistan; Guatemala. In all these examples, the legacy of colonialism as well as Cold War rivalries, and more recently their strategic importance with reference to oil access, is essential to understanding the specifics contours of the conflicts they have endured.

Colonialism may be held partially responsible for political instability in many regions; this is very pronounced in Africa, which has suffered most from conflict in recent decades. The legacy of the Berlin Conference (1884–1885) is pertinent to consider as it attempted to resolve the European 'Scramble for Africa'. At this Conference, the European colonial powers divided Africa into spheres of influence and established country boundaries. Unfortunately, these boundaries, still extant, were completely at odds with any social or cultural realities 'on the ground', which helps explain the continuing instability of governments in that continent.

The post-colonial states in Africa have proved extremely unstable, and hunger and conflict are highly correlated in that region. Some regions emerge as particularly unstable over time. The Greater Horn of Africa, parts of West Africa and Central Africa appear in conflict statistics depressingly regularly. European colonial history is also implicated in conflicts in Southeast Asia and the Middle-East. Legacies of Russian, and later Soviet, empire building in Central Asia are also essential to explanations of contemporary instability in Central Asia, where Russia competed with Britain for influence, particularly in the late nineteenth century. Massive migrations, in various manifestations, slavery, labour transfers (Chinese and Indian) and white settler communities were enthusiastically encouraged within the British Empire. These help explain the emergence of social tensions, which sometimes erupted into civil wars, across former colonial possessions from Sri Lanka to Kenya, Uganda, Malaysia and Zimbabwe. The ethnic composition of these places was transformed and often resulted in indigenous resentment and various forms of social, economic and political discontent on all sides.

More recently, the Cold War (1946–1991) between the West and the Soviet Union fuelled 'proxy wars' in the developing world. Some of the most notorious were in Central America, Southeast Asia and Africa. 'Conflicts in the 1970s and 1980s were fuelled in large part by Cold War policies that encouraged spending on arms and used food as a political tool. In 1989, hunger was being used as a weapon or existed as a consequence of earlier wars in 20 countries (excluding the Eastern Bloc): Afghanistan, Angola, Burma, Cambodia, Chad, El Salvador, Ethiopia, Guatemala, Indonesia–East Timor, Iraq, Iran, Mozambique, Nicaragua, Peru, the Philippines, South Africa, Sri Lanka, Sudan, Uganda, and Viet Nam' (Messer, 2000).

Millions of people in these places became victims of super powers and their surrogates in struggles for supremacy. The prices paid by these people were high by any measure – humanitarian, financial, constraints on development, and were most serious in those regions which experienced the fighting. The West and the Soviet Union exploited regional and local disagreements to their own ends during these years; conflicts, initiated then, often continue today, Afghanistan being an excellent example. Hence, historical and geopolitical factors are important to the explanation of most contemporary conflict situations.

While the tensions between the West and Russia are less, unfortunately conflict has prevailed in many regions, although it may have declined considerably. Ironically, as super power support for authoritarian regimes relaxed, an upsurge of civil unrest, previously contained by draconian state apparatuses, appeared. Colonial and Cold War competition was about access to resources and, in the post-Cold War era, similar objectives are implicated in most conflict zones, the Middle-East, Central Asia and Africa, for example. It is interesting to review the map about conflicts since 1990, available at: www.unhcr.org/refworld/category,REFERENCE,UNEP,,,49a6687f2,0.html; accessed 16 February 2011)

The legacy of European colonialism is illustrated by the dominance of African states in contemporary conflict situations.

> Armed conflict has resulted in a huge loss of life in sub-Saharan Africa. The Millennium Development Goals Report 2005 estimated that between 1994 and 2003, 9.21 million people in the region lost their lives as a result of armed conflict – this represents 70 per cent of all conflict-related casualties around the world. Ongoing conflicts continue to impact on large parts of the population, and their effects are magnified by the legacy of past conflict and the resulting culture and institutions of violence.
>
> (Abbott and Phipps, 2009: 4)

African states dominate the list of countries suffering from 'protracted crises', in some cases for over 19 years. A map illustrating the location and duration of these crises is available at: www.fao.org/economic/es-policybriefs/multimedia0/protracted-crises-map/en (accessed 10 August 2011).

A 'protracted crisis' is the name given to a region when 'a significant proportion of the population is acutely vulnerable to death, disease and disruption to livelihoods over a prolonged period of time' (Box 8.1). This box illustrates the links between conflict (human induced), natural disasters and food crises between the years 1996 and 2010. Attempting to disentangle the links between these causes is problematic; natural disasters usually only prove catastrophic if they occur in regions already desperately poor or politically instable. However, even if we allow for such a distinction, those classified as human induced far outnumber those deemed solely due to natural disasters.

Box 8.1

Protracted crisis situations

There is no simple definition of a country in protracted crisis. Protracted crises have been defined as 'those environments in which a significant proportion of the population is acutely vulnerable to death, disease and disruption of livelihoods over a prolonged period of time'. The governance of these environments is usually very weak, with the state having a limited capacity to respond to, and mitigate, the threats to the population, or provide adequate levels of protection. Food insecurity is the most common manifestation of protracted crises.

Protracted crisis situations are not all alike, but they may share some (not necessarily all) of the following characteristics.

- *Duration or longevity*. Afghanistan, Somalia and Sudan, for example, have all been in one sort of crisis or another since the 1980s – nearly three decades.
- *Conflict*. Conflict is a common characteristic, but conflict alone does not make for a protracted crisis, and there are some countries in protracted crisis where overt, militarized conflict is not a significant factor or is a factor in only one part of the country (e.g. Ethiopia or Uganda).

- *Weak governance or public administration*. This may simply be a lack of capacity in the face of overwhelming constraints, but may also reflect lack of political will to accord rights to all citizens.
- *Unsustainable livelihood systems and poor food security causes increased mortality rates*. Both transitory and chronic food insecurity tend to increase in protracted crisis situations. However, unsustainable livelihood systems are not just a symptom of protracted crises; deterioration in the sustainability of livelihood systems can be a contributing factor to conflict, which may in turn trigger a protracted crisis.
- *Breakdown of local institutions*. This is often exacerbated by state fragility. Relatively sustainable customary institutional systems often break down under conditions of protracted crisis, but state-managed alternatives are rarely available to fill the gap.

Source: www.fao.org/docrep/013/i1683e/i1683e.pdf (p. 12; accessed 9 February 2011)

Tellingly, food insecurity is the most common manifestation of protracted crises (Table 8.1) (Messer and Cohen, 2006; State of Food Insecurity in the World, 2010: 12). Table 8.2 details various indices of food insecurity for the same countries.

The scale of the problem is daunting; undernourishment in these places is almost three times as high as in other developing countries. More than 166 million undernourished people are in countries in protracted crises – roughly 20 per cent of the world's undernourished people, or more than a third of the total if large countries such as China and India are excluded from the calculation (FAO, 2010). Understanding the nature of these conflicts, a prerequisite for generating solutions is similarly daunting, rooted as they are in complex colonial legacies and contemporary political and economic struggles over legitimacy and resources. Typically, places suffering protracted crisis have weak governments with limited resources and capacities. Sometimes, conflict is a symptom of government weakness, sometimes a consequence, but it always means that resolving the conflict is more challenging.

'Protracted crises' situations are a category employed in humanitarian discourse; an allied term is 'complex emergency'. This is defined as a 'humanitarian crisis in a country, region or society where there is total

Table 8.1 ***Countries in protracted crisis***

Country	*Natural disaster only*	*Human-induced disaster only*	*Combined natural and human-induced disaster*	*Total disasters*	*Humanitarian aid/ total ODA (2000–2008) (%)*
Afghanistan		5	10	15	20
Angola	1	11	–	12	30
Burundi		14	1	15	32
Central African Republic	–	8	–	8	13
Chad	2	4	3	9	23
Congo	–	13	–	13	22
Cote d'Ivoire	–	9	–	9	15
Democratic People's Republic of Korea	6	3	6	15	47
Democratic Republic of the Congo	–	15	–	15	27
Eritrea	2	3	10	15	30
Ethiopia	2	2	11	15	21
Guinea	–	10	–	10	16
Haiti	11	1	3	15	11
Iraq	–	4	11	15	14
Kenya	9	–	3	12	14
Liberia	–	14	1	15	33
Sierra Leone	–	15	–	15	19
Somalia	–	–	15	15	64
Sudan	–	5	10	15	62
Tajikistan	3	–	8	11	13
Uganda	–	4	10	14	10
Zimbabwe	2	3	5	10	31
	38	143	107		

Source: www.fao.org/publications/sofi/en; p. 13

or considerable breakdown of authority resulting from internal or external conflict and which requires an international response that goes beyond the mandate or capacity of any single agency and/or the ongoing United Nations country program' (www.ifrc.org/en/what-we-do/disaster.../complex-emergencies). The distinction between

Table 8.2 ***Measures of food insecurity***

Country	*Percentage undernourished (2005–2007)*	*Percentage under 5 underweight for age (2002–2007)*	*Percentage under 5 Mortality rate*	*Percentage Global Hunger Index*	*Stunting (%)*
Afghanistan (Asia)	NA	33	26	Na	59
Angola (Africa)	41	14	16	25	51
Burundi (Africa)	62	35	18	39	63
Central African Republic	40	24	17	28	45
Chad (Africa)	37	34	21	31	45
Congo (Africa)	15	12	12	15	31
Cote d'Ivoire (Africa)	14	17	13	14	40
Democratic People's Republic of Korea (Asia)	33	18	5	18	45
Democratic Republic of Congo (Africa)	69	25	16	39	46
Eritrea (Africa)	64	35	7	36	44
Ethiopia (Africa)	41	35	12	31	51
Guinea (Africa)	17	22	15	18	39
Haiti (Americas)	57	19	8	28	30
Iraq (Asia)	NA	7	4	NA	27
Kenya (Africa)	31	16	12	20	36
Liberia (Africa)	33	20	13	25	39
Sierra Leone (Africa)	35	28	26	34	47
Somalia (Africa)	NA	33	14	NA	42
Sudan (Africa)	22	27	11	20	38
Tajikistan (Asia)	30	15	7	18	33
Uganda (Africa)	21	16	13	15	39
Zimbabwe (Africa)	30	14	9	21	36

Source: adapted from www.fao.org/publications/sofi/en; p. 13

these two terms is a fine one, as suggested by the characteristics of 'complex emergency' situations:

- extensive violence and loss of life; massive displacements of people; widespread damage to societies and economies
- the need for large-scale, multi-faceted humanitarian assistance
- the hindrance or prevention of humanitarian assistance by political and military constraints

- significant security risks for humanitarian relief workers in some areas

(Source: www.reliefweb.int/library/documents/ocha__orientation__handbook_on__.htm; accessed on 16 February 2011)

Both protracted crises and complex emergencies precipitate humanitarian crises and are multi-causal and associated with armed conflict. The term 'food wars' has been applied to many of these conflicts to emphasize the centrality of food to their dynamics (Messer and Cohen, 2006). In many ways, these conflicts confirm a major thesis of this text, that there is never anything that is accidental, inevitable or 'natural' about hunger in the contemporary world. If these conflicts and their associated humanitarian crises were simple, then relief efforts would be easy and successful – 'we have the technology'. However, relief efforts are never simple and rarely as successful as we expect because these conflicts are complex and politically inspired. Naïve interventions, based on simplistic analyses, may even exacerbate the crises rather than relieve its victims. That is not to suggest that interventions are never helpful, but to warn that they may sometimes inadvertently intensify the problem.

Conflicts normally have an objective and are inspired by political ambitions; there are beneficiaries as well as victims of conflicts and their associated food crises and dislocations. It should be noted that some of the largest beneficiaries are external agents. During the Cold War, arms and military equipment were channelled to the 'proxy war' zones in the developing world. Since the end of the Cold War, the flow of arms has continued, with arms companies and transnational criminal organizations (or individual arms dealers operating somewhere between the two) earning fortunes by trading the surplus arms supplies of ex-Warsaw Pact countries with rebel groups and, in many cases, the governments with which they fight.

The United States, United Kingdom and France accounted for 60 per cent of the legal international arms trade to the African continent in 2008, and China is playing an increasingly controversial role (Abbott and Phipps, 2009: 4). Many of these conflicts become self-generating because vested interests emerge which have a stake in continued fighting. Fragile political regimes, often with limited legitimacy among large sections of the population they govern, may have little to lose and a lot to gain from food crises and its attendant dislocations. Oppositional forces too operate in a 'conflict space', where there is a breakdown in civil institutions, official accountability, and law and order, which offers opportunities as well as costs. For the principal antagonists then, profiteering may become the fuel of conflict rather than ideologies or injustices.

Conflict situations and their associated food insecurity may actually be the main political objective in some cases; food is a weapon of war, and starving the opposition is a time-honoured strategy in war (O'Grada, 2009: 229–258). Analyses of recent conflicts confirm that there are beneficiaries of crises who maintain a volatile coalition to exploit the spoils of conflict. Both government and oppositional forces require food to sustain their armies and money to support their military campaigns. When humanitarian interventions deliver food aid and relief supplies, and relatively massive cash injections to distribute relief, all or some of these may be diverted from their intended target population and be requisitioned to sustain military operations. In most conflict situations, parallel economies emerge which are impossible to quantify but are judged by experts to be very lucrative.

There are a number of additional ways to exploit the situation. These alternative modes of income generation can be understood as entitlement packages which, if not unique to conflict zones, are certainly more important in such circumstances. Various types of prostitution, sexual exploitation and 'high-risk' activities usually increase in conflict zones (O'Grada, 2009: 45–68). Armed people, whether rebels or government-based, have a power which those without arms have not. They can normally negotiate very favourable terms of trade for food, shelter and gifts to support their cause. The 'war economies' which are created may dwarf the formal economy which prevailed before the conflict or which is likely to emerge from its ashes. In sum, conflict situations often create novel forms of entitlement for some members of a community. Sustained by violence or illegality, these may be worth the risk for an otherwise impoverished population. In many parts of the world, the prospects for lasting peace depend on weaning various groups away from violent strategies that have proved lucrative. Sustainable alternative entitlements must be provided before the conflict can be resolved.

Sometimes conflicts remain largely ignored by international actors and even the media; an example is the situation with the horrors associated with the activities of the Lord's Resistance Army. The lack of international alarm at this particular conflict has been explained because the region is of no particular significance for international powers (Box 8.2). In contrast, some conflicts have a great deal of

Box 8.2

A bad situation becomes worse: the case of the Democratic Republic of the Congo

The Democratic Republic of the Congo (DRC) is often represented as a typical example of the natural resource curse: while its abundant natural wealth makes it potentially one of the richest nations on Earth, its gross domestic per head is among the lowest in the world. Decades of economic mismanagement and patrimonial rule, the conversion of economic resources into political resources and profit-seeking activities by the ruling class have caused a total collapse of the Congolese economy.

Already, before the war, the Congolese population was faced with very low national income, limited access to health and education, and the total disintegration of economic and transport infrastructure. In 2002, about 80 per cent of the population lived below the poverty line of US$0.2 per day (AfDB/OECD, 2006). While life expectancy in Africa in 2002 was 51 years, in the DRC it was only 42 years.

Then, into this poverty-ridden situation arrived conflict. The war in eastern DRC has generated one of the most severe humanitarian crises since the Second World War. In a conflict that has involved six African nations and more than a dozen rebel groups, more than 3 million Congolese have died either as direct or indirect consequence of armed confrontations, according to the International Rescue Committee. Many have lost their physical and financial belongings, as well as access to arable land and health services.

Traditional livelihood resources have been destroyed, livestock killed or pillaged and hundreds of thousands of people have been displaced due to continuous insecurity in many rural areas. Interventions by international aid agencies and local development associations have tried to alleviate the most acute consequences of the war by addressing the most critical food security constraints. Recent analysis has revealed that most of these interventions have had limited success because they dealt with only the symptoms of food insecurity. The interventions were generally based on a very narrow range of actions and were not built on necessary assessments. They also focused too much on food production and relied too much on food aid.

Occasionally, the violence flares up, and Congo is thrust under the media spotlight for a short while. However, the rapes and killings have been going on for so long now that it's hardly 'news' any more. The country is so vast and remote that it's hard to access many of the worst-affected areas, and the sheer scale of the problem makes it seem so hopeless that it feels like the rest of the world has almost given up on the people of Congo.

Source: extracted from 'Food security responses to the protracted crisis context of the Democratic Republic of the Congo' at: ftp://ftp.fao.org/docrep/fao/009/ag307e/ag307e00.pdf (accessed 9 February 2011); War Child at: www.warchild.org.uk/issues/conflict-in-the-democratic-republic-of-congo-DRC?_kk=war%20in%20congo&_kt=dc4693ad-5414-442c-b85b-48438a3740e9&gclid=CPfCm4uzOaoCFVQLfAodUFma1A (accessed 15 August 2011)

salience for important international interests, especially corporate interests. This is the case when vital international resources are found in conflict zones; sometimes, this association is not coincidental, but a major part of the problem.

> In 2008, exports of oil and minerals from Africa were worth roughly US$393 billion – over ten times the value of exported farm products (US$38 billion) and nearly nine times the value of international aid (US$44 billion). If used properly, this wealth could lift millions of people out of poverty. However, more often than not, the main benefits of resource extraction go to political, military and business elites in producer countries, and oil, mining, timber and other companies overseas.
>
> (Source: www.globalwitness.org/our-campaigns; accessed on 8 August 2011)

Recent years have seen the emergence of a number of 'war economies' located in regions rich in natural resources. Where one might hope such rich endowments (enjoyed by many sub-Saharan nations) might provide wealth and development, they are often instead correlated with a tragic decline into political corruption, criminality and social and environmental devastation. This situation has been called the 'resource curse' or the 'paradox of plenty'. The money these resources generate, allied to the ready availability of arms in parts of the developing world, helps initiate or perpetuate conflicts. These are essentially struggles over the wealth created by exploiting a variety of resources. Civil wars in both Angola and Liberia were funded to a large extent by the diamond trade, while conflict in the DRC is sustained through the wealth generated by a variety of resources, ranging from aluminium and cadmium to timber, copper and zinc. Messer and Cohen (2006), in their excellent review of conflict, food insecurity and globalization, detail some of the connections. In addition to the obvious case of petroleum, they state that 'trade in gems, minerals and timber' finance arms and mercenaries for many current African and Asia hostilities' (Messer and Cohen, 2006: 18). They also use case studies of the place of agricultural exports in conflict creation (sugar, cotton and coffee). One of the most tragic examples is the nightmare that is the Niger Delta (Box 8.3).

Conflict and hunger: direct impacts

The direct impact of conflict is obvious and massive in some places. The most obvious way a conflict causes hunger is through the

Box 8.3

Oil and conflict in the Niger Delta: a human rights tragedy

The Niger Delta is one of the ten most important wetland and coastal marine ecosystems in the world, and is home to some 31 million people (Nigeria total 155 million, 2009 estimate). The Niger Delta is also the location of massive oil deposits, which have been extracted for decades (since 1956) by the government of Nigeria and by multinational oil companies (such as Shell, Chevron, Total and ExxonMobil). The relationship between the State and the international oil companies has been appropriately called the 'slick alliance'.

The oil and gas sector represents 97 per cent of Nigeria's foreign exchange revenues and contributes 79.5 per cent of government revenues.

Oil has generated an estimated US$600 billion since the 1960s. Despite this, the majority of the Niger Delta's population lives in poverty. The United Nations Development Programme (UNDP, 2006) describes the region as suffering from 'administrative neglect, crumbling social infrastructure and services, high unemployment, social deprivation, abject poverty, filth and squalor, and endemic conflict'. It also details that development indices actually fell in the core oil-producing states between 1996 and 2002. The majority of the people of the Niger Delta do not have adequate access to clean water or health-care. Their poverty, and its contrast with the wealth generated by oil, has become one of the world's starkest and most disturbing examples of the 'resource curse'.

Oil spills, waste dumping and gas flaring (gas is separated from oil and, in Nigeria, most of it is burnt as waste) are endemic in the Niger Delta. This pollution, which has affected the area for decades, has damaged the soil, water and air quality. Hundreds of thousands of people are affected, particularly the poorest and those who rely on traditional livelihoods such as fishing and agriculture. The human rights implications are serious, under-reported and have received little attention from the government of Nigeria or the oil companies. Only when locally based violent resistance effectively closes down oil production do corporations and the federal government begin to show some interest. Such violence has intensified since 2007.

Oil exploration in the Niger Delta has long been marked by protests by local communities about the negative impact of the oil industry, corruption and the failure of oil wealth to be translated into better living conditions. More recently, armed groups and criminal gangs have explicitly sought resource control on behalf of the oil-producing areas, and have engaged in theft of oil and in acts of violence which are sometimes claimed as retribution for the treatment of the people of the Niger Delta by the oil industry.

The people of the oil-producing areas of the Niger Delta have watched for more than half a century while oil companies, politicians and government officials get rich from the 'black gold' extracted from their land. Meanwhile, they have seen few benefits, if any. Even basic services, such as water and sanitation, are lacking in many areas. Many of the development initiatives that have been established have been marred by corruption and bad planning, leaving behind a trail of half-finished or non-functioning projects.

Discontent and anger at the lack of benefits from oil extraction is exacerbated by the damage that the oil industry has done in many communities. Widespread environmental damage associated with oil extraction has destroyed livelihoods, polluted water and undermined health. The same oil extraction that is generating wealth for the few is deepening the poverty of many.

The way in which some oil companies engage with communities is a central part of the problem. A lack of transparency in the award of compensation and clean-up contracts has fed inter- and intra-community tensions and conflict. Communities are often seen and treated as a 'risk' to be pacified, rather than as stakeholders with critical concerns about the impact of oil operations. The risk-based approach to communities underpins several damaging strategies in the Niger Delta. Some companies have effectively paid communities and youths off, hoping to prevent protests. This has underlined that threats, protests and violence are ways to access oil money.

Another strategy has been the deployment, by government, of heavily armed security forces. Protests by local communities about the oil industry (including peaceful protests) and attacks on oil installations by armed groups are frequently met with reprisals characterized by excessive use of force and serious human rights violations. Action has rarely been taken to bring to justice members of the security forces who are suspected of being responsible for grave human rights violations in the region. For many communities, the contrast between the government's actions to protect the oil industry and the almost total lack of action to protect their human rights reinforces the perception that the government is on the side of the oil companies, regardless of the damage they may do.

While poverty and the unchecked actions of the security forces are two of the factors that contribute to making the Niger Delta one of the most unsafe oil production areas in the world, armed groups have emerged in recent years as a serious threat. Since the end of 2005, armed groups and gangs have increasingly engaged in the kidnapping of oil workers and their relatives, including children, and attacking oil installations. Armed groups increasingly engage in battles with the Nigerian security forces. In response, the security forces have used excessive force without regard to the impact on the local population. The delta's armed groups and criminal gangs have emerged from and feed on local frustrations. They have also emerged as a result of political encouragement of armed 'youth' gangs in the run up to elections, and a context where they can engage in organized criminal activities such as illegal oil bunkering. The organized theft of oil by illegal bunkering or hot tapping is lucrative and widespread. There are persistent reports that current and former employees of some oil companies, as well as state officials and politicians, may be involved in illegal bunkering. The stolen oil is transported by barge or road tanker to the ports for sale on the international market, reportedly through refineries in West African countries such as Côte d'Ivoire and beyond.

The Niger Delta thus provides a stark case study of the lack of accountability of a government to its people and of multinational companies' almost total lack of accountability when it comes to the impact of their operations on human rights. A recent Amnesty International Report (2009) identified the following examples of serious human rights abuses in the Niger Delta:

- Violations of the right to an adequate standard of living, including the right to food – as a consequence of the impact of oil-related pollution and environmental damage on agriculture and fisheries, which are the main sources of food for many people in the Niger Delta.
- Violations of the right to gain a living through work – also as a consequence of widespread damage to agriculture and fisheries, because these are also the main sources of livelihood for many people in the Niger Delta.
- Violations of the right to water – which occur when oil spills and waste materials pollute water used for drinking and other domestic purposes.
- Violations of the right to health – which arise from failure to secure the underlying determinants of health, including a healthy environment, and failure to enforce laws to protect the environment and prevent pollution.
- The absence of any adequate monitoring of the human impacts of oil-related pollution – despite the fact that the oil industry in the Niger Delta is operating in a relatively densely populated area characterized by high levels of poverty and vulnerability.
- Failure to provide affected communities with adequate information or ensure consultation on the impacts of oil operations on their human rights.
- Failure to ensure access to effective remedy for people whose human rights have been violated.

Source: Amnesty International Pollution and Poverty in the Niger Delta (2009) at: www.amnesty.org/en/library/asset/AFR44/017/2009/en/e2415061-da5c-44f8-a73c-a7a4766ee21d/afr440172009en.pdf (accessed 18 February 2011); Watts, M. (2009)

deliberate use of hunger as a weapon. This has several manifestations; the most blatant is simply seizing food and livestock, cutting off food supplies or diverting food to the armed forces. Scorched earth tactics may also be used to terrorize and debilitate the opposition. Essential agricultural infrastructure, irrigation systems, grain stores, agricultural support systems and stocks are all easy targets which have immediate consequences for food availability. Another popular strategy is to pepper the rural area with land mines. The impact of these is horrific and concentrated on the civilian community. Long after a conflict situation has been resolved, land mines can impact food security by making large areas unsafe to farm. They are a major problem in countries across the globe, but are particularly serious in Somalia, Mozambique, Angola, Cambodia, Afghanistan, Iran, Egypt and Yemen (Box 8.4). Disabilities associated with land mines have tragic consequences both during and after conflicts, with implications for the individuals and their families (A fascinating small NGO is helping eliminate the treats of land mines by employing rats to help locate them. Read about these 'Hero Rats' and their role in land mine clearance operations at: www.apopo.org/cms.php?cmsid=107; accessed 16 December 2011).

Box 8.4

Yemen: the land mine problem

Yemen is located on the southern tip of the Arabian Peninsula, with a population of approximately 25 million people. It is the poorest nation in the Arab world, and ranks 140th of 182 countries in the 2009 Human Development Index. Infant mortality rates are at 8.5–9 per 100 live births and maternal mortality rates are at 1.4 per 100 births, ranking among the world's highest. Over 70 per cent of Yemen's rapidly growing population live in rural areas where agricultural work is the predominant occupation. The bulk of the population is concentrated in the mountainous highland region. Yemen currently faces an economic crisis as oil exports diminish and Yemen's capacity to generate foreign exchange fails. International assistance to Yemen is increasing, and in an attempt to address burgeoning food security issues, it is also subtly shifting from development to humanitarian assistance.

Politically, it also faces a range of challenges which generate instability. These include Zaydi secessionist factions in the north, Al Quaeda in the centre of the country and separatists in the south. The needs of a growing population alongside a diminishing economic base exacerbate these issues.

Yemen's contamination problem regarding landmines and explosive remnants of war (ERW) stems from a series of internal conflicts that have been ongoing since 1962. The great majority of the formal minefields were laid along the border of the Yemen Arab Republic (YAR) and the Peoples Democratic Republic of Yemen (PDRY), which were also known as North and South Yemen, respectively, before 1990 when the two were united. More recently, there have been unsubstantiated reports about the use of mines in the intermittent 6-year Saada conflict, which has been running intermittently since June 2004 to the present (2011).

All landmines were laid by hand, without markings or fencing and in no specific pattern. There are also several areas that contain deep-buried mines in sand dunes and wadi floors. The 2000 Landmine Impact Survey (LIS) found that 89 per cent of communities in Yemen reported having access to communal grazing land blocked because of ERW. This compares with 25 per cent reporting blocked access to rain-fed farms and 6 per cent reporting problems in accessing irrigated farms.

Source: extracted from www.reliefweb.int/rw/RWFiles2010.nsf/FilesByRWDocUnidFilename/JARR-8DYELQ-full_report.pdf/$File/full_report.pdf (accessed 17 February 2011)

A direct loss of entitlements, in the shape of access or ownership of land, is suffered by some in conflict situations; massive out-migrations may result and people then join the ranks of internally displaced people or refugees (Box. 8.5). People may be 'removed' by a variety of strategies – fear and intimidation and evictions, allied to larger ethnic cleansing objectives, are most frequent. Often, the vacuums created are short-lived as allies of the armed elements gain access and ownership. Such asset transfer is common in ethnic conflict situations and has been recorded in the recent past or present in places as varied as Somalia, Sudan, Angola, Mozambique, Rwanda, Afghanistan, Guatemala, Cambodia and Burma.

A common misconception about conflict situations is that most of the casualties are combatants, whereas, in fact, civilians often suffer almost as much, sometimes even more, and women and children may bear a disproportionate share of these costs. Conflict brings in its wake awful dislocations of whole communities as it undermines their social, economic and political systems, especially their food provisioning systems (Box 8.6). Markets and transport infrastructure are damaged; roads, ports and railways are often targeted; and networks linking producers to consumers are dismantled or decimated. As sanitation and water-provisioning systems collapse, some of the killer diseases appear, such as typhus, cholera, diarrhoea, etc. Then health infrastructure is usually damaged too, which exacerbates the problem of disease, and health care workers are required to divert their energies and efforts to caring for the seriously wounded.

Movements by people and animals may be seriously curtailed with implications for food security. One of the first indications of a food

Box 8.5

Colombia: armed conflict generates hunger, violence in the cities

BOGOTÁ, June 21, 2006 (IPS) – Ciudad Bolívar and Altos de Cazuca (Cazuca Heights) form an endless landscape of humble dwellings along mud or cement streets and sidewalks on the barren hills south of the Colombian capital, where thousands of displaced people continue to suffer the effects of the civil war.

But the displaced people from rural areas of the country are not the only ones who have settled in the most populous suburb of the capital. Members of rightwing paramilitary groups have also come. Their networks in Cazuca have now mutated into violent gangs that settle old scores, steal or extract forced payments from small shopkeepers, bus drivers and taxi drivers. The displaced people who flock to Bogotá swell virtually every poverty indicator, since they escaped with their lives leaving home, crops, land and loved ones behind.

'I'm not going back. Of course you miss your home, but the government can't guarantee safe conditions for returning. Back there in Montes de María, Sucre (in the northwest of the country), they killed my wife, and that's why I've come here with my two children', said Germán Luna, 40, a shoemaker when he gets the opportunity, and president of the displaced people's association 'Seeds of Hope'. 'The government doesn't support us, and when a person registers as a displaced person, their emergency aid takes two months to arrive', Luna added. Although people have lived there for 13 years, the settlement 'hasn't been legalised', that is, there is no official recognition yet that the land has been occupied by settlers.

The World Food Programme helps the 'Seeds of Hope' effort, sending provisions for a daily meal for 417 children. Luna's helpers from among the association's 350 families cook the meals on a three-burner stove. While they wait for lunch in a room with unplastered walls but a panoramic view reaching almost to the bustling centre of Bogotá, Luna, Alvarado and other displaced people from faraway departments tell all-too-similar stories of their dead relatives and sons who were forcibly recruited and then abandoned their homesteads. They also lament their situation here in the city, where they are hungry, homeless and jobless. Unfortunately, some displaced people continue their fight in the slums of the city where desperation and poverty mean that they are not short of recruits.

Source: extracts from http://ipsnews.net/news.asp?idnews=33705 (accessed 15 February 2011)

Box 8.6

Somalia: conflict and crisis continues – the tragic chronology

Somalia has been a country in crisis for decades. The country has no central authority and is ravaged by regional divisions. Conflict led to a major famine in south–central Somalia in 1992–1993. Since 2000, there have been localized food-security crises in various parts of the country. Fierce fighting in Mogadishu in 2006 led some half a million residents of the city to flee.

In 2009, some 3.2 million people in Somalia required immediate food assistance. Over half of these were internally displaced people; the remainder were affected either by the conflict, by drought and an underlying livelihoods crisis, or both. As of early 2010 and despite a good harvest in 2009, the food security situation for much of the population of south–central and central Somalia appeared increasingly worrying, while the security situation has forced almost all international agencies to withdraw from these areas.

In July 2011, famine has again caused the Horn of Africa make the headlines. The region is experiencing the worst drought in 60 years. The United Nations says 3.7 million people are 'now in crisis' and more than 10 million are affected by worst drought in decades.

The Horn of Africa is in a classic zone of vulnerability where a series of factors, natural and social, interact to create a region of extreme vulnerability. Many people in the region are pastoralists, and when the rains fail in two consecutive years, their animals die and their livelihoods collapse. Thousands of people are fleeing the drought to seek food and water, walking miles to reach refugee camps. The Dadaab refugee camp in eastern Kenya is now the fourth largest settlement in Kenya.

A massive global effort is now required to prevent a humanitarian disaster but interventions are fraught. In early 2010, the World Food Programme (WFP) suspended its aid operations in Al-Shabab-held areas. Al-Shabab is an armed movement that seeks to overthrow Somalia's Transitional Federal Government and impose Sharia law; it controls the majority of the country, including pockets of the capital Mogadishu. The situation illustrates some of the complications that surround humanitarian interventions in a region of extreme political instability and conflict. Recent reports have described human rights abuses on all sides of the conflict.

Source: FAO (2010) 'Countries in Protracted Crisis' at: www.fao.org/docrep/013/i1683e/i1683e03.pdf (accessed 26 December 2011). View some photographs from the refugee camps at: www.careinternational.org.uk/news-and-press/latest-news-features/1808-slideshow-more-pictures-from-east-africa (accessed 22 July 2011); FAO (2010) 'State of Food Insecurity' at: www.fao.org/docrep/013/i1683e/i1683e.pdf page 14, also: www.guardian.co.uk/global/2011/jul/20/famine-africa-result-modern-parctice

crisis is that people migrate in search of food, work and markets. Losing access to these because travel is disrupted means people become destitute more rapidly. Hence, restricting mobility may exacerbate the victims' plight, but so too can the forced migrations that often occur in conflict settings. Such forced reallocations have very heavy social and economic costs for the populations involved. Traumas associated with having to leave ancestral lands and places may be experienced through successive generations, as is the case with Palestinians. Some of the most trying of stories emerge from such forced migrations.

Conflict and hunger: indirect impacts

Indirect costs of war have various ramifications for food production and peoples' entitlements. Perhaps the most serious is that governments or communities divert resources to military expansion instead of more productive investments. Money spent on improving military capacity is money not available for economic and social investments. Wherever conflict exists, resources are deployed to enhance security or improve military capacity. Some interesting comparisons are highlighted in Table 8.3. Although some of these calculations may be flawed, they suggest the government's priorities.

At the international level too, resources are devoted to the relief of conflict-induced problems, rather than to more productive investments. Globally, around 10 per cent of total Official Development Assistance (ODA) is in the form of humanitarian assistance, while in protracted crisis countries, the share is much higher. In Somalia, for example, 64 per cent of assistance is in the form of humanitarian aid, and in the Sudan it is 62 per cent. At the global level, these countries receive close to 60 per cent of the total humanitarian assistance (FAO, 2010).

Another indirect cost of conflict is diversion of labour. In most conflict situations, young males are the main combatants, although in some notorious examples, young children are forced to fight too. The loss of this labour from agricultural areas has implications for food security. This represents a loss of entitlement for families and communities. During and after conflict, rural communities are very often left reliant on older people who are required to sustain young dependent populations. This has long-term implications for

Table 8.3 ***Military expenditure***

State	*Military Expenditure as Percentage of GDP*
United States	5.2
China	4.3
India	2.5
Russia	3.9
Saudi Arabia	10.0
Pakistan	3.0
Angola	3.6
Sudan	3.0
Sri Lanka	2.6
Kenya	2.8
Equatorial Guinea	6.3
Democratic Republic of the Congo	2.5
Burundi	5.9
Zimbabwe	3.8
Djibouti	3.8

Source: selected from www.globalsecurity.org/military/world/spending.htm (accessed 16 February 2011)

productivity as the maintenance of essential infrastructure becomes neglected. Some positive improvements in the status of females may be observed, but this is scant consolation for traumatized individuals and communities. Some of those most traumatized may become displaced people or refugees.

Refugees, displaced people and stateless populations

A stark indication of the extent of conflict is the number of refugees and displaced people in the world. A refugee is a person who has fled across an international boundary to escape conflict, while an internally displaced person (IDP) has fled their home but not crossed any international boundary. At the last count, the United Nations High Commission for Refugees (UNHCR, 2009) estimated that there were 36 million people of concern to them, approximately 10.4 million refugees and about 26 million IDP. In addition to these populations, we can add those who are deemed 'stateless' whose numbers are estimated to be 12 million. Stateless people often exist in precarious circumstances on the margins of society; they usually lack any form of documentation and may be subject to discrimination. The great majority of the people in these categories live in the developing world, in countries where resources are already scarce. They are frequently allocated temporary 'camps' where legal rights are tenuous or non-existent. Increasingly, they are found in urban areas in the developing world, with restricted access to all basic needs as well as employment opportunities. The danger is that, in such circumstances, their presence may exacerbate social tensions and political instability (Box 8.7).

Box 8.7

Refugees: Kenya's Dadaab refugee camp is a haunting place

Most refugee camps are sad places, but few are as tragic as Dadaab, on the border between Somalia and Kenya. Dadaab has the dubious distinction of being the world's largest refugee camp; it actually consists of three different sub-camps. Initially established in the 1990s for 90,000 people, it now houses over 370,000, and waves of desperate people are still arriving. Some estimates suggest that the numbers could reach 5,000,000 before the end of next year (2012). Literally thousands of people fleeing the conflict in Somalia arrive in this camp every day. Years of conflict and 2 years of drought help explain the mass exodus from Somalia. People are trying to survive by seeking help and shelter in the Kenyan camp. What they find upon arrival is pitifully insufficient; shelter, food, water and medical care, the basics, are scarce.

After long journeys by foot over poor terrain and with meagre provisions, if any, people arrive very weak and malnourished. The hardships of the journey, the long wait for food, their limited access to water and desperate living conditions mean that they are vulnerable to disease and infections. As in all such communities, respiratory tract infections and diarrhoea are rife. These conditions become endemic in the high temperatures of the region where water is very scarce. The camp is in a semi-desert, and the shelter is make-shift.

The UNHCR is funding the camp and its inhabitants, and Médecins Sans Frontières MSF (Doctors Without Borders) is helping to alleviate illness. All resources are stretched, and urgent help is required if death and disease is to be checked. As with all refugee circumstances, the best solution would be if the situation in Somalia improved and families and communities could be re-established in their traditional homelands; prospects for this are bleak for Somali refugees.

Source: based on material at: www.msf.org.uk/Dadaab_malnutrition_20110714.news?gclid=CMW29Ji90aoCFcRO4Qod8ALfOA (accessed 15 August 2011); www.guardian.co.uk/commentisfree/2011/jul/14/dadaab-camp-kenya-drought (accessed 8 August 2011)

Food aid

Food aid is a complex area, and I have included it here because most readers will assume it is allied to conflict and emergency food crises, and so it is, in part. However, food aid is also one part of the much larger picture which includes international trade, development and geopolitics. It comes in three major forms: Project Food Aid, Program Food Aid and Relief or Emergency Food Aid. Since the 1950s, the first two forms of aid have dominated global flows.

Food aid is dispersed by numerous agencies and countries too, through multilateral agencies (WFP) and bilateral arrangements between two governments. These variables mean that it is a difficult subject to summarize. It is timely, however, because this year (2011) the Food Aid Convention (FAC), which has been guiding the process since 1967, is under review. Therefore, there have been some serious efforts in the last few years to evaluate its effectiveness and consider how it might be reformed. Most experts accept that reform is well overdue and its effectiveness for recipients is partial always and negative often (Crow, 1990; Friedmann, 1990; Millstone and Lang, 2003: 28–29).

Its reform, however, will be the occasion for considerable tensions because 'political interests and legal obligations, such as those enshrined in the FAC, continue to be more powerful drivers of food aid than recipient needs' (Gaus *et al.*, 2011: 1). FAC arrangements are reminiscent of those of the WTO; they lack transparency and reflect the interests of a narrow donor base rather than the requirements of those actually receiving the aid. Indeed, most of the institutions established to govern international food have close ties to the donor countries rather than the recipient states.

From the beginning, just after the Second World War, food aid has been controversial and, although improvements have been made to its operations, serious reservations remain about its operations and impacts in some circumstances. The project and program of food aid suffer serious flaws. They have often served as vehicles to 'dump' cereal surpluses, initially from the United States and Canada, later the European Union and to a smaller extent Australia, China, Korea and Japan. Special legislation has permitted the disposal of surplus grains in the developing world, in the name of aid, but in effect such disposals are an instrument of foreign policy intended to modify trading regimes and promote geopolitical interests. Humanitarian concerns are rarely the only motivation of food aid. Food aid may be manipulated and, in the past, it has been used for various not-so-philanthropic ends.

The United States has been open about their objectives, namely, it is designed to:

- Combat world hunger and malnutrition and their causes
- Promote broad-based, equitable and sustainable development, including agricultural development
- Expand international trade
- Develop and expand export markets for US agricultural commodities
- Foster and encourage the development of private enterprise and democratic participation in developing countries

(Source: Mousseau, 2005: 5)

In addition, there are important domestic objectives in the US food aid efforts. The farming lobby is very important in the United States, and their interests have been obvious since the outset, since food aid extents the market for their outputs. However, shipping interests were included too as food shipments were required to be carried by US boats. Foreign policy objectives have followed the dynamics of global politics, determined for many years by the dictates of Cold War when India, Pakistan and Indonesia were large recipients, and then later by trade-related incentives. Large food aid donations from the United States were employed to extend market opportunities, particularly in South Korea, the Philippines and Indonesia. A further limitation of food aid was that it could be used as a lever to make recipient countries implement economic reforms.

It would be remiss, however, not to recognize the contribution that food aid has made to saving lives in emergency situations since the 1950s. In its emergency relief form, food aid has been vital; it can save lives and can protect livelihoods and promote recovery, if delivered appropriately. Just how aid is delivered – its speed, effectiveness and motives – have also come under scrutiny (Mousseau, 2005: 23, 25). Mixed strategies are obvious in the largest food aid programmes.

The WFP is an arm of the United Nations and is the world's largest multilateral channel for food aid. Some of its projects support environmental improvements and sustainable farming solutions in some of the world's most fragile ecosystems and among some of the world's poorest populations. Their Managing Environmental Resources to Enable Transitions to More Sustainable Livelihoods (MERET) scheme is one such example which is proving successful in one of the poorest districts of Ethiopia. Learn about this scheme at: www.wfp.org/countries/Ethiopia/News/Ethiopian-Growers-Turn-Barren-Land-Into-Booming-Farms (accessed on 7 March 2011). Another interesting WFP programme is in Ghana; see: www.wfp.org/students-and-teachers/teachers/blog/country-spotlight-ghana (accessed 7 March 2011).

Although the WFP is involved in long-term relief, recovery and development projects, its main activity is relief operations in emergencies. Most of the financial resources of the WFP are sourced from a few countries. The donations from the United States dwarf any other country's contributions, at approximately 50 per cent. It is important to confirm the very vital role that food aid has in supporting people suffering extreme food crises; it certainly saves lives. In 2009, the WFP provided food for 101.8 million people in 75 countries. The number of hungry people reached in 2009 included: 84.1 million women and children; 20.7 million children reached through school meals projects; 14.1 million internally displaced persons (IDPs); 2 million refugees; and 2.6 million people affected by HIV/AIDS. In 2010, it spent US$965 million on buying food (WFP.org), and each year, on average, the WFP feeds more than 90 million people in more than 70 countries. In disaster situations, their help is indispensible, whether it is in conflict zones or when a natural disaster exacerbates an already tenuous food security, such as in Haiti.

At present (March 2011), a new emergency situation has emerged in North Africa, and the WPF is involved in delivering emergency food supplies. The WFP has launched a regional emergency operation that aims to provide food assistance to vulnerable populations in Libya, and to people fleeing across borders into Tunisia and Egypt. A total of 1.06 million people are targeted over a 3-month period. Many of the people caught up in this crisis are poor migrant workers from the rest of North Africa or other poor countries.

> 'Standing on the Libya border with tens of thousands fleeing violence I realized that unless the world acts we may be facing an historic human tragedy', said WFP Executive Director Josette Sheeran who travelled to the

> Tunisia–Libya border last week to meet local authorities, aid organisations, and people who have fled violence.
>
> (WFP website; accessed 8 March 2011)

Until recently, this element of food aid was only a small fraction of the total. In recent years (the past 10 years), the nature and type of food aid has shifted in relatively positive directions although, as yet, reform is limited. There has emerged a stronger focus on priority countries, so that food aid has been directed to countries in need rather than countries that are useful. One of the most significant changes has involved the origins of the food delivered. Recognizing the serious problems associated with dumping subsidized grain in developing markets (and thereby damaging local food production), more food is now sourced locally or from neighbouring countries. This has several advantages; it supports and develops local food production and markets, and it helps build more resilience into food systems. This introduces another important issue: in the past, food aid created dependencies and almost ensured continued food insecurity.

This is where food aid has been most seriously wanting. To eliminate hunger, developing countries must be encouraged and allowed to establish productive farming sectors. This means challenging unfair international trading regimes and inequitable subsidies and regulating international agri-business. Ideally, food aid should become unnecessary; development policies should promote food security by increasing food production and national food sovereignty in the developing countries. Although emergency food aid will always have a role, development assistance should be designed to promote food sovereignty and economic expansion to provide more diverse opportunities for those in the industrial and manufacturing sectors. A better focus on creating sustainable farming systems in the developing world must be effected across development policy. Some effective programmes are being initiated by some NGOs and others.

Conclusion

Although some positive trends are discernible in the statistics concerning conflict, its victims still represent a significant number of people, and their incidence is marked in sub-Saharan Africa. As the ‘protracted crisis’ category suggests, many of these show little sign of being resolved. Most of the conflicts have their origins in historical circumstances, but are fuelled by contemporary poverty and injustice. International relief efforts have enjoyed some successes but that they are still so frequently required is depressing (Box 8.8). Humanitarian aid is vital in emergencies to prevent loss of life but more financial investments are needed to eliminate poverty, which is so often a precondition to conflict. Important international interventions are required to ensure that trade and investment decisions are made that support progressive development in agriculture and social infrastructure that supports human development.

Box 8.8

Hope for a troubled Sudan?

Sudan is the largest country in Africa, bordering Egypt and the Red Sea. Sudan suffered from regional rivalries throughout the nineteenth century. In 1898, Egypt and Britain ruled Sudan jointly although the British administration dominated. In 1956, Sudan gained independence. Since independence, protracted civil wars, rooted in the deep cultural and religious differences, have constrained Sudan's economic and political development and forced massive internal migrations. The civil war has roots in colonial times. As in the rest of Africa, Sudan's borders were drawn up by colonial powers with little regard to cultural realities on the ground. Most northerners are Arabic-speaking Muslims, while the south is made up of numerous different ethnic groups who are mostly Christian or follow traditional religions. The presence of extensive oil reserves in the South added fuel to this already volatile situation.

With the government based in the north, many southerners felt discriminated against, and north–south conflict has coloured most of the country's history. Southerners felt that the revenues from oil reserves in the South were monopolized by northern interests. Southerners were also angered at attempts to impose Islamic law on the whole country. During more than 20 years of conflict (which started in 1983) between the northern and southern groups, violence, famine, and disease have killed more than 2 million people and forced an estimated 600,000 people to seek refuge in neighbouring countries. An estimated 4 million are internally displaced, creating the world's largest population of internally displaced people.

A separate conflict, which broke out in the western region of Darfur in 2003, has displaced nearly 2 million people and caused an estimated 200,000–400,000 deaths. Sudan continues to cope with the countrywide effects of conflict, displacement and insecurity. Decades of fighting have left Sudan's infrastructure in tatters, and there is a pressing need for reconstruction. On the plus side, Sudan is endowed with considerable natural resources, and its oil reserves are ripe for further exploitation.

Elections were held in Southern Sudan in January 2011, and the people voted overwhelmingly to separate from the North and establish an independent state. Africa's newest country came into being on 9 July 2011. Numerous contentious issues will continue to complicate relationships between the north and south, however, and governance and infrastructure in both places is weak.

Source: www.countrywatch.com/country_profile.aspx?vcountry=162 (accessed 21 January 2011) and www.bbc.co.uk/news/world-africa-12111730 (accessed 15 February 2011)

Summary

- **Conflict is directly and indirectly responsible for food insecurity in many regions.**
- **Conflict often has structural causes allied to historical and geopolitical circumstances that are politically complex.**
- **Refugees from conflict situations usually suffer from limited entitlements.**
- **Food aid may occasionally be essential to relieve emergency hunger situations, but it has often been employed to further the donor's economic and political objectives.**
- **It is important to reform food aid policy to support food sovereignty and security rather than reduce them, as has frequently been the case.**
- **Food has long been employed as a weapon of war.**

Discussion questions

1. Discuss the relevance of historical and geopolitical circumstances to contemporary conflict situations.
2. Evaluate the direct and indirect implications of conflict for food security.
3. Who benefits from food aid?
4. Discuss the food security of refugees with reference to specific case studies.
5. Evaluate how food aid may compromise development.

Further reading

Keen, D. (2008) *Complex Emergencies*, London, Polity Press. This book offers an insightful analysis that investigates the causes of conflict and its beneficiaries as well as its multiple victims; an excellent and relevant read.

Messer, E. and Cohen, M. J. (2004) 'Breaking the Links between Conflict and Hunger in Africa', International Food Policy Research Institute, IFPRI. Accessible consideration of the themes in this chapter with useful evaluation of positive interventions.

Messer, E. and Cohen, M. J. (2006) 'Conflict, Food Security and Globalization' International Food Policy Research Institute (IFPRI) at: www.fao.org/docrep/007/ae044e/ae044e00.htm (accessed 8 August 2011). This is an excellent study that considers the relationships between conflict and food security, conflict and globalization and globalization and food security; very useful case studies are included.

Teodosijevi, Slobodanka B. (2003) 'Armed Conflicts and Food Security', published by the Food and Agriculture Organization at: www.fao.org/righttofood/kc/downloads/vl/docs/armed%20conflicts%20and%20FS.pdf (accessed 8 August 2011). A helpful review of the causes and consequences of conflict, especially their impacts on food security and strategies employed by communities caught in conflict situations.

Mousseau, F. (2005) *Food Aid or Food Sovereignty? Ending World Hunger in Our Time,* Oakland, The Oakland Institute. This book exposes how food aid has been manipulated in the past and suggests how the system could be reformed to promote food sovereignty rather than food dependency.

Useful websites

www.worldbank.org The 2011 World Development Report 'Conflict, Security and Development' examines the links between conflict and development. Read the report at: http://wdr2011.worldbank.org

http://unu.edu This United Nations University website has material of relevance to this and other chapters of the text. Its content changes but is worth reviewing regularly.

www.oxfam.org.uk/resources/issues/conflict/introduction.html The Oxfam site has an excellent number of in-depth case studies of conflict and disaster situations; often, these two circumstances interact to exacerbate hunger and distress.

www.unhcr.org.uk A great deal of relevant material is available at this site, including latest news and publications. One of the most interesting for insights into life as a refugee is the Refugee Magazine at: www.unhcr.org/cgi-bin/texis/vtx/search?page=&comid=4b66b4656&cid=49aea93ae2&scid=49aea93a6e (accessed 8 August 2011).

www.peacedirect.org/?gclid=CKO-k6T9v6oCFQEf4QodnSVA4w Peace Direct is an NGO dedicated to supporting peace makers in conflict situations. Their website has profiles from many conflict regions and details about how people are working to overcome past and present hostilities.

www.globalwitness.org An NGO devoted to finding solutions to the 'resource curse', so that citizens of resource-rich countries can get a fair share of their country's wealth.

Chapter 8

Follow Up

1. Visit the World Hunger map produced by the WFP at: http://documents.wfp.org/stellent/groups/public/documents/communications/wfp229328.pdf
2. Learn more about the activities of the WFP at: www.wfp.org/ (accessed 7 March 2011).
3. Visit the WPF site and learn about their successful programme in Ghana at: www.wfp.org/students-and-teachers/teachers/blog/country-spotlight-ghana (accessed 7 March 2011).
4. Learn about the WFP's MERET scheme and its impacts at: www.wfp.org/countries/Ethiopia/News/Ethiopian-Growers-Turn-Barren-Land-Into-Booming-Farms (accessed on 7 March 2011).
5. Visit this address and listen to a talk about peace and security: www.ted.com/talks/jody_williams_a_realistic_vision_for_world_peace.html. What relevance has this talk to food security?
6. Visit this address and watch the video 'Stained with Blood and Oil: The Niger Delta' by Ed Kashi about conditions in the Niger Delta: http://geography.berkeley.edu/ProjectsResources/ND%20Website/NigerDelta/VIDEO/KashiNigerDelta.mov (accessed 17 February 2011). What evidence of social and environmental destruction is obvious in the video?
7. Now, visit the Amnesty International Report (2009) and read about the environmental pollution in the Niger Delta and how it has compromised people's livelihoods in that region: www.amnesty.org/en/library/asset/AFR44/017/2009/en/e2415061-da5c-44f8-a73c-a7a4766ee21d/afr440172009en.pdf (accessed 17 February 2011).
8. It is interesting to review the map about conflicts since 1990. Examine the location of these conflicts and regions where they are concentrated. Available at: www.unhcr.org/refworld/category,REFERENCE,UNEP,,,49a6687f2,0.html (accessed 16 February 2011).

9. Visit the map at this address to view where land mines remain a serious threat to civilians: www.canadianlandmine.org/LessonPlans/Landmine_Problem%20map.pdf (accessed 17 February 2011).
10. Rats and their contribution to food security and development. Visit the Hero Rats charity and learn about its work at: www.herorat.org (accessed 7 March 2011).

9 Alternative visions

Learning outcomes

At the end of this chapter, the reader should be able to:

- **Review some of the reasons why the global industrial food system has become hegemonic**
- **Review why the same system has been challenged in recent decades**
- **Appreciate the diversity and extent of challenges to 'business as usual'**
- **Evaluate the political character of these oppositional forces, and their strengths and weaknesses**
- **Evaluate the potential of some of the oppositional forces to challenge the hegemony of the contemporary food system, singularly or in political alliances**
- **Consider how each of us are implicated in the food system and the complex politics of our individual consumption habits**

Key concepts

Neoliberal perspectives; reformist perspectives; progressive food movements; radical food movements; externalities; agricultural efficiency; oppositional movements; transnational movements

Introduction

The previous chapters reviewed some of the most obvious failings of our food system, specifically its inability to deliver a decent diet to millions, and its negative environmental consequences. This chapter summarizes why the system has proved so resilient and then considers some more positive trends. Given the problems identified with the contemporary food system, why has it not been more effectively challenged and changed? Corporate power has driven the agendas in food and agricultural policy since the 1950s with one ambition, profit maximization. Since the 1980s, as globalization has intensified, corporate power has enjoyed ever greater access to previously protected markets, both to produce raw materials and to sell processed goods through internationally owned retail outlets. Robinson argues:

> [I]n the last three decades both food production and distribution have been radically restructured in favour of a more global scope and character, with TNCs (Transnational Corporations) playing an increasingly important role, especially in activities 'upstream' and 'downstream' from farms.
>
> (Robinson, 2004: 53)

Although not solely responsible of the problems identified, the power of corporate houses to influence government policies and international financial institutions (IFIs) must be understood as a major element of any explanation. Corporate power is of course facilitated by many other actors and agents, so it is important to consider who helps them realize their agendas. Globalization of food and agriculture, intensified by economic liberalization in the 1980s initiated by structural adjustments policies (SAPs), was boosted by the signing of the Agreement on Agriculture (AoA) as part of ongoing WTO negotiations in Marrakesh in 1994. These negotiations were supported by a number of powerful Western governments as well as some major new agricultural exporting countries.

The consequences have been very mixed and simple distinctions between developed and developing interests are far from satisfactory. However, it is important to recognize that liberalization policies have been implemented more thoroughly in the developing world than the developed; this is significant. It reflects the importance that countries in the developed world, especially the United States, Japan, and those in the European Union, grant their agricultural sectors and signals their refusal to concede their food sovereignty. Such countries have found ways and means to subsidize their agricultural sectors and promote corporate interests while insisting, through the IMF and WTO, that less powerful players remove all forms of agricultural support. Many countries in the global south have effectively been required to reduce their food sovereignty in recent decades.

Trade liberalization associated with SAPs has promoted corporate power and reach in food production across the developing world. Evidence is mounting that unless agricultural policies are fundamentally refocused, the changes wrought by contemporary patterns of globalization are fated to intensify all of the problems and contradictions identified in this book. Corporate influence has shaped all aspects of the food system since the 1980s, not just agricultural production and trade. As detailed earlier in this text, corporate food processor and retailers, through marketing strategies and pricing policies, are implicated in the obesity pandemic.

A caveat is required here: emphasizing the dominance of corporate power exercised through globalization, while accurate, must be used carefully. The process is neither linear nor inevitable. The following discussion appreciates the significance of corporate power but also identifies spaces that have escaped its embrace and some emergent political challenges to the dominant paradigm. This is important because understanding the corporate sector as all powerful is both inaccurate and politically defeatist. The global food system is fragile, fragmented and contested; its future is not assured and evidence of oppositional efforts is multifaceted and increasing. However, corporate power and associated trends are undeniable.

Globalization has been central to development strategies since the 1970s, and these have promoted industrial farming, which is allied to corporate expansion. There are many business opportunities that come in the wake of the corporate penetration of emerging markets. A great many benefit from the current state of affairs, employees and shareholders are unlikely to complain as the food sector retains its profit margins despite the global economic recession which began in 2008.

Millions of consumers in India and China, as well as the United States and United Kingdom, benefit from the status quo; they enjoy a diet divorced from seasons or major fluctuations in price and spend a smaller percentage of their income on food than ever before in human history. Nutritional improvements in some classes and regions are marked and reflected in dramatic improvements in life expectancy and infant mortality rates. Farmers with large landholdings have benefitted from the collapse of marginal producers, whether in the global north or south, and have enjoyed greater incomes as the market for agricultural commodities increased. In addition to vested interests in the food system, the story of agricultural success and its attendant ideologies have very been convincingly promulgated.

The narrative of food production since the 1950s has been celebrated as a triumph of science and technology, and it is understandable that optimism prevailed as 'productivity' (narrowly conceived by its advocates) seemed to grow exponentially. Indeed, as stated by Weis, 'the productive efficiency of industrial capitalist agriculture was an important foundation of dominant development narratives' (2010: 315). In the last 40 years, the revolution in agriculture has been profound and has been effectively portrayed as an unqualified success. The 'myriad upheavals wrought by industrial capitalist agriculture, and its enduring inequalities and tensions' were justified by promises 'of more, cheaper and better food (understood as more animal-derived protein) along with an end to the drudgery of farm work' (Weis, 2010: 316). Until recently, its externalities were understood as a price worth paying for 'progress'.

However, the system's social and environmental failings, its negative externalities (see Weis, 2010), have become increasingly obvious and in some places so politically destabilizing that food, its production and distribution, are back on international and national agendas.

Perspectives on the politics of change

Exploitation is endemic in the global food system, of some people, animals and the environment, which helps explain the appearance of multiple sites of vociferous resistance. There are the general anti-globalization, anti-corporatist networks that work to redress the social polarization and environmental despoliation that have accompanied corporate globalization. Then, there are progressive political parties, numerous international and national NGOs and other transnational networks devoted to challenging the worst externalities of the globalization project. Some of these are clearly focused and dedicated to a specific theme, such as the anti-land-grab campaign or opposition to GMOs; others have more extensive remits that may include advocacy, lobbying and activism. The disparate nature of these various challenges reflects the multifaceted character of the problems associated with contemporary food production; those concerned with human rights, food security and social justice; academics and activists concerned about addressing global poverty; environmental groups; public health officials; animal rights activists; bird lovers, etc. Understanding their various and diverse capacities to promote significant transformations of the food system is therefore problematic; their diversity also may constrain their political effectiveness. Exploring their capacities and potential as agents of social and political change requires politically informed analyses.

A recent article (Holt-Giménez and Shattuck, 2011) provides us with a helpful overview; it reviews the nature of the contemporary food regime and examines various interpretations of its problems and solutions. In addition, they suggest a typology within which we may begin to place oppositional efforts and their politics before considering their potential effectiveness. Their conclusions are guardedly optimistic as they suggest that a global alliance of forces based in the north (what they label 'progressive food movements') and the global south (which they label 'radical food movements') present an opportunity to transform some of its most glaring anomalies. They stress that such an alliance is not inevitable, but understand the current food crisis as presenting an opening within which progressive political change may be possible (see Weis, 2010, for another guardedly optimistic scenario). The following discussion draws heavily on their analysis while recognizing some limitations and partialities in their perspectives.

The categorizations employed by Holt-Giménez and Shattuck (2011) are too simplistic. One problem is in their use of the north–south dichotomy, but they are certainly not alone in finding this dualism difficult to escape. This book, while recognizing the limitations of north–south categories, continues to employ them; alternatives are also problematic. The radical–progressive dichotomy is similarly unsatisfactory. It ignores the phenomenal diversity of movements within each of these categories and denies that radical and progressive movements exist in both the north and the south. Their reading supposes that northern-based oppositional efforts are, by definition, only progressive in nature, while those in the global south are granted a more radical character. This too is very problematic (Box 9.1). Radicalism is not limited to southern perspectives, and some of the 'radical' forces in the global south may not always be as politically pure, as this discussion suggests.

However, I have found their paper pedagogically helpful when introducing students to the myriad oppositional forces and their political tendencies. Most students, in my experience, come to the themes in this text as remarkably naïve participants. It is important to encourage them to understand interventions as theoretically and practically embedded political discourses. The following discussion is designed to suggest the diverse politics at work within the multitude of suggested policies, at the global, national and local scales. Given the contestations in food literature (see one recent example in the following text, which considers the literature about alternative

Box 9.1

Radical Homemakers: how radical is this northern-based resistance to global corporatism?

As I looked more closely at the role that homemaking could play in revitalizing our local food system, I saw that this was a linchpin position, capable of more than just making use of garden produce and chicken carcasses. Individuals who had taken this path in life were building a great bridge from our existing extractive economy – where corporate wealth was regarded as the foundation of economic health, where mining our Earth's resources and exploiting our international neighbours was accepted as simply the cost of doing business – to a life-serving economy, where the goal is, in the words of David Korten, to generate a living for all, rather than a killing for a few, where our resources are sustained, our waters are kept clean, our air pure, and families can lead meaningful and joyful lives.

Radical Homemakers are men and women who have chosen to make family, community, social justice and the health of the planet the governing principles of their lives. They reject any form of labour or the expenditure of any resource that does not honour these tenets. For about 5,000 years, our culture has been hostage to a form of organization by domination that fails to honour our living systems, where 'he who holds the gold makes the rules'. By contrast, Radical Homemakers use life skills and relationships as a replacement for gold, on the premise that he or she who doesn't *need* the gold can *change* the rules. The greater our domestic skills, be they to plant a garden, grow tomatoes on an apartment balcony, mend a shirt, repair an appliance, provide for our own entertainment, cook and preserve a local harvest or care for our children and loved ones, the less dependent we are on the gold.

These thoughts led me to wonder if salvation from our global woes – the rampant social injustices, climate change, peak oil – was going to be dependent upon the women, upon considering all the hard-fought battles of both the first and second waves of feminism that have swept this country.

Source: extracted from 'Introduction to 'Radical Homemakers: Reclaiming Domesticity from a Consumer Culture by Shannon Hayes http://radicalhomemakers.com/ (accessed 28 July 2011)

food networks), delivering a comprehensive review of the politics of food provisioning is daunting and not in the remit of this text. All that is attempted here is to review some significant interventions, at different sites and scales, and their potential for significantly changing the injustices in the current food system.

Table 9.1 summarizes the various perspectives and their characteristics and key documents. Two broad categories are

Table 9.1 ***A food regime/food movement's framework***

	Corporate food regime food movements			
Politics	←-------------------- *Spectrum* --------------------→ *Neoliberal reformist progressive radical*			
Discourse	*Food enterprise*	*Food security*	*Food justice*	*Food sovereignty*
Main institutions	IFIs; UDAID; USDA; CGIAR; GAFSP part of WB initiative; Global Harvest; Gates Foundation; Agri-business TNCs	IFIs; FAO; CGIAR; UN High Level Task Force on Food Security (HLTF); mainstream Fair Trade; Slow Food movement; Worldwatch; some major INGOs, Oxfam; CARE etc.; food aid programmes	CFS; some slow food chapters; some fair trade initiatives; community food security initiatives; food policy councils; some unions and labour organizations; urban food justice movements	Via Campesina and other agrarian-based farmers' movements; ROPPA; EAFF; International Planning Committee on Food Sovereignty; many food justice- and rights-based movements; PCFS; Institute for Food and Development Policy, Food First

Orientation	*Corporate/global*	*Development/aid*	*Empowerment*	*Entitlement/redistribution*
Model	Productionist, promoting industrial agriculture; corporate concentration; monocultures; technical fixes, GMOs etc. Increasingly large farms	Mainstreaming/certification of niche markets (organic, fair, local, sustainable); 'sustainable' roundtables for agro fuels, soy and forest products; market-led land reform; micro credit initiatives	Agroecologically produced local food; investment in marginal farming communities; cooperative approaches; new business models; land access; regulated markets	Dismantle corporate agri-business monopoly power and trade distortions; democratization of food system; redistributive land reform; community rights to water and seed; regionally based food systems; sustainable livelihoods; support/protect peasant agriculture
Approaches to food crisis	Increased production; corporate control; liberalized markets; increased technical inputs	Modified neo-liberal perspectives; more agricultural aid; promotion of local food and sustainable agriculture; no serious challenge to patterns of power	Right to food; better safety nets; sustainably produced, locally sourced food; agro ecologically based agricultural development; CFS	Human right to food; locally sourced, sustainably produced food; culturally appropriate, democratically controlled food; focus on UN/FAO negotiations
Main documents	World Bank Development Report 2008; HLTS Comprehensive Framework for Action (2008)	World Development Report (2008); FAO and the Rome Declaration (1996)	IAASTD (2009)	Declaration of Nyelení (2007); UN Declaration of Peasant Rights; IAASTD

Source: based on Table 1 in Holt-Giménez and Shattuck (2011: 17–18); HLTS at: www.un.org/issues/food/taskforce; CGIAR at: www.cgiar.org; GAFSP at: www.gafspfund.org/gafsp; Gates Foundation at: www.gatesfoundation.org/Pages/home.aspx; Global Harvest Initiative at: www.globalharvestinitiative.org/index.php/about-us; International Federation of Agricultural Producers at: www.ifap.org; FAO, see the Rome Declaration on World Food Security and Plan of Action at: www.fao.org/docrep/003/w3613e/w3613e00.htm; Committee of Food Security (CFS) at: www.fao.org/cfs/en; see their revised more inclusive vision (2009) at: ftp://ftp.fao.org/docrep/fao/meeting/017/k3023e3.pdf; International Assessment of Agriculture, Knowledge, Science and Technology for Development (IAASTD), see summary of their Report 'Agriculture at the Crossroads' (2009) at: www.agassessment.org/reports/IAASTD/EN/Agriculture%20at%20a%20Crossroads_Synthesis%20Report%20(English).pdf; the network of Peasant organizations and Producers in West Africa (ROPPA) established in 2000, see goals and principles at: www.roppa.info/spip.php?rubrique37&lang=en; European and African Family Farms NGO (EAFF), see www.europafrica.info/en/chi-siamo; read 'Food Sovereignty: a common challenge in Africa and Europe (2010) at: www.europafrica.info/en/pubblicazioni; Peoples Coalition on Food Security at: www.foodsov.org/index.html; Declaration of Nyelení (2007) at: www.viacampesina.org/en/index.php?option=com_content&task=view&id=282&Itemid=38; Institute for Food and Development (Food First) at: www.foodfirst.org/en/about/programs/development

employed: corporate food regime approaches and food movements. The main distinctions between these categories are that corporate perspectives understand the contemporary food provisioning system to be essentially working but in need of a few adjustments or modifications. The food movements perspectives, on the contrary, recognize the system to be fundamentally flawed and in need of serious reconstruction, from the bottom up and from the local to global level. The distinctions are not always very sharp, as the appearance of the same groups and institutions in different boxes indicate. Hence, I have merged some boxes and suggested that the character of most interventions and actions have tendencies and are better considered as lying on a spectrum. As mentioned in the preceding text, the distinction between progressive and radical perspectives is not always convincing.

Corporate food regime: neoliberal and reformist

Evaluating the corporate perspectives first, these are further divided into neoliberal and reformist camps. Again, the categories are best understood as being on a spectrum, with an extreme corporatist vision at one extreme and market-friendly perspective at the other which accepts that the current regime is unjust. The neoliberal approach calls for a deepening and further extension of the industrial model, and corporate control is unchallenged and assumed to be the most efficient. This approach allows that some non-business actors have an important place, especially philanthropic donors and bilateral aid, which can subsidize the diffusion of industrial models of food production (the Gates Foundation and Golden Harvest initiatives are good examples). Since this text considers this approach to be the cause of the contemporary food crisis, it is not considered in any more detail here. Its limitations I hope have been adequately detailed in earlier chapters. It is, however, worth evaluating the reformist approach because it represents some very powerful players and is, in all probability, likely to be the dominant paradigm for the next few years, certainly among some of the most influential international institutions.

The reformist agenda has become more powerful since the food crisis of 2008 and the associated political unrest which threatens to destabilize the status quo. This perspective accepts that the system has flaws and that modifications are essential to alleviate some of its most chronic failures, namely severe food insecurity and environmental collapse. Interventions are seen as necessary to contain or placate resistance and to preserve corporate dominance; they are in effect adjustments necessary to preserve the status quo and to ensure that the political climate facilitates business. Even some elements within the IFIs accept that past prescriptions have failed and that new policies are required. They do not advocate drastically new policies, however; rather, their focus is on policies designed to eliminate or reduce the worst impacts of the current system. This perspective argues for more liberalization of trade and agriculture and more long-term investment in agriculture in the global south. They understand food insecurity as a temporary problem that must be alleviated in the short term until the system is improved and it 'works'. Alleviation efforts have assumed a number of guises, from safety nets and emergency relief to food-for-work programmes and state subsidies. This approach accepts that access to food is a human right and that national governments have a legal obligation to secure food security for their citizens.

The food movements: progressive and radical approaches

It is within the food movement's perspectives that we find the most compelling challenges to the status quo. Holt-Giménez and Shattuck (2011) employ another distinction here between progressive movements and radical activism. As mentioned earlier, this distinction is an interesting assertion rather than a convincing political reality. Radical thought and action is not limited to actors and activists in the global south, as their typology would suggest. The progressive trend, they assert, represent those movements that are probably most familiar to readers of this text in the north where 'it is based primarily in the middle and working classes … and has particular appeal to the young' (p. 124); again, a rather unsafe assumption. This category represents some very diverse groups and initiatives, including those concerned with social justice, the alternative food networks (AFNs), Slow Food Movement, Fair Trade initiatives, Sustain vegetarianism, environmental activism of all sorts, public health campaigners, as well as animal rights advocates, etc. The debate that surrounds the AFNs illustrates the complexity of oppositional efforts and efforts to determine their politics (Box 9.2).

Box 9.2

Alternative food networks: what's the politics?

Resistance to globalization can take form in different types of activities (Cohen *et al.*, 2005), including anti-consumerism campaigns (for example, 'Adbuster') and anti-logo initiatives (Klein, 2000), but also as promotion of alternative forms of consumption and lifestyles (Micheletti, 2003). As suggested by Pretty (2005), 'eating is the most political act we do on a daily basis because of its effect on farms, landscapes and food businesses'. It is in this context that initiatives focusing on local food have acquired new relevance (Roos, Terragni and Torjusen, 2009).

AFNs come in many forms: quality food schemes; organic foods; community-supported agriculture; farmers' markets; direct farm sales; box schemes; urban food gardens; Slow Food Movement; Fair Trade, etc.

Debates that surround AFNs illustrate the difficulties involved in evaluating the politics of resistance to globalization. AFNs refer to a wide range of food production, distribution and retail activities presented as alternatives to the corporate global food system. Most of the literature on AFNs has come from the United Kingdom and United States, which reflects their particular salience in these markets. Their emergence in the 1990s reflects 'the confluence of longer-term developments and more conjunctural factors' (Goodman, 2009: 3). To some extent, they were a response by concerned consumers to the increasingly obvious problems with industrial food production (food scares; awareness of environmental issues; animal welfare issues; food miles, etc.). The ethics of food consumption acquired a political significance as the social and environmental impacts of industrial food production become more glaring. Concern about provenance and traceability also became a market opportunity for small-scale producers with access to affluent consumers. In the EU, modifications to the Common Agricultural Policy also help explain their emergence as farmers found greater opportunities to diversity their livelihoods and enjoy profits from 'unique' products sold in local markets or speciality shops.

A significant attraction of AFNs was their presumed oppositional nature; participation in these networks was regarded as a political statement. Consumers and producers were consciously rejecting the 'placeless and faceless' homogenized foods delivered by the corporate sector. Farmers and artisan producers found a market for their 'quality' products that evoked specific localities or traditional or organic production processes that were diametrically opposed to the standardized fare from the supermarkets. Latterly, however, the corporate sector has appreciated the commercial opportunities of such products too. These 'alternative' foods are increasingly available on supermarket shelves, but this is a double-edged sword for producers and processors in the AFNs. Competitive pressures in the large retail chains squeeze down prices, and the original premium profits for producers are diluted or eroded. Here is one dilemma: should alternative food producers seek larger markets by compromising an original core principle by selling through supermarkets? Other questions emerge: is a product still alternative if it has a bar code and goes through the cash machine at a major food retailer? Do such moves signal more democratic access for 'quality' foods or the demise of the alternative ideal? Clearly, the 'interface between "alternative" and "conventional" food provisioning is increasingly permeable and a highly contested terrain' (Goodman and Goodman, 2007: 5).

Some have seriously challenged the assertion that AFNs are what they seem; most agree that they are rarely radical, but are they even progressive? The most problematic issue concerns the social relations of consumption; at present, these are very unequal. Goodman (2009) describes two examples of 'foodie gentrification' in London, namely Marylebone 'Village' and Northcote Road in Battersea. He describes the transformation of these areas and comments how they 'emphatically underline, even to the point of caricature, the huge social and spatial inequalities that generally are typical of the consumption relations of organic and artisanal quality foods' (25). He also quotes Slocum (2006), who drew attention to the connections between AFNs and privilege and paler skins.

In the global north, only highly privileged consumers are in a position to join this 'flight to quality', leaving others as the 'missing guests at the table' (Goodman, 2004: 12–130). Low-income families are often excluded from the goods sold in AFNs. How many low-income, minority families are enthusiastic about foods from the AFNs, and how can their participation be encouraged? Low incomes help explain their low participation rates, but some analysts argue that exclusion is a result, not of income, but because of a knowledge and skills gap. Processed food is often cheap, but unprocessed raw, wholesome food is often cheap too, but traditional cooking skills are required to turn this food into tasty meals. All sorts of social and cultural changes are implicated in the decline of traditional cooking skills. Despite the popularity of celebrity chefs and their calls to pick up our kitchen tools, there is scant evidence of a popular revival of traditional cooking. Only in affluent households can money and time be released to indulge in what are rapidly becoming arcane activities. However, anyone living in or visiting the developing world will have eaten delicious meals conjured up from very modest ingredients at a very low cost. Are AFNs not alive and well in many places in the global south (see Abrahams, 2007)? Corporate incorporation has not completely obliterated traditional food production or retail outlets. Just how resilient these skills and habits will prove is uncertain, but recent trends are not reassuring.

Elitism emerges in other debates about AFNs. Its most serious critics argued that they may actually reproduce the neoliberal forms they purport to challenge (DuPuis and Goodman, 2005). They observe that naïve assumptions which equate the 'local' with 'good' serve to obscure numerous class, race and gendered contradictions in local social relations that sustain local food systems. These critics perceive real dangers in uncritical assumptions about the character of 'local'. Some of these discourses deny the heterogeneity of local communities and may further establish reactionary and elitist tendencies. However, while acknowledging some of these reservations, most analysts agree that AFNs represent tentative moves towards a more positive sustainable food system (Harris *et al.*, 2009; Goodman, 2009).

Another issue is the Eurocentric character of much of the debate about AFNs. Most of the literature is based on Anglo-American experience. As mentioned earlier, AFNs exist and thrive in the global south, especially in urban areas where poor communities, or ethnic and religious minorities, represent a significant percentage of the participants. What a contrasting picture appears in these contexts! The papers by Abrahams (2006, 2007) provide a fascinating discussion about the nature of AFNs in the global south where they have a very different profile and purpose. In Johannesburg, she describes how:

> Alternative food supply systems for poor people are particularly noteworthy; surrounding farms sell poultry and beef off-cuts to residents in the township who re-sell them, while others times give away the offal, head, skin and feet of their slaughtered livestock to the poorer residents at no cost. Informalised alternative food supply systems that cater to poorer members of the community are preferred by poorer people even though the food may not always be cheaper or safer. Face-to-face interaction of producers and consumers are typical, and relationships of trust are increasingly the *modus operandi* of small peri-urban producers to gain access to speciality markets, even in formal retailing structures.
>
> (Abrahams 2006: 13)

The discussion about the importance of urban gardens in the global south (Chapter 7) also suggests that definitions of AFNs, based on affluent consumers in the global north, is probably a very small part of a much more significant model of food provisioning. In the urban areas of the global south, AFNs serve a vital role in livelihoods and survival strategies for producers and consumers; there, it is not a lifestyle choice.

The debates that surround the politics of AFNs suggest how contested and complex the politics of social movements can be in theory and in practice. They invite us to examine the nature of our theoretical constructs and concepts, their relevance in the north and global south, as well as illustrate the complexities of consumer choice and their political associations. We may conclude that affluent consumers are but one of the factors driving AFNs, and this raises another question, namely, how vital are they as agents of change in the global food system? More generally, where and how is change in the global food system likely to emerge?

Sources: based on material in: Goodman, D (2009), also available at: www.kcl.ac.uk/sspp/departments/geography/research/epd/GoodmanWP21.pdf (accessed 10 August 2011); Harris, E (2010), available at: http://edmundharris.com/wp/wp-content/uploads/2011/06/Harris-2010-Eat-Local.pdf; Caryn N. Abrahams (2007), available at: www.era.lib.ed.ac.uk/bitstream/1842/1465/1/cabrahams001.pdf (accessed 10 August 2011). Consult this book for an in-depth review of the issues considered in this box: Goodman, D. DuPuis, E.M. and Goodman, M (2011). Read about the state of AFNs in the United States at this website: USDA (2010) Local Food Systems: Concepts, Impacts, and Issues. Available at: http://permanent.access.gpo.gov/lps125302/ERR97.pdf (accessed 10 August 2011). For examples of local food initiatives in Europe, see: www.faanweb.eu/sites/faanweb.eu/files/FAAN_Booklet_PRINT.pdf (accessed 13 August 2011). For an interesting case study from Norway, see: Terragni, L. Torjusen, H and Vittersø, G (2009), available at: http://aof.revues.org/index6400.html (accessed 10 August 2011). For a consideration of ethical consumption and case studies of fair trade, slow food, and community-supported agriculture, see: Roos, G. Terragni, L. and Torjusen, H. (2009), available at: http://aof.revues.org/index489.html (accessed 10 August 2011)

The potential of consumers to construct a fairer food system has generated a great deal of interest and activism. The most significant example is ethical trading, where attention is paid to environmental issues, human rights concerns, animal welfare and other social issues. 'Fair trade' is a general term used to describe trade that helps disadvantaged producers access global markets and to obtain a 'fair' price for their goods. Fair trade is described at the Fair Trade Resource Network website as:

> Fair Trade is a lot of things: a social justice movement, an alternative business model, a system of global commerce, a tool for international development, a faith-based activity. It means different things to different people. There is no single, regulatory, authoritative body. So, individuals need to explore various models and concepts. Fair Trade's many definitions do always center around the exchange of goods based on principles of economic and social justice.
>
> (Source: www.fairtraderesource.org/wp/wp-content/uploads/2007/10/The-New-Conscious-Consumer.pdf; accessed 10 August 2011)

As with the AFNs, it is an alternative trading system in opposition to conventional 'free' trade. It offers consumers a way to support more ethical production systems than conventional trade operating under the dictates of the WTO.

Table 9.2 ***Free trade is not fair trade***

	Free trade	*Fair trade*
Objective: presumed to promote	National economic growth	Empowerment, especially of poor rural producers
Focus	International trade between independent countries	Trade between individual producers and individual consumers and businesses
Primarily beneficiaries	TNCs, powerful business interests	Vulnerable farmers, artisans and workers in less industrialized countries
Problems	Often exploits poor communities and their environments. Normally excess profits enjoyed at middle and top of the commodity chain	Interferes with free market, inefficient, too small scale for impact
Major actions	Countries liberalize markets: lower tariffs, quotas and labour and environmental standards. Loss of national control over production and markets	Businesses offer producers favourable financing, long-term relationships, reasonable prices and encourage higher labour and environmental standards. Often, support is available for small producers in the initial stages

(Continued)

Table 9.2 ***Cont'd***

	Free trade	*Fair trade*
Producer compensation determined by	Market and government policies, often prejudiced by long-established biases in trading relationships	Living wage and community improvements. More of value captured at the producer end of the commodity chain; this can be used to invest in local community
Supply chain	Includes many parties between producer and consumer. Often, large power asymmetries obvious	Includes fewer parties, more direct trade and more value is captured at producer end of chain
Key advocate organizations	WTO, WB and IMF	Fairtrade Labelling Organizations International, World Fair Trade Organization

Source: adapted from www.fairtraderesource.org/wp/wp-content/uploads/2007/10/The-New-Conscious-Consumer.pdf (accessed 10 August 2011)

The Fair Trade movement is an example of a transnational effort to challenge the unfair impacts of trade – in this case, the exploitation of small-scale producers and their environments who export goods to the international community. Most successful campaigns have included those associated with agricultural products such as coffee, tea and herbs, bananas, cocoa and chocolate, wine, olive oil and some fresh fruit and flowers, sugar, rice and vanilla. Fair Trade initiatives are global and represent a limited intervention by affluent consumers to make trade more just. It has helped improve the livelihoods and entitlements of some of the poorest producers who sell in the global arena (Box 9.3).

Box 9.3

Cocoa cooperative with 9,000 members in the Dominican Republic builds capacity for sustainable business

Many of the Dominican Republic's small farmers still struggle to survive. Founded in 1988 and a participant in the Fair Trade system since 1995, the Conacado cooperative aims to generate work and income for disadvantaged groups. Conacado has about 9,000 members who receive the majority of their income from cacao. Fair Trade provides added support for farming methods that are safe for the environment and public health.

Fair Trade also ensures a higher minimum price, long-term contracts and access to credit. Most farmers do not have savings and face extreme difficulty in securing loans, while the government has very limited means for aiding these farmers.

With the higher price from Fair Trade, sales have provided the cooperative with enough income to meet basic expenses and invest in the future. In the wake of the massive destruction of Hurricane George in 1998, the importance of Fair Trade revenues was clear, as the communities were able to rebuild.

Fair Trade has helped the cooperative to organize workshops to teach farmers how to improve fermentation techniques, expand sustainable growing methods, increase productivity and participate more actively in the cooperative.

Source: www.fairtraderesource.org/wp/wp-content/uploads/2007/10/The-New-Conscious-Consumer.pdf (accessed 10 August 2011)

While admirable in their ambitions, the impact of these initiatives is limited and partial. The most serious limitation is that registration procedures can be both complex and expensive, and the percentage of such products in world trade remains very small. Some research also suggests the corporate involvement, claimed as fulfilling their social responsibility obligations, may in fact be strategies to secure commodity supplies. As with AFNs, the market opportunities associated with Fair Trade products have not be missed by the corporate players. The involvement of corporate giants such as Cadbury's and Nestle in Fair Trade products further complicates the picture, as an ethical concept has now become a 'brand'; is it in danger of becoming just another corporate logo (Box 9.4)? Lastly, it is unsatisfactory to consider that our participation as ethical global citizens is limited to our patterns of consumption; this represents only a small part of our potential political participation.

Another transnational movement that challenges conventional trading systems is the Trade Justice Movement (TJM). This is a large coalition of organizations concerned with trade justice and includes trade unions, aid agencies, environment and human rights campaigns, Fair Trade organizations, and faith and consumer groups. They base their activism on the following:

> We believe that everyone has the right to feed their families, make a decent living and protect their environment. But the rich and powerful are pursuing trade policies that put profits before the needs of people and the planet. To end poverty and protect the environment, we need Trade Justice, not free trade.
>
> (Source: www.tjm.org.uk/about-us.html; accessed 10 August 2011)

Box 9.4

Corporate social responsibility: a positive trend?

Many global companies are involved in the CSR agenda, either actively or at least at a rhetorical level: for example, Nestlé now has a line of Fair Trade coffee; Starbucks aims to have 100 per cent of its coffee '... responsibly grown, ethically traded...' by 2015; McDonalds 'envisions... engaging in equitable trade practices...'; and Chiquita 'cares about the people who live and work on the farms'. Coca-Cola, working in collaboration with an NGO, TechnoServe, and the Bill and Melinda Gates Foundation, has launched a partnership with over 50,000 small mango and passion fruit farmers in Uganda and Kenya, which is intended to create new market opportunities for them to supply the fruits for Coca-Cola's locally produced juice. CSR is also being institutionalized through numerous standards and codes for food products. Some are business-led (e.g. the Ethical Tea Partnership, Business for Social Responsibility, World Cocoa Foundation), while others are multi-stakeholder-driven (e.g. the Ethical Trading Initiative, the International Cocoa Initiative, the Common Code for the Coffee Community, the Rainforest Alliance agricultural certification and the Sustainable Agriculture Initiative). There can be significant differences between company-led and multistakeholder initiatives in terms of substance, credibility and application, and the impact of each still has to be evaluated.

Source: IFAD, Rural Poverty Report, 2010: 135, available at: www.ifad.org/rpr2011/report/e/rpr2011.pdf (accessed 11 August 2011)

There are a number of interesting policy and briefing papers on their website. The list of their membership illustrates the limitations of the categorization employed by Holt-Giménez and Shattuck (2011), where oppositional movements are understood to be 'northern' or 'southern' and 'radical' or 'progressive', respectively. The membership of TJM is transnational and of various political types. One of the most important issues raised by this type of movement is that it suggests the alternative possibilities that globalization presents; globalization does not have to be a corporate creation. The process has generated progressive transnational spaces where activism may be more effective than if it is nationally based.

It is in the radical category of the food movement that we encounter some of the most politically charged and therefore, perhaps, the most potentially significant forms of resistance to industrial food production and the status quo. Central to these efforts are notions

Plate 9.1 Fair trade coffee (Source: Wikimedia Commons, http://en.wikipedia.org/wiki/File:Coffee_beans_being_sorted_and_pulped.jpg).

'of entitlement and redistribution of wealth and power' of the membership of these groups (Holt-Giménez and Shattuck, 2011: 128).

Patel observed that '[T]he most systematic and comprehensive organic and living alternative to existing hegemonies comes not from the ivory towers or the factories but from the fields' (2007: 90). Indeed, the Via Campesina has become a major radical movement seeking to reduce global inequalities in the food system. The concept of food sovereignty is a core element of their perspective. This transnational agrarian social movement is campaigning to have the United Nations negotiate and implement a declaration, and eventually an international convention, on peasants' rights. Peasant organizations hope to follow in the footsteps of indigenous peoples' movements that participated in the negotiations preceding the 2007 UN Declaration on the Rights of Indigenous Peoples. This grassroots movement promotes food sovereignty and sustainable agriculture. Although originating in Latin America, it has now established links with small producers across the globe (Box 9.5). They claim that peasant and

Box 9.5

La Via Campesina: the birth and evolution of a transnational social movement

Martínez-Torres and Rosset (2010) examine the rise and changing ambitions of one of the most successful social movements devoted to improving food sovereignty across the globe. They describe its origins in Latin America and its evolution as a 'transnational peasant movement', and detail five evolutionary stages. In the 1980s, the withdrawal of the state from rural areas simultaneously weakened corporate and client control over rural organizations, even as conditions worsened in the countryside. This gave rise to a new generation of more autonomous peasant organizations, who saw the origins of their similar problems as largely coming from beyond the national borders of weakened nation-states. A transnational social movement defending peasant life, La Via Campesina, emerged out of these autonomous organizations, first in Latin America, and then at a global scale, during the 1980s and early 1990s (phase 1).

Subsequent stages saw leaders of peasant organizations take their place at the table in international debates (1992–1999, phase 2), muscling aside other actors who sought to speak on their behalf, taking on a leadership role in global struggles (2000–2003, phase 3) and engaging in internal strengthening (2004–2008, phase 4).

More recently (late 2008–present, phase 5), the movement has taken on gender issues more squarely and defined itself more clearly in opposition to transnational corporations. Particular emphasis is given to La Via Campesina's fight to gain legitimacy for the food sovereignty paradigm, to its internal structure and to the ways in which the (re)construction of a shared peasant identity is a key glue that holds the struggle together despite widely different internal cultures, creating a true peasant internationalism.

Source: Martínez-Torres, M. E. and Rosset, P. (2010), available at: www.landaction.org/IMG/pdf/Rosset-Martinez_ViaCampesina-movement.pdf (accessed 10 August 2011)

small farmers hold the secret of a more ethical agricultural future, where people are fed without extracting such a high price from the environment. It also supports a number of educational programmes and has produced free online teaching material: 'Food for Thought Curriculum' at: www.grassrootsonline.org/publications/educational-resources/food-thought-action-a-food-sovereignty-curriculum (accessed 11 August 2011).

The Declaration at Nyelení in 2006 in Mali (see Chapter 1) is a key document of the radical food movement. At the heart of these approaches is a call for the complete transformation of the food system and the power relations at its centre. Corporate control over food and its production is rejected in favour of local control by small-scale producers and local consumers. It is also a green movement that rejects the chemical model of food production, which has become completely dependent upon products from the petrochemical industries. It wants to reassert democratic control over food systems, wresting back ownership of food from corporate actors that have become so dominant at every stage of the food chain, from seed to plate.

This organization is addressing 'the immediate problems of hunger, malnutrition, food insecurity and environmental degradation, while working steadily towards the structural changes needed for sustainable, equitable and democratic food systems' (Holt-Giménez and Shattuck, 2011: 132).

The analysis in this text argues that obesity is another manifestation of the broken global food system. If this problem is correctly diagnosed, by medical and health campaigners in the north and global south, then their combined influence would represent a powerful addition to the food movement and would increase its political clout. It is naive to assume that medical opinion would necessarily become an ally, however. Professional ethics do not always prove able to resist the corporate embrace, and are often enamoured by costly 'high-tech' solutions to socially embedded health problems (Young, 2008). At present, debates about obesity in the medical profession are largely dominated by behavioural and biomedical discourses that emphasize proximate causation and individual responsibility, rather than structural factors. Such analyses open more market opportunities for corporate food companies as they develop diet foods and functional foods, the former already a massively profitable business and the latter recognized as the new frontier:

> The revolution in functional foods – commodities manipulated by science and marketing to promise health and wellness – is the modern battleground where industry titans are fighting for dominance in the global food business.
>
> (Source: www.bbc.co.uk/programmes/b00wmvck; accessed 3 March 2011)

However, some medical opinion certainly recognizes the role of corporate profits to the obesity problem. The global obesity pandemic threatens to be an economic crisis for the liberal welfare state as well as a crisis with individual consequences. This fact will become increasingly politically salient, although it may not guarantee progressive political solutions. Experts agree that governments can no longer ignore the problem as they begin to recognize the health consequences and costs from obesity. The costs of health-related problems are estimated to be as serious as those associated with smoking. These costs are of a magnitude at least as serious as tobacco.

What needs to change?

As identified by the recent Food Ethics Council Report (2010), the dimensions of the problem are daunting:

> We are faced with social injustice across all aspects of the production and consumption of our food, and the interrelationships within and between social justice and sustainability problems make the picture even more complicated. … to a great extent, these problems are rooted in structural features of 'how the world works'. Trade liberalisation, the role of global corporations, the influence of the financial sector, deregulation, socio-economic policy orthodoxy, consumption-led growth … these are the factors that the committee has found to underlie the unfairness and unsustainability of our food system.
>
> (Source: Food Justice: The Report of the Food and Fairness Inquiry', page 79, at: www.foodethicscouncil.org/system/files/FoodJustice_reportweb.pdf; accessed 10 August 2011)

This text has employed different scales to analyse the problems of the global food system, and here we employ the same scales to evaluate some of the changes required for a more sustainable and equitable food system to emerge. Transforming food production policies means contesting long-established ideologies and powerful vested interests; this requires '[D]ense, diverse, organised collective action' (Evans, 2008: 57) in multiple political spaces. Political challenges are obvious at traditional sites of contestation as well as in more recent transnational spaces facilitated by new communications technologies.

Global-level perspectives

Clearly, change is required at the global level, especially where development policies and trading regimes are constructed and enshrined. Many factions of the global food movement explicitly target the WTO, and this institution has become the bête noir of most anti-globalization movements. The distortions embedded in the international trading system with regard to food and agricultural commodities have been established in earlier chapters; without these being addressed, serious inequities in food security will prevail. An associated issue is how development policy might extend livelihoods in the global south.

Many countries in the global south are still very dependent on the export of primary products and have a very limited industrial base. In this area too, the IFIs need to adjust their policies. Even if agricultural opportunities are supported, millions will still need to find employment in the urban areas. The Asian Tiger economies are often hailed as exemplifying what is possible by expanding industrial opportunities. Examining their success illustrates the importance of specific international and national contexts to success; in the current global economic climate, it is difficult to be optimistic about their success stories being repeated elsewhere. However, development policy must extend and support the multiple livelihood strategies in the global south to improve the poorest population's food entitlements. The global shift identified by Dicken (2011), while undoubtedly improving economic opportunities for millions in the global south, remains geographically limited; sub-Saharan Africa, for example, has failed to enjoy much benefit. Trade may help diversify entitlements but other initiatives will be vital too.

The fixation on trade can deflect attention from the other vital institutions of global governance, namely the IMF and WB and their affiliates:

> The international financial institutions (IMF and WB) should pay greater attention to the protection of the right to food in their leading policies and credit agreements and in international measures to deal with the debt crisis.
>
> (General Comments of the UN Committee on Economic and Social Rights, 2009)

It is crucial to recognize the role of the IMF and the WB and their place in designing and implementing development programmes across the developing world. These have proved disastrous in some contexts and have singularly failed to deliver development for the majority. There is evidence to suggest that they are encouraging more investment in agriculture but, if this mirrors past policies, it will not benefit small producers or local food sustainability but large commercial producers. Development in small-scale agriculture can

(a)

(b)

Plate 9.2a,b Tourists present a market opportunity for rural populations in Kenya (Courtesy: Geography Department, Staffordshire University).

deliver a series of benefits to national and local economies and is necessary to limit massive rural–urban migrations.

Several experts have suggested the creation of an international food agency, akin to the International Energy Agency (IEA). Its mission would be to maintain buffer stocks to be used to stabilize global food prices. Evans (2009) argues that the WFP might be a suitable agency to operate and manage such an agency. He stresses, however, that it would be 'essential to specify that the role of any system of reserves would be limited to emergency assistance – not to act as a price support for producers or a permanent system for managing food aid' (p. 6). As reviewed in the previous chapter, food aid has in the past been used as a Trojan Horse to access markets and to create food dependency.

The other global players are, of course, the transnational corporations. Corporate profits are buoyant as the UN announces that world food prices have reached record highs (www.bbc.co.uk/news/business-12638085; accessed 3 March 2011). Market concentration is high and increasing in all sections of the food industry, and a few 'Titians' are remarkably powerful, both economically and politically. This monopolistic situation helps explain the extreme volatility of food prices and high corporate profits. Such circumstances are dangerous for consumers everywhere but are disastrous for poor populations vulnerable to food price fluctuations. Monitoring the activities of these corporations is vital to exposing some of the worst social and environmental impacts of global industrial agriculture; opposition to their activities has fuelled some of the most effective transnational resistance movements. Recent speculation in food commodities has caused concern, and it is important that these activities are regulated.

The disastrous part played by conflicts in explaining malnutrition was the theme of the last chapter. The international community has a vital part to play in trying to prevent conflicts, and if this fails, in intervening to limit the implications. Unfortunately, the incidence of resource-based conflicts will probably increase as pressure mounts on Earth's resources. The international media has a very fickle nature and a limited attention span. Until this week, the crisis in the Horn of Africa was in the headlines, millions were being forced to migrate in search of food, but riots in the streets of London soon replaced the Horn in the headlines. Interest in conflict situations are also prone to be fleeting. It is difficult not to see the business of war remaining a significant cause of hunger for the foreseeable future, but efforts must continue at the United Nations. In many respects, therefore, the global level is vital to the resolution of malnutrition in its several forms. Various battles must be waged at the global level, the site of considerable power, materially and ideologically. Transnational political movements, responding to this realization, have emerged to challenge the corporate nature of contemporary globalization.

The potential of such transnational oppositional movements to effect significant social change is widely debated. Certainly, their profile has increased in the media and in social science research since the 'Battle of Seattle' (Evans, 2008: 272). They, as with corporate actors, are creatures of globalization. All the factors that enable global corporate integration and success may be employed by global social movements with different ambitions, namely to build more equitable and democratic global politics. Global communications infrastructure facilitates international alliances and helps coordinates their activities. We can all appreciate the enormity of the challenge and the ideological differences that may divide, rather than unite, a broad-based transnational alliance to resist corporate control. Will a broad-based alliance between the numerous radical and progressive movements, from the global north and south, present a political potent force? Their combined activism and advocacy could form an impressive transnational coalition which would impress international, national and local policy-makers.

The annual World Social Forum (WSF) meeting represents a platform for many such dissenting voices. The most recent gathering was in Senegal during 6–11 February, 2011. 'Its mantra of social change "another world is possible" proved especially resonant as anti-government protests rocked Egypt in the north. The eleventh forum, an alternative to the elite World Economic Forum held in Davos, brought together civil society groups from around the world'. This was an assembly of people from a great many backgrounds and with various agendas, and academics, activists and field workers all contributed to the discussions. Santos (2008) argues that 'the WSF may be said to represent today, in organizational terms, the most consistent manifestation of counter-hegemonic globalization'. This is an accurate description, and readers are urged to visit the WSF website to assess its diverse constituencies. The prominence of food politics (of several sorts) at the heart of their efforts indicates the potential of the issue to 'capture dreams and spark the energies of ordinary citizens' which is the *sine qua non* of building oppositional power (Evans, 2008: 295). While transnational global oppositional politics are increasingly evident, most analysts still accept that nationally based political movements remain most salient for progressive politics.

National perspectives

The policies of national governments will be among the most crucial to positive change; policies need to be implemented to increase entitlements of the poor and to support sustainable agriculture and healthy diets. Most of these are most effectively implemented at the national level; the Cuban example is a particularly interesting illustration of what a state can do when it must (Box 9.6). Earlier in this text, we established that states are obliged under the Human Rights Act to ensure that their citizens have adequate diets; pressure must be applied to ensure that nations fulfil this obligation. Ironically, Sen's work throughout the 1980s and 1990s established the essential role of an 'activist' state in securing entitlements, just as international economic ideologies were making it more difficult for developing states to establish progressive development policies in the realms of public health and agriculture, and decreases in public sector spending was the norm.

Explicit state intervention has long been recognized as necessary to increase entitlements of poor populations, through land reform, subsidized agricultural inputs and food support payments of various kinds. The role of agricultural extension services, state-led marketing boards and some state buffer food stocks have played important roles in the past and should be revisited. Investments in these declined as liberalization deepened though the 1980s and 1990s. All these interventions have decreased through the liberalizing decades and must be reintroduced to reduce traditional malnutrition. It is important that national governments begin to intervene to secure food security and sovereignty, the case being urgent in numerous countries. As noted by Evans (2009), investment in agriculture by international and national governments has fallen drastically in recent years. He cites the example of the African continent: '[I]n Africa, for example, governments spend on average only 4.5 per cent of their budgets on agriculture – despite an African Union target of allocating 10 per cent of public spending to agriculture by 2008' (2009: 8). He also

Box 9.6

Cuba: mixed results?

Almost overnight, following 1989, US$6 billion in Soviet subsidies to Cuba disappeared. At the same time, the US trade embargo tightened, and Cuba was plunged into an economic crisis. Cuba radically changed the state sector in 1993; about 80 per cent of the farmland was then held by the state and over half was turned over to workers in the form of cooperatives. Farmers lease state land rent-free in perpetuity, in exchange for meeting production quotas. They may even bequeath the land, as long as it continues to be farmed. A 1994 reform permitted farmers to sell their excess production at farmers' markets.

The reforms emphasized five basic principles. First, environmentally friendly farming was promoted, supported by the state. A second included land reform; state farms were transformed to cooperatives or broken into smaller private units, and anyone wishing to farm could do so rent-free. In effect, a right-to-farm policy was implemented. A third principle of the reform was fair prices to farmers: farmers can sell their excess production at farmers' markets; average incomes of farmers are three times that of other workers in Cuba. A fourth principle of reform is an emphasis on local production in order to reduce transportation costs, and urban agriculture was key. The fifth principle of reform is farmer-to-farmer training as the backbone of the extension system.

What were the results of these reforms? Production of many of the basic food crops, vegetable and cereals improved dramatically. One of the most interesting elements of the changes was that Cuba was required to seek environmentally friendly options. The conversion of Cuba's agriculture to more sustainable practices has focused on urban agriculture and domestic crops. Urban agricultural production climbed from negligible in 1994 to more than 600,000 metric tons in 2000.

Cuba continues to rely on food imports, especially rice, beans, milk products, feed grains, soybeans, some chicken and pork, as well as substantial amounts of cooking oil, soybean meal, meat and malt.

Cuba's shift to organic agriculture is not taking place without controversy. One common point of view holds that what is taking place is not a process of conversion, but rather a temporary substitution during a period of crisis. The opposite point of view holds that the present change is long overdue and that further transformations are needed to develop truly rational production systems.

Source: extracts from Zepeda, L. (2003) 'Cuban Agriculture: A Green and Red Revolution' in *Choices* at: www.choicesmagazine.org/2003-4/2003-4-01.htm (accessed 10 August 2011); Rosset, P. 'Organic Farming in Cuba' at: http://multinationalmonitor.org/hyper/issues/1994/11/mm1194_06.html (accessed 10 August 2011)

identifies the five basic resources that are essential to support small farmers; these are: access to (1) assets (such as land, machinery and renewable resources such as water); (2) markets (e.g., adequate infrastructure, communication networks that give farmers access to up-to-date price information and the capacity to meet supplier standards for supermarkets); (3) credit (to prevent small farmers from falling prey to predatory lending, and to improve access to inputs such as fertilizers); (4) knowledge (where there is an urgent need to invest in agricultural extension services to help disseminate R&D findings in the field); and (5) risk management tools (e.g., through social protection systems, mechanisms for hedging against bad weather and improved crop storage systems). Developing-country governments and donors alike need to focus on supporting these outcomes (Evans, 2009: 8).

Sometimes relatively small initiatives can produce considerable improvements in livelihoods. Read about some successful initiatives to improve small-scale food production at: www.worldwatch.org/system/files/NtP-Innovations-in-Action.pdf (accessed 26 February 2011).

With reference to obesity too, as I have maintained elsewhere (2004), government 'activism' is essential. Patterns of obesity represent new inequalities in nutrition at global and national levels, and national governments remain the most likely and legitimate institution able to address the inequalities identified. The list of potentially useful interventions is extensive and includes action in most areas of national policy. Some of the most obvious include investments in public health; a more aggressive preventive health agenda is necessary to tackle childhood obesity in particular. Robust action to alter corporate behaviour is essential and sadly lacking. Ruthless marketing of unhealthy foods, especially to the young, is obvious everywhere, as are supine government interventions in health debates which fail to address the problem of corporate interests.

Redirecting national agricultural policy and subsidies to help support the low-cost delivery of sustainable, locally produced food would be a useful place to start. Massive subsidies currently support the production of unnecessary and/or unhealthy foods which are attractive to poor households. Multiple benefits may be secured by using national public procurement programmes to promote better diets, as well as by supporting environment objectives. The excellent analysis by Morgan and Sonnino (2008) examines the obstacles to, and potential of, sustainable school food in a great variety of national contexts. One fears that, in the current economic climate, the continuation of some really innovative programmes in public procurement and local food systems may suffer, at least temporarily. Morgan describes one such threat at: http://purefoodlinks.eu/files/2011/08/Morgan-2011_let-them-eat-pickles-_-budget-cuts-school-dinner-revolution.pdf (accessed 11 August 2011).

Alternatively, as climate changes, energy security and water scarcity intensify the merits of local food systems, and their social, economic and environmental effects become even more obvious.

Rethinking urban planning models to promote green cities where physical activity is encouraged rather than constrained should be prioritized, especially in the emerging world economies as they design their new megacities. The multiple benefits of urban garden initiatives are recognized in the north and south (see Chapters 6 and 7). New visions about urban planning are obvious across the north where the food issue is now beginning to receive the attention it deserves; perhaps, urban planners in the global south will emulate these innovations (Box 9.7). All of this calls for joined-up initiatives, so that changes in one element of government policy supports and enhances progressive changes in other areas. Health, environment and food policy are obviously all allied, but their respective policies are seldom designed to reinforce their respective objectives (Lang, Barling and Caraher, 2009).

Box 9.7

Feeding the City: the challenge of urban food planning

Another sign of the popular resonance of food planning is the growth of Food Policy Councils in North America, where there are now more than 100 in various cities and counties. To give a flavour of the food planning movement at a local level, it is worth highlighting the following three examples:

- *Baltimore*: In the first update to its comprehensive master plan in 45 years, city planners discovered that residents were concerned about poor access to healthy, affordable food, triggering a process that eventually led the city to adopt the

Baltimore Sustainability Plan, which explicitly states the need for a food system that supports public health, quality of life and environmental stewardship.

- *King County, Washington*: The comprehensive plan update adopted in King County in 2008 is thought to be one of the few plans in the United States that systematically addresses food system issues. Though not a traditional planning theme, the plan says that food is as important to health and well-being as air and water, and therefore a food dimension has been woven into the other planning policies.
- *Marin County, California*: In 2007, Marin County adopted an innovative countywide plan for sustainability, one of the key elements of which was 'agriculture and food', which addresses not only the preservation of agricultural lands and resources, but many other facets of the food system, such as sustainable farming practices and community food security (Hodgson, 2009).

To these food planning pioneers in the United States, we might add New York and San Francisco, and in Europe, too, the food planning movement is beginning to register its presence. Two of the most prominent examples are London, which launched a sustainable food strategy in 2006, and Amsterdam, where the food strategy has multiple objectives, one of which is to help the city to re-connect to its regional hinterland for both economic and ecological reasons.

Source: extracts from Morgan, K. (2010), available at: www.tandfonline.com/doi/pdf/10.1080/13563471003642852 (accessed 12 August 2011)

Only concerted collective action can challenge the massive vested interests in the global food and agricultural sectors. Large business interests are involved in both promoting sedentary behaviour and the passive over consumption of food. The changes necessary to create a more equitable and rational food system, which provides more people with a sufficient, safe and nutritious diet, will not be served up by magnanimous agricultural corporations. It is not in the nature of corporate sector to behave in public interest. They find it better to present the changes in patterns of production and consumption as 'efficient' and as a 'natural' consequence of individual choice, and then market expensive technological 'solutions' to alleviate the resultant environmental and/or health problems. For example, diabetes is set to become very big business. Diets already are big money, of course; a conservative estimate for expenditure on specialty diet foods in the European Union is €15 billion per year, while the 'functional food' sector already accounts for 2.4 per cent of the global food and drink market.

Sub-national perspectives

What power might be exerted at the sub-national scale? Many of the most powerful social movements are based on ethnic-, indigenous-, gender- or peasant-based activism. Many of these people are among those most directly threatened by food insecurity and who suffer when food prices increase. Change may be most effective if launched by those most seriously burdened by the system at present – women in the developing world, the rural poor, small peasant producers, the urban poor, etc. These vulnerable populations and poor consumers are more likely to challenge their politicians and the status quo than those who are sheltered from major food shocks. It is from within these populations that some of the most serious challenges are obvious; and it is from these communities too that some of the most inspirational figures emerge. As discussed in the preceding text, building effective alliances between these communities and northern-based movements may be the critical variable if a more humane food system is to emerge.

Across the globe, the fight to reduce hunger continues, and inspirational individuals emerge to promote social justice (visit www.fian.org/resources/photo-gallery/face-it-act-now-exhibition/gallery_slide_view?b_start:int=3&-C to read about a few examples; accessed 12 January 2011). However, most of us will play more humble parts. There is a heighted awareness of the ethical implications of food consumption, and affluent consumers are beginning to drive forward some innovative policies, nationally and globally. Most analysts welcome increased awareness of the ethics of consumption in a global world, and their educational impact is certainly positive. Such efforts are limited, as government action is ultimately required to alter the patterns of trade, pricing and consumption – the structural components of the food system. As Monbiot (2009) notes, changing consumption habits are limited 'unless it is backed up by government action'. He added that 'our power comes from acting as citizens, demanding political change, not acting as consumers'. However, probably most people who start to shop ethically become more committed with time, and may eventually become more active politically. Similar individual histories are associated with allied issues such as animal rights; sometimes, an initial vague disquiet can mature into constructive resistance.

Some of the most vociferous criticism of the global food system emanates from animal rights activists. An initial concern about animal well-being may cause people to learn more about how people as well as animals are exploited in food production. Among affluent Western consumers, the image of caged chickens or crated calves can be very politically potent (Nierenberg, 2005). Animal welfare groups, from the mild to the radical, are increasingly effective at challenging the worst examples of animal exploitation in farming systems. Given the number of animal-based disease scares in recent years, animal rights campaigners can now add the dangers of animal diseases and their potential to infect humans in their war on animal cruelty.

Conclusion

This text examines two forms of malnutrition, food insecurity and obesity. Some may judge the discussion of problems of want (insufficient food) and overconsumption (too much) in the same analysis contentious, so a word of justification. Food is about politics and power. At every site of its transformation from seed to supermarket, power is inherent if often obfuscated. Globalization has shifted the balance of power in the global food factory, and many have lost and a few have gained.

This text exposes some of these shifts, at global, national and local scales, and tries to identify where power might be asserted to reshape the contours of food provisioning. It finds that globalization has generated new problems and is a legitimate target for oppositional efforts. However, the analysis recognizes the continued relevance of the state for ensuring food security and food sovereignty – that is, to ensure sufficient, safe and nutritious food 'at all times' for its inhabitants. It would be foolish for any government to ignore this imperative, although in the West the abundance of apparently 'cheap' food since the 1960s might encourage them to do so. It is much easier for some states to succeed than others, but it remains an obligation for all.

This text argues that economic liberalization has not proved to be the correct strategy for enhancing food security, in any respect, quite the contrary. This is not to argue for national self-sufficiency, but it is to recognize that every government is ultimately responsible for the provision of the public's health, and that is still inherently tied to the provision of nutritious foods and healthy environments. Food has proved a force for revolutionary change in the past. The evidence of the global food and agricultural crisis is mounting, and perhaps its transformative capacity is about to reappear (Magdoff and Tokar, 2010).

The strategic significance of oil may be crucial (Evans, 2009: 6). It is impossible to overestimate the importance of oil to the global food system described in this analysis. Oil and the products of the petrochemical industry are simply essential to our industrial food model; a decline in its availability or an increase in its price will destabilize the way we feed ourselves. Evidence of this instability is already obvious; its appearance has proved politically potent in North Africa in early 2011. Ensuring the West's access to petroleum products helps explain the geopolitical significance of the Middle-East, and Western countries' involvement in conflicts in that region and others.

The global food system is one of the clearest manifestations of hydrocarbon capitalism – we are simply hooked. Wall Street is getting nervous because oil prices are increasing and threatening the survival of capitalism as we have known it for the last few centuries. While capitalism may evolve, the transition to alternative energy sources is likely to be long and difficult. Some of the greatest changes and some of the most significant pain will be those associated with the transition in our food system. The optimistic scenario is that an oil crisis will precipitate a food crisis which may occasion the birth of a more socially equitable and environmentally benign food system.

Summary

- **Many vested interests benefit from the global industrial food provisioning system.**
- **The narrative of the industrial food system and its triumphs has been dominant since the 1950s.**
- **Challenges to this dominant narrative have emerged and are growing in intensity as its social and environmental limitations have become more obvious.**
- **Radical and progressive opposition is obvious in the global north and south.**
- **We are all implicated in the global food system and can influence its direction, as consumers and as global and national citizens.**

Discussion questions

1. Debate the assertion that 'the global food system is fragile, fragmented and contested'.
2. Evaluate the narrative of global food production since the 1950s.
3. Why is the 'food question' back on international and national political agendas?
4. Describe and evaluate some of the multiple sites of vociferous resistance to the global food system.
5. Compare and contrast neoliberal and reformist politics with reference to the food system.
6. Have individuals any role to play in challenging the injustices in the global food system?

Further reading

Magdoff, F. and Tokar, B. (eds) (2010) *Agriculture and Food in Crisis*, London, Monthly Review Press. A very timely discussion about the failure of the capitalist food system which includes a review of some positive changes too.

Holt-Giménez, E. and Patel, R. (2009) *Food Rebellions: Crisis and the Hunger for Justice*, Oakland, Food First Books. The authors detail the failures of the current food system and review some more promising alternatives.

Evans, A. (2009) *The Feeding of the Nine Billion: Global Food Security for the 21st Century*, London, Chatham House. This provides a very measured overview of the main issues and suggests some reforms at the international and national levels to reduce hunger: www.cic.nyu.edu/staff/docs/alx_evans/evans_9_billion.pdf (accessed 13 August 2011).

Motta, S. and Nilsen, A. G. (eds) (2011) *Social Movements in the Global South: Dispossession, Development and Resistance,* London, Palgrave Macmillan. A very interesting collection of chapters about resistance to globalization in the global south; interesting case studies as well.

Steel, C. (2008) *Hungry City*, London, Chatto and Windus. Fascinating reflections of food and cities.

Useful websites

http://viacampesina.org/en Via Campesina organizational material relevant to all chapters of this text (agrarian reform; food sovereignty; human rights; female empowerment, etc).

www.foodethicscouncil.org The website of the Food Ethics Council (United Kingdom). Excellent resources: see their recent report at: www.foodethicscouncil.org/system/files/FoodJustice_reportweb.pdf.

www2.ohchr.org/english/bodies/hrcouncil/docs/16session/A-HRC-16-40.pdf United Nations: Human Rights Council (2011) Study of the Human Rights Council Advisory Committee on Discrimination in the context of the right to food (accessed 13 August 2011).

http://uk.oneworld.net/guides/globalisation Very useful website with themed sections on issues addressed in this text.

www.grassrootsonline.org Grassroots International is a northern-based NGO that works to create a just and sustainable world by building alliance with progressive movements. Their primary focus is on land, water and food as human rights.

Chapter 9

Follow Up

1. A supplement to the recent WorldWatch Report 'Nourishing the Planet' (2011) reviews some initiatives to improve small-scale food production at: www.worldwatch.org/system/files/NtP-Innovations-in-Action.pdf (accessed 26 February 2011).
2. Select three from those detailed in the preceding Report and summarize their advantages for small-scale farmers and their potential contribution to reducing hunger in the developing world.
3. Read the article about the history and evolution of the Via Campesina Grassroots Peasant Movement by Marínez-Torres and Rosset (2010) at: www.informaworld.com/smpp/content~db=all~content=a918802179~frm=titlelink (accessed 1 March 2011) and then visit the online curriculum 'Food for Thought and Action', available at: www.grassrootsonline.org/publications/educational-resources/download-food-thought-action-a-food-sovereignty-curriculum. Then read about their alternative vision for feeding the world at: www.viacampesina.org/downloads/pdf/en/paper6-EN-FINAL.pdf (accessed 1 March 2011).
4. Learn how an NGO called Bridges to Understanding uses digital technology and the art of storytelling to empower and unite youth worldwide, enhance cross-cultural understanding and build global citizenship at: www.bridgesweb.org
5. Visit the International Planning Committee for Food Sovereignty at: www.foodsovereignty.org/Aboutus/WhatisIPC.aspx (accessed 15 August 2011). They identify five core issues. Select one of these core issues and summarize its main objectives to present to a seminar group.
6. Compare and contrast some examples of AFN from the United States and Europe at: USDA (2010) Local Food Systems: Concepts, Impacts, and Issues. Available at: http://permanent.access.gpo.gov/lps125302/ERR97.pdf

(accessed 10 August 2011). And, for examples of local food initiatives in Europe, see: www.faanweb.eu/sites/faanweb.eu/files/FAAN_Booklet_PRINT.pdf (accessed 13 August 2011).

7. Some effective and amusing material is available at this website. Adbusters are: a global network of culture jammers and creatives working to change the way information flows, the way corporations wield power and the way meaning is produced in our society'. Their spoof ads about food are very pertinent, see them at: www.google.co.uk/search?q=adbusters+spoof+ads&hl=en&source=hp&biw=1090&bih=762&gs_sm=c&gs_upl=250l192l1l0l4l185l9l9l0l3l3l0l234l1l06l2l0.4.2l6l0&oq=adbusters+&aq=5&aqi=g10&aql=

Bibliography

Abbott, C. and Phipps, T. (2009) Beyond Dependence and Legacy: Sustainable Security in Sub-Saharan Africa, at: www.oxfordresearchgroup.org.uk/sites/default/files/dependenceandlegacy.pdf (accessed 17 February 2011).

Abbott, E. (2010) *Sugar: A Bittersweet History*, New York, Overlook Press.

Abrahams, C. (2006) Globally Useful Conceptions of Alternative Food Networks: The Case of Johannesburg's Urban Food Supply System, available at: www.era.lib.ed.ac.uk/bitstream/1842/1465/1/cabrahams001.pdf.

Abrahams, C. (2007) 'Globally Useful Conceptions of Alternative Food Networks: The Case of Johannesburg's Urban Food Supply System' in Maye *et al.* ibid, pp. 95–114.

Action Against Hunger (2011) Undernutrition: What Works, at: www.actionagainsthunger.org.uk/fileadmin/contribution/0_accueil/pdf/Undernutrition%20What%20Works.pdf (accessed 25 July 2011).

Action Aid (2010) 'Her Mile: Women's Rights and Access to Land: The Last Stretch of Road to Eradicate Hunger', www.actionaid.it/filemanager/cms_actionaid/images/DOWNLOAD/Rapporti_DONNE_pdf/HerMile_AAItaly.pdf (accessed 12 January 2011).

Action Aid (2010) Meals Per Gallon: The Impact of Industrial Bio Fuels on People and Global Hunger, at: www.actionaid.org.uk/doc_lib/meals_per_gallon_final.pdf (accessed 26 October 2010).

Adams, W. M. (1990) *Green Development: Environment and Sustainability in a Developing World*, London, Routledge.

Agrawal, A. (2001) 'Common Property Institutions and Sustainable Governance of Resources', *World Development Report* 29(10): 1649–72.

Alpert, E., Smale, M. and Hauser, K. (2009) Investing in Poor Farmers Pays, at: www.oxfam.org/sites/www.oxfam.org/files/bp-129-investing-in-poor-farmers-summary.pdf (accessed 13 November 2009).

Altieri, M. (2001) Modern Agriculture: Ecological impacts and the Possibility of Truly Sustainable Farming, at: www.cnr.berkeley.edu/~agroeco3/modern_agriculture.html (accessed 29 January 2010).

Altieri, M. (2004) *Genetic Engineering in Agriculture: The Myths, Environmental Risks, and Alternatives*, Oakland, Food First Books.

Ambler-Edwards, S. *et al.* (2009) *Food Futures: Rethinking UK Strategy*, London: Chatham House, Institute of Foreign Affairs, available at: www.chathamhouse.org.uk/files/13248_r0109foodfutures.pdf (accessed on 13 November 2009).

Amin, S. (1997) *Capitalism in the Age of Globalization*, Zed Books, London.

Amnesty International (2009) 'Petroleum, Pollution and Poverty in the Niger Delta' London, Amnesty International, at: www.amnesty.org/en/library/asset/AFR44/017/2009/en/e2415061-da5c-44f8-a73c-a7a4766ee21d/afr440172009en.pdf (accessed 18 February 2011).

Anderson, K. and Valenzuela, E. (2010) Agricultural and Trade Policy Reforms in Latin America CEPAL Review 100 April 2010 online, at: www.eclac.org/publicaciones/xml/4/40534/RVI100Andersonetal.pdf (accessed 26 October 2010).

Anderson, S. (ed.) (2000) *Views from the South: Effects of Globalization and the WTO on Third World Countries*, Oakland, Calif., Food First Books and International Forum on Globalization.

Angus, I. (2008) 'Food Crisis: The Greatest Demonstration of the Historical Failure of the Capitalist Model', Global Research, at: http://globalresearch.ca/index.php?context=va&aid=8836 (accessed 24 September 2010).

Appelbaum, R. P. and Robinson, W. I. (eds) (2005) *Critical Gobalization Studies*. Routledge, Oxford, United Kingdom.

Atkins, P. and Bowler, I. (2001) *Food in Society*, London, Arnold.

Bacchetta, M. (2009) Globalization and Informal Jobs in Developing Countries, at www.rynekpracy.pl/pliki/pdf/34.pdf (accessed 16 January 2011).

Bailey, R. (2011) Growing a Better Future: Food Justice in a Resource-Constrained World, at www.oxfam.org.uk/resources/papers/downloads/growing-a-better-future-010611-en.pdf (accessed 21 May 2011).

Bank for International Settlements (2010) www.marketwatch.com/story/daily-currency-trading-turnover-hits-4-trillion-2010-09-01 (accessed 12 September 2010).

Baranyi, S., Deere, C. D. and Morales, M. (2004) Scoping Study on Land Policy Research in Latin America Online, at: www.landaction.org/gallery/LAND%20in%20LA.pdf (accessed 26 October 2010).

Barker, D. (2007) The Rise and Predictable Fall of Industrial Agriculture, at www.ifg.org/pdf/ag%20report.pdf (accessed on 29 January 2010).

Barraclough, S. (1991) *An End to Hunger?* London, Zed Press.

Barraclough, S. (1999) 'Land Reform in Developing Countries: The Role of the State and Other Actors', United Nations Research Institute for Social Development, Discussion Paper No. 101, at: www.unrisd.org/80256B3C005BCCF9/(httpAuxPages)/9B503BAF4856E96980256B66003E0622/$file/dp101.pdf (accessed 22 July 2011).

Barrett, C. and Maxwell, D. (2005) Food Aid After Fifty Years: Recasting Its Role, London, Routledge.

Barrientos, S. (2011) 'Labour Chains': Analysing the Role of Labour Contractors in Global Production Networks', Brooks World Poverty Institute, University of Manchester, available online at: www.capturingthegains.org/pdf/bwpi-wp-15311.pdf (accessed 13 July 2011).

Barrow, G. (2010) quoted at 'The F-word: Hunger in the Media' London Seminar Tackles Media's Hunger Coverage at: http://www.wfp.org/stories/london-seminar-tackles-media%E2%80%99s-hunger-coverag (accessed 5 January 2012).

Becker, G. S. (2008) 'Food and Agricultural Imports from China', CRS Report for Congress, at: www.fas.org/sgp/crs/row/RL34080.pdf (accessed 18 November 2010).

Bellamy Foster, J., Clark, B. and York, R. (2010) *The Ecological Rift: Capitalism's War on the Earth*, New York, Monthly Review Books.

Bernstein, H. (2000) 'Colonialism, Capitalism, Development', in T. Allen and A. Thomas (eds) *Poverty and Development into the 21 Century*, Oxford, OUP, pp. 241–70.

Bernstein, H., Crow, B. and Johnson, H. (eds) (1992) Rural Livelihoods: Crises and Responses, Oxford, Open University Press.

Blay-Palmer, A. (ed.) (2010) *Imagining Sustainable Food Systems*, London, Ashgate.

Borras, Jr. S. M. and Franco, J. (2009) *Transnational Agrarian Movements Struggling for Land and Citizenship Rights*, Institute for Development Studies Working Paper 323 at: www.ntd.co.uk/idsbookshop/details.asp?id=1105 (accessed 8 November 2010).

Borras, Jr. S. M. and Franco, J. (2010) *Towards a Broader View of the Politics of Global Land Grab: Rethinking the Issues, Reframing Resistance*, ICAS Working Paper Series No. 001 at: www.tni.org/sites/www.tni.org/files/Borras%20Franco%20Politics%20of%20Land%20Grab%20v3.pdf (accessed 5 October 2010).

Bramall, C. (2009) *Chinese Economic Development*, London, Routledge.

Broadhurst, T. (2011) 'Biofuels and Sustainability; A Case Study from Tanzania' in *PISCES (Policy Innovation Systems for Clean Energy Security*), Edinburgh, University of Edinburgh, available online at: www.pisces.or.ke/sites/default/files/Pisces%20Working%20Brief%20No.%203.pdf.

Brooks, R. and Maklakov, A. (2010) 'Sex Differences in Obesity Associated with Total Fertility Rate', *PLoS ONE* 5(5): e10587. doi:10.1371/journal.pone.0010587 at: www.plosone.org/article/info%3Adoi%2F10.1371%2Fjournal.pone.0010587 (accessed 15 January 2010).

Bujra, J. (2000) 'Diversity in pre-capitalist societies' in T. Allen and A. Thomas (eds) *Poverty and Development into the 21 Century*, Oxford, Oxford University Press, pp. 219–40.

Burns, J. and Thomson, N. (2008) Review of Nutrition and Growth among Indigenous Peoples, at: www.healthinfonet.ecu.edu.au/health-risks/nutrition/reviews/our-review (accessed 30 January 2011).

Cambio (Canada–Mexico Battling Childhood Obesity) (2010) Unraveling the Emerging Childhood Obesity Epidemic in Mexico: The Nutritional Transition, at: www.cambio-red.net/ENGLISH/EN-Project.php (accessed 12 December 2011).

Campbell, W. (2009) 'A Passion for Pungent Sauce', *Cultural Survival* 33(2), at: www.culturalsurvival.org/publications/cultural-survival-quarterly/uganda/passion-pungent-paste (accessed 30 January 2011).

Cappelle, J. (2009) Towards a Sustainable Cocoa Chain, available at: www.oxfam.org/sites/www.oxfam.org/files/towards-a-sustainable-cocoa-chain-0901.pdf (accessed 13 November 2009).

Carson, R. (1962) *Silent Spring*, Boston, Houghton Mifflin.

Caryn, N. A. (2006) 'Globally Useful Conceptions of Alternative Food Networks in the Developing South: The Case of Johannesburg's Urban Food Supply System', online papers archived by the Institute of Geography, School of Geosciences, University of Edinburgh, at: www.era.lib.ed.ac.uk/bitstream/1842/1465/1/cabrahams001.pdf (accessed 10 August 2011).

Case, A. and Menendez, A. 'Sex Differences in Obesity Rates in Poor Countries: Evidence from South Africa', *Economics and Human Biology*, pg: 271–82 at: www.princeton.edu/rpds/papers/Case_and_Menendez_EHB_Dec_2009.pdf (accessed 15 January 2011).

Chang, Ha-Joon (2007) *Bad Samaritans: The Guilty Secrets of Rich Nations and the Threat to Global Prosperity*, London, Random House.

Chossudovsky, M. (1997) *The Globalisation of Poverty: Impacts of IMF and World Bank Reforms*, London, Zed Books.

Christian Aid (2009) Growing Pains: The Possibilities and Problems of Bio Fuels, at: http://www.christianaid.org.uk/images/biofuels-report-09.pdf (accessed 5 January 2012).

Christodoulou, D. (1990) *The Unpromised Land: Agrarian Reform and Conflict Worldwide*, London, Zed Books.

Clay, J. (2008) Aquaculture: Greening the Blue Revolution, at: www.worldwildlife.org/what/globalmarkets/aquaculture/greeningthebluerevolution.html (accessed on 12 December 2011).

Cohen, M., Comrov, A. and Hoffner, B. (2005) 'The New Politics of Consumption: Promoting Sustainability in the American Marketplace', *Sustainability: Science, Practice & Policy* 1: 58–76.

Cole, D. C., Lee-Smith, D. and Nasinyama G. W. (eds) (2008) *Healthy City Harvests: Generating Evidence to Guide Policy on Urban Agriculture*, CIP/Urban Harvest and Makerere University Press, Kampala, Uganda, and Lima, Peru.

Colls, R. and Evans, B. (2010) 'Re-thinking 'the Obesity Problem', *Geography* 95(2): 99–105.

Corbridge, S. (2002) 'Third World Debt', in V. Desai and R. B. Potter (eds) *The Companion to Development Studies*, London, Hodder Education, pp. 477–80.

Cotula, L. (2009) Land Grab or Development Opportunity? Agricultural Investment and International Land Deals in Africa, at: www.ifad.org/pub/land/land_grab.pdf

Country Fact Sheets about Aquaculture, at: www.fao.org/fishery/naso-maps/fact-sheets/en/ (accessed 21 November 2010)

Cox, K. R. (2002) *Political Geography*, Oxford, Blackwell.

Craig, D. and Porter, D. (2006) *Development beyond Neoliberalism: Governance, Poverty and Political Economy*, London, Routledge.

Cray, C. (2010) 'ADM's New Frontiers: Palm Oil Deforestation and Child Labour', online at: www.corpwatch.org/article.php?id=15587 (accessed 5 January 2012)

Cray, C. (2010) ADM's New Frontiers: Palm Oil Deforestation and Child Labor, at: www.corpwatch.org/article.php?id=15587 (accessed 17 November 2010).

Crisis Group at: www.crisisgroup.org/en.aspx (accessed 15 February 2011).

Croll, E. (2001) 'Amartya Sen's 100 Million Missing Women', *Oxford Development Studies, 1469–9966*, 29(3): 225–44.

Crow, B. (1990) 'Moving the Lever: A New Food Aid Imperialism', in H. Bernstein, B. Crow and C. Martin (eds) *The Food Question: Profit versus People*, London, Earthscan.

Cuevas, A., Alvarez, V. and Olivos, C. (2009) 'The Emerging Obesity Problem in Latin America' *Expert Review of Cardiovascular Therapy* 7(3): 281–8, view summary at: www.expert-reviews.com/doi/abs/10.1586/14779072.7.3.281 (accessed 18 November 2010).

Dalle Mulle, E. (2010) Exploring the Global Food Supply Chain: Markets, Companies, Systems, available at: www.3dthree.org/pdf_3D/3D_ExploringtheGlobalFoodSupplyChain.pdf (accessed 27 July 2011).

Dalle Mulle, E. and Ruppanner, V. (2010) Exploring the Global Food Supply Chain Markets, Companies, Systems, at: www.3dthree.org/pdf_3D/3D_ExploringtheGlobalFoodSupplyChain.pdf (accessed 12 December 2011).

Daly, M, E. (1986) *The Famine in Ireland*, Dublin, Dublin Historical Association by Dundalgan Press.

Delpeuch, F., Maire, B., Monnier, E. and Holdsworth, M. (2009) *Globesity: A Planet Out of Control?* London, Earthscan.

Démurger, S., Sachs, J. D., Woo, W. T., Bao, S., Chang, G. and Mellinger, A. (2002) Geography, Economic Policy and Regional Development in China, at: http://web.cenet.org.cn/upfile/115169.pdf (accessed 7 November 2010).

Desmaris, A. A. (2007) *Globalization and the Power of Peasants: La Vía Campesina*, London, Pluto Ptress.

Devereux, S. (1993) *Theories of Famine*, London and New York, Harvester Wheatsheaf.

Devereux, S. (2000) 'Famines in the Twentieth Century', Sussex, Institute of Development Studies, available to download, at: www.ntd.co.uk/idsbookshop/details.asp?id=541 (accessed on 7 July 2010).

Devereux, S. (2001) 'Sen's Entitlement Approach: Critiques and Counter-critiques', *Oxford Development Studies* 29(3): 245–63, Taylor and Francis.

Devereux, S. (2007) (ed.) *The New Famines: Why Famines Persist in the Era of Globalization*, London, Routledge.

Devereux, S. (2007) *The New Famines: Why Famines Persist in an Era of Gobalization*, London, Routledge.

Diamond, L., Linz, J. and Lipset, S. (eds) (1988) *Democracy in Developing Countries: Africa*, London, Adamantine Press.

Dicken, P. (2003) '*Fabric-ating Fashion' in Global Shift*, London, Sage, pp. 317–54.

Dicken, P. (2007) '*We Are What We Eat' in Global Shift*, London, Sage, pp. 347–78.

Dicken, P. (2011) *Global Shift*, 6th edition, London, Sage.

Dower, N. and Williams, J. (2002) *Global Citizenship: A Critical Reader*, Cambridge, CUP.

Downes, G. (2004) 'TRIPs and Food Security: Implications of the WTO's TRIPs Agreement for Food Security in the Developing World', *British Food Journal* 106(5): 366–79, available at: www.emeraldinsight.com/journals.htm?articleid=870756&show=abstract (accessed 7 October 2010).

Dreze, J, and Sen, A. (1989) *Hunger and Public Action*, Oxford: Oxford University Press.

DuPuis, E. M. and Goodman, D. (2005) 'Should We Go ''Home'' to Eat?: Toward a Reflexive Politics of Localism', *Journal of Rural Studies* 21: 359–71, available at: http://cgirs.ucsc.edu/conferences/whitefood/foodx/papers/dupuis.pdf (accessed 16 December 2011).

Eagleton, D. (2005) Power Hungry: Six Reasons to Regulate Global Food Corporations, at: www.actionaid.org.uk/_content/documents/power_hungry.pdf (accessed 20 September 2010).

Ellis, F. (2000) 'The Determinants of Rural Livelihood Diversification in Developing Countries', *Journal of Agricultural Economics* 51: p. 289–302 at: http://onlinelibrary.wiley.com/doi/10.1111/j.1477-9552.2000.tb01229.x/pdf (accessed 28 July 2011).

El-Ojeili, C. and Hayden, P. (2006) *Critical Theories of Globalization*, Basingstoke, Palgrave Macmillian.

European Union (2007) Peru: Country Strategy Paper at www.eeas.europa.eu/peru/csp/07_13_en.pdf (accessed 21 January 2011).

Evans, A. (2009) *The Feeding of the Nine Billion: Global Food Security for the 21st Century*, London, Chatham House.

Evans, P. (2008) 'Is an Alternative Globalization Possible?' *Politics and Society* 36: pp. 271–305 at: http://pas.sagepub.com/content/36/2/271.abstract (accessed 21 May 2011).

Everington, S. (2010) The Dispossessed: I See Undernourished Children with Bloated Stomachs… Like in Africa' London Evening Standard, 3 March 2010. www.thisislondon.co.uk/standard/article-23811532-the-dispossessed-i-see-undernourished-children-with-bloated-stomachs-and-8201-like-in-africa.do (accessed 9 July 2010).

Fan, Shenggen, Omilola, B., Rhoe, V. and Salau, S. (2008) 'Towards a Pro-poor Agricultural Growth Strategy in Nigeria', International Food Policy Research Institute, at: www.ifpri.org/publication/towards-pro-poor-agricultural-growth-strategy-nigeria (accessed 10 January 2011).

FAO (2005) Rural Women and Food Security in Asia and the Pacific: Prospects and Paradoxes, at: ftp://ftp.fao.org/docrep/fao/008/af348e/af348e00.pdf (accessed 15 December 2011).

FAO (2006) The State of World Aquaculture, at: ftp://ftp.fao.org/docrep/fao/009/a0874e/a0874e04.pdf (accessed 25 November 2010).

FAO (2010) Countries in Protracted Crisis, at: www.fao.org/docrep/013/i1683e/i1683e03.pdf (accessed 26 December 2011).

Feagan, R. (2007) 'The Place of Food: Mapping Out the 'Local' in Local Food Systems', *Progress in Human Geography* 31: 23–42.

Ferguson, N. (2004) *Empire: How Britain Made the Modern World*, London, Penguin Books.

Food and Agricultural Organisation (2003) The Role of Aquaculture in Improving Food Security and Nutrition, at: www.fao.org/DOCREP/MEETING/006/Y8871e.HTM#P43_4335 (accessed 25 November 2010).

Food and Agricultural Organisation (2008) The State of World Fisheries and Aquaculture, at: www.fao.org/docrep/011/i0250e/i0250e00.htm (accessed 21 November 2010).

Food and Agricultural Organization (2001) Gender and Nutrition: Fact Sheet (2001) at: www.fao.org/sd/2001/pe0703a_en.htm (accessed 6 January 2011).

Food and Agricultural Organization (2009) The State of Food Insecurity www.fao.org/docrep/012/i0876e/i0876e00.htm.

Food and Agricultural Organization (FAO) (2008) The State of World Fisheries and Aquaculture, at: www.fao.org/docrep/011/i0250e/i0250e00.htm (accessed 21 November 2010).

Food and Agricultural Organization (FAO) (2010) 'Investing in Rural Women Contributes to Food Security' Gender, Insight, at: www.fao.org/gender/gender-home/gender-insight/gender-insightdet/en/?dyna_fef%5Buid%5D=46508 (accessed 15 January 2011).

Food and Agricultural Organization (FAO) (2010) The State of Food Insecurity in the World: Addressing Food Insecurity in Protracted Crises, at: www.fao.org/docrep/013/i1683e/i1683e.pdf (accessed 13 January 2011).

Food First Foundation Action Network (FIAN) (2010) 'Land Grabbing in Kenya and Mozambique: A Report on Two Research Missions and a Human Rights Analysis of Land Grabbing www.fian.org/resources/documents/others/land-grabbing-in-kenya-and-mozambique/pdf (accessed 12 January 2011).

Franco, J. (2008) 'Rural Democratisation: (Re) Framing Rural Poor Political Action', Transnational Institute, at: www.tni.org/paper/rural-democratisation-reframing-rural-poor-political-action (accessed 8 November 2010).

Fraser, A. (2009) *Harnessing Agriculture for Development*, Oxford, Oxfam, at: www.oxfam.org/sites/www.oxfam.org/files/bp-harnessing-agriculture-250909.pdf (accessed 21 January 2011).

French, H. (2000) Vanishing Borders: Protecting the Planet in an Era of Globalization [New York, W.W. Norton and Company, Worldwatch Institute]

French, P. and Crabbe, M. (2010) *Fat China: How Expanding Waistlines are Changing a Nation*, London: Anthem Press, summary online at: http://adage.com/china/article?article_id=145320 (accessed 18 November 2010).

Friedland, W. H. (1994) 'The New Globalization: The Case of Fresh Produce', in A. Bonanno, L. Busch, W. Friedland, L. Gouveia and E. Mingione (eds)

From Columbus to ConAgra: The Globalization of Agriculture and Food, Lawrence: University Press of Kansas, pp. 210–31.

Friedmann, H. (1990) 'The Origins of Third World Food Dependence', in H. Bernstein, B. Crow and C. Martin (eds) *The Food Question: Profit versus People*, London, Earthscan.

Friends of the Earth (2010) Africa; Up for Grabs. The Scale and Impact of Land Grabbing for Agro Fuels, at: www.foeeurope.org/agrofuels/FoEE_Africa_up_for_grabs_2010.pdf (accessed 12 July 2011).

Gamberoni, E. (2009) Trade Protection: Incipient but Worrisome Trends, at: www.voxeu.org/index.php?q=node/3183 (accessed 11 October 2010).

Gandhi, M. (1999) 'Factory Farming and the Meat Industry in India', in G. Tansey, and J. D'Silva (eds) *The Meat Business: Devouring a Hungry Planet*, London, Earthscan, pp. 92–100.

Gatrell, A. C. and Elliott, S. J. (2009) *Geographies of Health: An Introduction*, 2nd edition, Oxford, Wiley-Blackwell.

Gaus, A., Steets, J., Binder, A., Barrett, C. and lentz, E. (2011) How to Reform the Outdated Food Aid Convention, at: http://reliefweb.int/sites/reliefweb.int/files/resources/3A5B6357AA95D861C125785700480195-Full_Report.pdf.

George, S. (1999) A Short History of Neo-liberalism, at: www.globalexchange.org/campaigns/econ101/neoliberalism.html (accessed 9 September 2010).

Ghimire, K. B. (ed.) (2001) *Land Reform and Peasant Livelihoods: The Social Dynamics of Rural Poverty and Agrarian Reform in Developing Countries*, London, ITDG Publishing.

Ghosh, A. (2000) *The Glass Palace*, New York, HarperCollins.

Ghosh, J. (2011) Poverty Matters Blog, Guardian.co.uk, at: www.guardian.co.uk/global-development/poverty-matters/2011/may/03/food-insecurity-death-doha-trade-talks (accessed 27 June 2011).

Giddens, A. (1990) *The Consequences of Modernity*, Stanford: Stanford University Press.

Gist, G. (2008) *The History of Development: From Western Origins to Global Faith*, 3rd Edition, London, Zed Books.

Goodman, D. (2004) 'Rural Europe Redux? Reflections on Alternative Agro-food Networks and Paradigm Change', *Sociologia Ruralis* 44(1): 3–16.

Goodman, D. (2009) 'Place and Space in Alternative Food Networks: Connecting Production and Consumption', Environment, Politics and Development Working Paper Series, Department of Geography, King's College London paper no 21 at: www.kcl.ac.uk/sspp/departments/geography/research/epd/GoodmanWP21.pdf (accessed 10 August 2011).

Goodman, D. and Goodman, M. K. (2007) Alternative Food Networks' Entry for the Encyclopaedia of Human Geography, available at: www.kcl.ac.uk/content/1/c4/98/59/DGMKGAFNJulyFinal.pdf (accessed 9 August 2011).

Goodman, D., DuPuis, E. M. and Goodman, M. (2011) *Alternative Food Movements: Knowledge, Practice and Power*, London, Routledge.

Goodman, M. (2008) 'Towards Visceral Entanglements: Knowing and Growing Economic Geographies of Food', Environment, Politics and Development

Working Paper Series, Paper 5 Department of Geography, King's College London, at: www.kcl.ac.uk/sspp/departments/geography/research/epd/GoodmanWP5.pdf (accessed 12 July 2011).

Goodman, Z. (2009) Seeds of Hunger: Intellectual Property Rights on Seeds and the Human Rights Response, at: www.3dthree.org/pdf_3D/3D_THREAD2seeds.pdf.

Gragnolati, M., Shekar, M., Das Gupta, M., Bredenkamp, C. and Lee, Yi-Kyoung (2005) 'India's Under Nourished Children: A Call for Reform and Action', World Bank Health, Nutrition and Population (HNP) Discussion Paper http://siteresources.worldbank.org/SOUTHASIAEXT/Resources/223546-1147272668285/IndiaUndernourishedChildrenFinal.pdf (accessed 24 January 2011).

Grain (2009) Corporate Candyland: The Looming GM Sugar Cane Invasion, at: www.grain.org/seedling/?id=589 (accessed 28 September 2010).

Grain (2010) 'Global Agribusiness: Two Decades of Plunder, at: www.grain.org/seedling/?id=693 (accessed 27 September 2010).

Grain (2010a) World Bank Report on Land Grabbing: Beyond the Smoke and Mirrors at www.grain.org/articles/?id=70 (accessed 28 September 2010).

Greenpeace (2008) Greenhouse Gas Emissions and Agriculture, at: www.greenpeace.org/canada/en/recent/agriculture-and-climate-change (accessed 13 November 2009).

Greenpeace (2009) Agriculture at a Crossroads, at: www.greenpeace.org/international/press/reports/agriculture-at-a-crossroads-report (accessed 13 November 2009).

Gudynas, E. (2008) 'The New Bonfire of Vanities: Soybean Cultivation and Globalization in South America', *Development* 51: 512–8, www.agropecuaria.org/publicaciones/GudynasBonfireVanitiesSoybean08.pdf (accessed 2 November 2010).

Gündüz, Zuhal Yeşilyurt (2011) 'Water – On Women's Burdens, Humans' Rights, and Companies' Profits', *Monthly Review* 62(8), available at: http://monthlyreview.org/2011/01/01/water-on-womens-burdens-humans-rights-and-companies-profits (accessed 15 December 2011).

Gurian-Sherman, D. (2008) 'CAFOs Uncovered: The Untold Costs of Confined Animal Feeding Operations' published by the Union of Concerned Scientists of United States of America (UNCUSA) at: www.ucsusa.org/assets/documents/food_and_agriculture/cafos-uncovered.pdf (accessed 12 November 2009).

Gwynne, R. N., Klak, T. and Shaw, D. J. B. (2003) *Alternative Capitalisms*, London, Arnold.

Halweil, B. and Gardiner, G. (2000) *Underfed and Overfed: The Global Epidemic of Malnutrition*, Washington, Worldwatch Institute.

Halweil, B. and Nierenberg, D. (2008) *Meat and Seafood: The Global Diet's Most Costly Ingredients' in State of the World*, Washington, World Watch Institute.

Harris, D., Moore, M. and Schmitz, H. (2009) 'Country Classifications for a Changing World' Institute for Development Studies (IDS, University of Sussex) Working Paper no 326 at: www.ntd.co.uk/idsbookshop/details.asp?id=1114 (accessed 8 November 2010).

Harris, E. (2010) 'Eat Local? Constructions of Place in Alternative Food Politics', *Geography Compass* 4(4): 355–69 at: http://edmundharris.com/wp/wp-content/uploads/2011/06/Harris-2010-Eat-Local.pdf.

Hartwick, E. (1998) 'Geographies of Consumption: A Commodity Chain Approach', *Environment and Planning D: Society and Space* 16: 423–37.

Harvey, D. (1990) 'Between Space and Time: Reflections on the Geographical Imagination', *Annals of the Association of American Geographers* 80: 418–34.

Harvey, D. (1990) *The Condition of Postmodernity: An Enquiry into the Origins of Cultural Change*, Cambridge, MA: Blackwell.

Hawkes, C. (2006) 'Uneven Dietary Development: Linking the Policies and Processes of Globalization with the Nutritional Transition, Obesity and Diet-related Chronic Diseases', in Globalization and Health, at: www.globalizationandhealth.com/content/2/1/4 (accessed 19 August 2010).

Held, D., McGrew, A., Goldblatt, D. and Perraton, J. (1999) *Global Transformations – Politics, Economics and Culture*, Cambridge: Polity Press.

Hermandez-Triana, M. (2010) 'Cuba: Gender and Geography Influence Childhood Obesity', *MEDICC Review*, 12(2).

Hoddinott, J. and Cohen, M. (2007) Renegotiating the Food Aid Convention: Background, Context and Issues. Washington D.C., IFPRI, Food Consumption and Nutrition Division.

Holmes, K. (2001) Carnivorous Cravings: Charting the World's Protein Shift, at: http://earthtrends.wri.org/features/view_feature.php?theme=8&fid=24 (accessed on 28 October 2009).

Holt-Gimenez, E. (2008) 'The World Food Crisis', Institute for Food and Development Policy, Policy Brief No. 16.

Holt-Gimenez, E. and Patel, R. (2009) *Food Rebellions: Crisis and the Hunger for Justice*, Oxford, Fatamu Books.

Holt-Giménez, E. and Shattuck, A. (2011) 'Food Crises, Food Regimes and Food Movements: Rumblings of Reform or Tides of Transformation?', *Journal of Peasant Studies* 38(1): 109–44 at: http://pdfserve.informaworld.com/760268__932368147.pdf (accessed 1 March 2011).

Hoogeveen, J. G. (2003) Measuring Welfare for small Vulnerable Groups Poverty and Disability in Uganda, at: www.tinbergen.nl/files/papers/Hoogeveen.pdf (accessed 12 December 2011).

Hoyos, C. and Blas, J. (2008) 'Security fears over food and fuel crisis', *Financial Times*, June 20. Available at: <http://www.ft.com/cms/s/0/29cff8ec-3ef4-11dd-8fd9-0000779fd2ac.html?nclick_check=1 (accessed 5 January 2012).

Icaza, R., Newell, P. and Saguier, M. (2009) 'Democratising Trade Politics in the America's: Insights from the Women's Environmental and Labour Movements', IDS Working Paper, No 328 (Institute of Development Studies, Sussex) at: www.ntd.co.uk/idsbookshop/details.asp?id=1119 (accessed 19 January 2011).

Ikelegbe, A. (2005) 'The Economy of Conflict in the Oil Rich Niger Delta Region of Nigeria', *Nordic Journal of African Studies* 14(2): 208–34, available at: www.sweetcrudemovie.com/pdf/njas2005.pdf (accessed 2 February 2011).

Intermón Oxfam (2005) Goliat Contra David. Quién gana y quién pierde con la PAC en Españya en los países pobres, at: https://doc.es.amnesty.org/cgi-bin/ai/BRSCGI/Informe?CMD=VEROBJ&MLKOB=29913365656 (accessed 12 December 2011).

International Assessment of Agricultural Knowledge, Science and Technology for Development (2009) Agriculture at the Crossroads, at: www.agassessment.org/reports/IAASTD/EN/Agriculture%20at%20a%20Crossroads_Synthesis%20Report%20(English).pdf (accessed 19 January 2011).

International Food Policy Research Institute [IFPRI] (2009) 'When Speculation Matters', Issue Brief 57 February, 2009, at: www.ifpri.org/sites/default/files/publications/ib57.pdf (accessed 6 December 2011).

International Fund for Agricultural Development (2011) Rural Poverty Report 2011: New Realities, New Challenges, New Opportunities, at: www.ifad.org/rpr2011/report/e/rpr2011.pdf (accessed 10 January 2011).

International Labor Office (ILO) and United Nations Development Program (2009) 'Decent Work in Latin America and the Caribbean: Work and Family: Towards a New Form of Reconciliation with Social Co-responsibility', Geneva, ILO and UNDP, www.undp.org/publications/pdf/undp_ilo.pdf (accessed 11 January 2011).

International NGO/CSO Planning Committee (IPC) at: www.foodsovereignty.org/new/index.php (accessed 13 November 2009).

International Obesity Task Force www.iotf.org/globalepidemic.asp (accessed 9 July 2010).

International River 'Ethiopia's Gibe 111 Dam' at: www.internationalrivers.org/files/Gibe3Factsheet2011.pdf (accessed 15 February 2011).

Jacobs, S. (2002) 'Land Reform: Still a Goal worth Pursuing for Rural Women?' *Journal of International Development Studies* 14: 887–98.

Jarosz, L. (2009) 'Energy, Climate Change, Meat, and Markets: Mapping the Coordinates of the Current World Food Crisis', *Geography Compass*, 3: 2065–83, online at: http://onlinelibrary.wiley.com/doi/10.1111/j.1749-8198.2009.00282.x/abstract;jsessionid=D8E4EC97BF3BFF8414FF083CC9A9F21F.d02t02 (accessed 25 October 2010).

Jensen, M. F. and Gibbon, P. (2007) 'Africa and the WTO Doha Round', *Development Policy Review* 25(1): 5–24.

Johnson, R. J., Taylor, P. J. and Watts, M. (2002) *Geographies of Global Change: Remapping the World*, London, Blackwell.

Josling, T. (2009) 'Agricultural Trade Disputes in the WTO' in Professor Hamid Beladi and Professor E. Kwan Choi (eds) '*Trade Disputes and the Dispute Settlement Understanding of the WTO: An Interdisciplinary Assessment', Frontiers of Economics and Globalization, Volume 6*, Emerald Group Publishing Limited, pp. 245–82 www.emeraldinsight.com/books.htm?chapterid=1775985&show=abstract (accessed 9 September 2010).

Kay, C. (2002) 'Why East Asia overtook Latin America: Agrarian Reform, Industrialisation and Development', *Third World Quarterly* 23(6): 1073–102, online at: www.alasru.org/textos/Kay-TWQ23(6)2002.pdf (accessed 26 October 2010).

Keen, D. (2008) *Complex Emergencies*, London, Polity.

Kelly, B., Halford, J. C., Boyland, E. J., Chapman, K., Bautista-Castaño, I., Berg, C., Caroli, M., Cook, B., Coutinho, J. G., Effertz, T., Grammatikaki, E., Keller, K., Leung, R., Manios, Y., Monteiro, R., Pedley, C., Prell, H., Raine, K., Recine, E., Serra-Majem, L., Singh, S. and Summerbell, C. (2010) 'Television Food Advertising to Children: A Global Perspective' *American Journal of Public Health* 100(9): 1730–6.

Khera, R. and Nayak, N. (2009) 'Women Workers and Perceptions of the National Rural Employment Guarantee Act', *Economic and Political Weekly* 44(43): 24–30; Oct., pp. 49–57.

Khor, M. (2001) Re-thinking Trips in the WTO – NGOs Demand Review and Reform of TRIPS at Doha Ministerial Conference, available at: www.twnside.org.sg/title/joint5.htm (accessed 7 October 2010).

Khor, M. Destructive Impacts and Resistance from Local Communities, at: www.twnside.org.sg/title/aqua-ch.htm (accessed 21 November 2010).

Kimbrell, A. (2002) *Fatal Harvest: The Tragedy of Industrial Farming*, Washington, DC: Island Press.

Ki-moon, B. (2009) Secretary-General's Message on the International Day of Rural Women, at: www.un.org/apps/sg/sgstats.asp?nid=4161 (accessed 16 December 2011).

Klein, N. (2000) *No Logo*, London, Flamingo.

Kneafsey, M., Cox, R., Holloway, L., Dowler, E., Venn, L. and Tuomainen, H. (2008) *Reconnecting Consumers, Producers and Food: Exploring Alternatives*, London, Berg.

Knox, J., Agnew, J. and McCarthy, L. (2003) *The Geography of the World Economy*, 3rd edition, London, Arnold.

Knox, P., Agnew, J. and McCarthy, L. (2008) *The Geography of the World Economy*, London, Hodder Arnold.

Knudsen, M. T., Halberg, N., Olesen, J. E., Byrne, J., Iyer, V. and Toly, N. (2006) 'Global trends in agriculture and food systems', in Halberg, Niels; Alrøe, Hugo Fjelsted; Knudsen, Marie Trydeman and Kristensen, Erik Steen (Eds.) *Global Development of Organic Agriculture—Challenges and Prospects*, Wallingford, Oxfordshire: CABI Publishing. p. 1–48.

Kurlansky, M. (1999) *Cod: A Biography of the Fish that Changed the World*, London, Vintage.

Lang, C. (2008) Plantations, Poverty and Power (World Rainforest Movement), available at: www.wrm.org.uy/publications/Plantations_Poverty_Power.pdf (accessed 2 November 2010).

Lang, T. (2008) Interview about World Food Crisis, available at: www.grain.org/seedling/?id=553.

Lang, T. (2010) 'Food Standards Agency: What a Carve Up' *The Guardian*, 21 July, 2010, at: www.guardian.co.uk/commentisfree/2010/jul/21/fsa-what-a-carve-up (accessed 27 July 2011).

Lang, T. and Heasman, M. (2004) Food Wars. The Global Battle for Mouths, Minds and Markets, London: Earthscan.

Lang, T. and Rayner, G. (2002) Why Health is the Key to the Future of Farming and Food, available at: www.agobservatory.org/library.cfm?refID=30300 (accessed 3 March 2011).

Lang, T. and Rayner, G. (2005) 'Obesity: A Growing Issue for European Policy?' *Journal of European Social Policy,* November 15(4): 301–27.

Lang, T., Barling, D. and Caraher, M. (2009) *Food Policy: Integrating Health, Environment and Society*, Oxford, Oxford University Press.

Lapierre, D. (1985) *City of Joy*, New York, Time Warner Books.

Lappé, F. M. and Collins, J. (1986) *World Hunger: Twelve Myths*, New York, Grove Press.

Larrea, C. and Friere, W. (2002) 'Social Inequality and Child Malnutrition in Four Andean Countries', *Pan American Journal of Public Health Public* 11(5/6), 2002 at: www.scielosp.org/pdf/rpsp/v11n5-6/10720.pdf (accessed 24 January 2011).

Lawrence, F. (2008) *Eat Your Heart Out*, London, Penguin.

Lee, R. (2007) Food Security and Food Sovereignty, online at: www.ncl.ac.uk/cre/publish/discussionpapers/pdfs/dp11%20Lee.pdf

Lee-Smith, D. (2010) Cities Feeding People: An Update on Urban Agriculture in Equatorial Africa, at: http://eau.sagepub.com/content/22/2/483.full.pdf+html (accessed 3 February 2011).

Leturque, H. and Wiggins, S. (2010) Ghana's Story: Ghana's Sustained Agricultural Growth: Putting Underused Resources to Work, Overseas Development Institute, at: http://www.developmentprogress.org/progress-stories/ghanas-sustained-agricultural-growth-putting-underused-resources-work?page=2 (accessed 5 January 2012).

Leturque, H. and Wiggins, S. (2010) Thailand's Story: Thailand's Progress in Agriculture: Transition and Sustained Productivity Growth, at www.developmentprogress.org/sites/default/files/thailand_agriculture.pdf (accessed 7 November 2010).

Levine, S. (2011) 'Here We Go Again: Famine in the Horn of Africa', see Overseas Development Institute Blog at: http://blogs.odi.org.uk/blogs/main/archive/2011/07/06/horn_of_africa_famine_2011_humanitarian_system.aspx (accessed 12 December 2011).

Liou, Tsan-Hon, Pi-Sunyer, Xavier, F., Laferre`re, Blandine (2005) Physical Disability and Obesity, at: http://onlinelibrary.wiley.com/doi/10.1111/j.1753-4887.2005.tb00110.x/pdf (accessed 1 February 2011).

LIAMBI, L. (1990) 'Transitions to and within Capitalism: Agrarian Transitions in Latin America', *Sociologia Ruralis* 30: 174–96, available online at: http://onlinelibrary.wiley.com/doi/10.1111/j.1467-9523.1990.tb00408.x/pdf (accessed 26 October 2010).

Lotter, D. (2009) 'The Genetic Engineering of Food and the Failure of Science – Part 1: The Development of a Flawed Enterprise', *International Journal of Sociology of Agriculture and Food*, available at: www.ijsaf.org/archive/16/1/lotter1.pdf (accessed 7 December 2011).

Lucas, L. (2010) 'Pet Food Sales Forecast to Rise by 2.5%', *Financial Times*, at: http://www.ft.com/cms/s/0/f0cd8b8c-da40-11df-bdd7-00144feabdc0.html#axzz1igBJHzKh (accessed 5 January 2012).

MacDonald, M. and Iyer, S. (2009) Skillful Means: The Challenge of China's Encounter with Factory Farming, at: http://www.brightergreen.org/files/china_bg_pp_2011.pdf (accessed 5 January 2012).

MacGregor, N. (2010) History of the World in 100 Objects: The Tea Set, at: www.bbc.co.uk/podcasts/series/ahow (accessed 28 July 2011).

Mackenzie, F. D. (2010) 'Gender, Land Tenure and Globalisation: Exploring the Conceptual Ground', in D. Tsikata and P. Golah (eds) *Land Tenure, Gender and Globalisation: Research and Analysis from Africa, Asia and Latin America*, Ottawa, International Development Research, ibid., pp. 35–69.

Madeley, J. (2000) *Hungry for Trade*, London, Zed Books.

Madeley, J. (2002) *Food for All*, London Zed Books.

Madeley, J. (ed.) (1999) Hungry for Power, available at: www.ukfg.org.uk/docs/Hungry%20For%20Power.pdf (accessed 20 September 2010).

Magalhaes, S. B. and Hernandez, F. (2009) Experts Panel Assesses Belo Monte Dam Viability, at: www.internationalrivers.org/files/EXEC%20SUMMARY%20ENGLISH_0.pdf (accessed 13 November 2009).

Magdoff, F. (2004) A Precarious Existence: The Fate of Billions? at: www.monthlyreview.org/0204magdoff.htm (accessed 12 November 2009).

Magdoff, F. (2008) 'The World Food Crisis: Sources and Solutions', *Monthly Review*, at: www.monthlyreview.org/080501magdoff.php (accessed 12 November 2009).

Magdoff, F. and Tokar, B. (2010) *Agriculture and Food in Crisis: Conflict, Resistance, and Renewal*, New York, Monthly Review Books.

Magdoff, F., Bellamy Foster, J. and Buttel, F. (eds) (2000) *Hungry for Profit*, New York, Monthly Review Press.

Malloch Brown, M. (2002) The Millennium Development Goals and Africa: A New Framework for a New Future, available at: content.undp.org/go/newsroom/2002/november/mmb-uganda.en;jsessionid=axbWzt8vXD9 (accessed 11 October 2010).

Markkanen, P. (2009) *Shoes, Glues and Homework: Dangerous Work in the Global Footwear Industry*, New York, Baywood Publishing Company.

Martinelli, L. and Filosos, S. (2008) 'Expansion of Sugarcane Ethanol Production in Brazil: Environmental and Social Challenges', *Ecological Applications* 18(4): 885–98, online at: www.wilsoncenter.org/news/docs/brazil.martinelli.filoso.sugarcane.production.pdf (accessed 25 October 2010).

Martínez, C. J. (2006) Study of the Problem of Discrimination Against Indigenous Populations, at: http://www.un.org/esa/socdev/unpfii/en/spdaip.html (accessed 12 December 2011).

Martinez-Alier (2000) Environmental Justice, Sustainability and Valuation http://ecoethics.net/hsev/200003txt.htm (accessed 21 November 2010).

Martínez-Torres, M. E. and Rosset, P. (2010) 'La Via Campesina: The Birth and Evolution of a Transnational Social Movement', *Journal of Peasant Studies* 37(1) at: www.stwr.org/food-security-agriculture/la-via-campesina-the-birth-and-evolution-of-a-transnational-social-movement.html (accessed 7 October 2010).

Matondi, P., Havnevik, K. and Beyene, A. (2011) *Biofuels, Land Grabbing and Food Security in Africa*, London, Zed Books.

Maxey, L. (2007) *From 'Alternative' to 'Sustainable'* in Maye *et al*., ibid, pp. 55–76.

Maxwell, S. and Slater, R. (2004) 'Food Policy Old and New', in Maxwell, S. and Slater, R. (eds) *Food Policy Old and New*, Overseas Development Institute, Oxford, Blackwell Publishing, pp. 1–19.

Maye, D., Holloway, L. and Kneafsey, M. (2007) *Alternative Food Geographies: Representation and Practice*, Oxford, Elsevier.

Mayhew, H. (2010) *London Labour and the London Poor*, Douglas-Fairhurst (ed.), Oxford, Oxford University Press.

McDonald, B. (2010) *Food Security: Addressing Challenges from Malnutrition, Food Safety and Environmental Change*, London, Polity Press.

Mehta, A. G. (2007) 'The Global Market for Agricultural Machinery and Equipment', *Business Economics*, (October 1, 2007) at: www.allbusiness.com/economy-economic-indicators/economic-indicators/5497022-1.html (accessed 4 October 2010).

Messer, E. (2000) Armed Conflict and Hunger – The Extent of Armed Conflict, at: www.worldhunger.org/articles/fall2000/messer2.htm (accessed 16 December 2011).

Messer, E. and Cohen, M. (2004) 'Breaking the links between Conflict and Hunger', *IFPRI Briefing Paper*, at: www.ifpri.org/sites/default/files/publications/ib26.pdf (accessed 15 February 2011).

Messer, E. and Cohen, M. (2006) 'Conflict, Food Insecurity and Globalization', *International Food Policy Research Institute*, at: www.ifpri.org/sites/default/files/publications/fcndp206.pdf (accessed 10 August 2011).

Micheletti, M. (2003) *Political Virtue and Shopping: Individuals, Consumerism, and Collective Action*, New York, Palgrave Macmillan.

Miller, A. (2010) New Wave of Agricultural Land Grab Reaches Canada http://farmlandgrab.org/15808 (accessed 28 September 2010).

Millstone, E. and Lang, T. (2003) *The Atlas of Food: Who Eats What, Where and Why*, London, Earthscan.

Mintz, S. W. (1986) *Sweetness and Power*, London, Penguin.

Minxin, Pei (2009) 'China's Not a Superpower', *The Diplomat*, at: www.carnegieendowment.org/publications/index.cfm?fa=view&id=24404 (accessed 19 November 2010).

Monbiot, G. (2009) 'We Cannot Fight Climate Change with Consumerism', *The Guardian*, 8 November, 2009 online at: www.countercurrents.org/monbiot081109.htm.

Moncrieffe, J. (2005) Beyond Categories: Power, Recognition and the Conditions of Equity', Background Paper for the World Development Report 2006.

Monteiro, C. A., Moura, E. C, Wolney, L. C. and Popkin, B. M. (2004) 'Socioeconomic Status and Obesity in Adult Populations of Developing Countries: A Review', *Bulletin World Health Organ* 82(12), Genebra December 2004 at: www.scielosp.org/scielo.php?pid=S0042-96862004001200011&script=sci_arttext&tlng=en (accessed 15 January 2011).

Montgomery, M. (2009) 'Urban Poverty and Health in the Developing Countries', at: *Population Reference Bulletin* 64(2) www.prb.org/pdf09/64.2urbanization.pdf (accessed 12 December 2011).

Morgan, K. (2010) Feeding the City: The Challenge of Urban Food Planning, at: api.ning.com/files/…/ipseditorial.doc (accessed 9 August 2011).

Morgan, K. and Sonnino, R. (2008) *The School Food Revolution: Public Food and the Challenge of Sustainable Development*, London, Earthscan.

Morgan, K., Marsden, T. and Murdoch, J. (2006) *Worlds of Food: Place, Power, and Provenance in the Food Chain*, Oxford, Oxford University Press.

Mousseau, F. (2005) Food Aid or Food Sovereignty: Ending Hunger in Our Time: at: http://media.oaklandinstitute.org/content/food-aid-or-food-sovereignty-ending-world-hunger-our-time-0 (accessed 27 July 2011).

Mulle, Dalle E. and Ruppanner, V. (2010) Exploring the Global Food Supply: Companies, Markets, Systems, at: www.3dthree.org/pdf_3D/3D_ExploringtheGlobalFoodSupplyChain.pdf.

Murphy, S. (2002) Europe's Double Standards; How the EU Should Reform Its Policies with the Developing World, at: www.oxfam.org.uk/resources/policy/trade/downloads/bp22_eutrade.pdf (accessed 11 October 2010).

Nally, D. (2011) 'The Biopolitics of Food Provisioning', *Transaction of British Geographers* 36: 37–53.

National Trust, UK Food and Farming 2009, at: http://www.nationaltrust.org.uk/main/w-chl/w-countryside_environment/w-food_farming.htm (accessed 13 November 2009).

Nederveen Pieterse (2010) *Development Theory*, 2nd Edition, London, Sage.

Nierenberg, D. (2005) Happier Meals: Rethinking the Global Meat Business, Worldwatch Paper, 171, available at: www.worldwatch.org/node/819.

Nissanke, M. and Thorbecke, E. (2010) *The Poor Under Globalization in Asia, Latin America, and Africa*, Oxford, OUP.

Nuetzenadel, A. and Trentmann, F. (eds) (2008) *Food and Globalization: Consumption, Markets and Politics in the Modern World*, London, Berg.

Nwanze, K. F. (2010) 'President's Foreword', in IFAD (2010) Annual Report, at: www.ifad.org/pub/ar/2010/e/index.htm (accessed 16 December 2011).

O'Grada, C. (2006) *Ireland's Great Famine: Interdisciplinary Perspectives*, Dublin, University College Dublin Press.

O'Grada, C. (2009) *Famines: A Short History*, Princeton, Princeton University Press.

Obi, C. and Rustad, S. A. (2011) *Oil and Insurgency in the Niger Delta*, London, Zed Books.

Oddy, D. J., Atkins, P. J. and Amilien, V. (2009) *The Rise of Obesity in Europe: A Twentieth Century Food History*, London, Ashgate.

Oniang'o, R. and Mukudi, E. (2002) 'Nutrition and Gender', in Nutrition: Foundation for Development, Geneva, ACC/SCN, at: www.ifpri.org/sites/default/files/pubs/pubs/books/intnut/intnut_07.pdf (accessed 6 January 2011).

Organisation for Economic Cooperation and Development (OECD) (2006) 'Promoting Pro-Poor Growth: Agriculture, at: www.oecd.org/dataoecd/9/60/37922155.pdf (accessed 10 January 2011).

Osterhammel, J. and Petersson, N. P. (2005) *Globalization: A Short History*, Princeton, University of Princeton Press.

Oxfam (2008) Another Inconvenient Truth: How Bio Fuel Policies Are Deepening Poverty and Accelerating Climate Change, at: www.oxfam.org.uk/resources/policy/climate_change/bp114_inconvenient_truth.html (accessed 6 January 2011).

Oxfam (2011) Grow, Life, Planet: Growing a Better Future: Food Justice in a Resource-Constrained World, at: www.oxfam.org.uk/resources/papers/downloads/cr-growing-better-future-170611-en.pdf (accessed 19 June 2011).

Palmer, R. (2010a) 'Would Cecil Rhodes have signed a Code of Conduct? Reflections on Global Land Grabbing and Land Rights in Africa, Past and Present', Paper Presented at African Studies Association of the United Kingdom, Biennial conference, Oxford, 16–19 September 2010, available at: www.oxfam.org.uk/resources/learning/landrights/downloads/would_cecil_rhodes_have_signed_a_code_of_conduct.pdf.

Palmer, R. (2010b) Annotated Bibliography on Land Grab, at: www.oxfam.org.uk/resources/learning/landrights/downloads/annotated_guide_to_bibliogs_biofuels_africanlandrights_global_land_grabbing_sept_2010.pdf (accessed 13 November 2009).

Panitchpakdi, S. (2010) UNCTAD Promotes Responsible Foreign Direct Investment in Agriculture, at: www.unctad.org/templates/webflyer.asp?docid=14032&intItemID=5267&lang=1 (accessed 18 July 2011).

Patel, R. (2007) *Stuffed and Starved*, London, Portobello Books.

Patel, R. (2010) Down with the Clown: Why Ronald McDonald Has No Business Talking to Children at http://globalpolicy.org/social-and-economic-policy/world-hunger/agribusiness-companies/48942.html (accessed 12 September 2010).

Pearson, R. (1992) 'Gender Matters in Development' in T. Allen and A. Thomas (eds) (1992) *Poverty and Development in the 1990s*, pp. 291–312.

Perry, K. E. G. (2010) 'Secrets and Lies: Tackling HIV among Sex Workers in India', *The Guardian*, 7 December, at: www.guardian.co.uk/global-development/2010/dec/07/india-prostitution-hiv (accessed 16 December 2011).

Philpott, T. (2010) Haiti, U.S. Agricultural Policy Reform, and Bill Clinton, at: www.energybulletin.net/node/52366 (accessed 24 September 2010).

Pikerman, A. (2002) The Iberian Golden Age, 1400–1650, online at: http://history-world.org/iberian_golden_age.htm.

Pollan, M. (2006) *The Omnivore's Dilemma*, London, Penguin.

Pollan, M. (2008) *In Defence of Food*, London, Penguin.

Popkin, B. (2004) 'The Nutritional Transition in the Developing World' in S. Maxwell and R. Slater (eds) *Food Policy Old and New*, London, Blackwell, pp. 43–56.

Popkin, B. (2009) *The World is Fat: The Fads, Trends, Policies and Products that Are Fattening the Human Race*, Avery, Penguin.

Potter, D. (2000) The Poer of Colonial States', in T. Allen and A. Thomas (eds) *Poverty and Development into the 21st Century*, Oxford, Oxford University Press, pp. 271–88.

Potter, R. B., Binns, T., Elliott, J. A. and Smith, D. (2004) *Geographies of Development*, 2nd Edition, Harlow, Pearson.

Potts, L. (1990) *The World Labour Market: A History of Migration*, London, Zed Books.

Prakask, A. (ed.) (2011) Safeguarding Food Security in Volatile Global Markets, at: www.fao.org/docrep/013/i2107e/i2107e.pdf (accessed 12 December 2011).

Pretty, J. (2005) Local Food 'Greener than Organic' at: http://news.bbc.co.uk/1/hi/sci/tech/4312591.stm (accessed 10 August 2011).

Pretty, J. (2005) *The Earthscan Reader in Sustainable Agriculture*, London, Earthscan.

Rachman, G. (2010) 'End of the World as We Knew It', *Financial Times*, 23–24 October, 2010, Life and Arts, p. 1.

Rainforest Action Network (2009) http://ran.org/campaigns/rainforest_agribusiness/ (accessed 13 November 2009).

Rapley, J. (1996) *Understanding Development: Theory and Practice in the Third World*, 3rd Edition (2007), Boulder Colorado, Lynne Rienner Publishers.

Rayner, G., Hawkes, C., Lang, T. and Bello, W. (2007) Trade Liberalisation and the Diet Transition: A Public Health Response', Health Promotion International, available at: http://heapro.oxfordjournals.org/cgi/content/full/21/suppl_1/67 (accessed 29 January 2010).

Reeves, H. and Baden, S. (2000) Gender and Development: Concepts and Definitions, BRIDGE (development – gender), BRIDGE Reports – 55, Institute of Development Studies, Brighton UK. www.bridge.ids.ac.uk// bridge/reports/re55.pdf.

Rigg, J. (1991) *Southeast Asia: A Region in Transition*, London, Unwin Hyman.

Roberts, I. and Edwards, P. (2010) *The Energy Glut: The Politrics of Fatness in an Overheating World*, London, Zed Books.

Roberts, P. (2008) *The End of Food: The Coming Crisis in the World Food Industry*, London, Houghton Mifflin.

Robinson, G. (2004) *Geographies of Agriculture: Globalisation, Restructuring and Sustainability*, London, Pearson.

Roos, G., Terragni, L. and Torjusen, H. (2009) 'The Local in the Global – Creating Ethical Relations between Producers and Consumers', Anthropology of Food, at: http://aof.revues.org/index489.html (accessed 10 August 2011).

Ross, E. B. (1998) *The Malthus Factor: Population, Poverty, and Politics in Capitalist Development*, London, Zed Books.

Rossett, P., Patel, R. and Courville, M. (eds) (2006) *Promised Land: Competing Visions of Land Reform*, California, Food First Books.

Rossillo-Calle, F. and Johnson, F. X. (2010) *Food versus Fuel: An Informed Introduction to Biofuels*, London, Zed Books.

Rostow, W. W. (1960) *The Stages of Economic Growth: A Non-Communist Manifesto*, Cambridge: Cambridge University Press.

Roy, A. (1999) The Greater Common Good, at: www.narmada.org/gcg/gcg.html (accessed on 29 October 2011).

Royal Society for the Protection of Birds (RSPB) Advice for Farmers, at: www.rspb.org.uk/ourwork/farming/advice/ (accessed 13 November 2011).

Ryan, O. (2011) *Chocolate Nations: Living and Dying for Cocoa in West Africa*, London, Zeb Books.

Sachs, J. (2007) 'The Promise of the Blue Revolution', Scientific American, July, at: www.scientificamerican.com/article.cfm?id=the-promise-of-the-blue-revolution-extended.

Sachs, W. (1999) *Planet Dialectics*, London, Zed Books.

Safran Foer, J. (2010) *Eating Animals*, London, Hamish Hamilton.

Salaman, R. N. and Hawkes, J. G. (1985) *The History and Social Influence of the Potato*, Cambridge, Cambridge University Press.

Santos, B. De Sousa (2008) 'The World Social Forum and the Global Left', *Politics & Society* 36(2, June): 247–70.

Schlosser, E. (2001) *Fast Food Nation*, London, Penguin.

Scholte, Jan Aart (1997) 'Global Capitalism and the State', *International Affairs* 73(3, July): 427–52.

Schoonover, H. and Muller, M. (2006) 'Food Without Thought: How U.S. Farm Policy Contributes to Obesity', Institute for Agricultural and Trade Policy, Minneapolis, Minnesota, United States, online at: www.iatp.org/iatp/publications.cfm?accountID=421&refID=80627 (accessed on 18 November 2010).

Scudder, T. (2006) *The Future of Large Dams: Dealing with the Social, Environmental, Instructional and Political Costs*, London, Earthscan.

Seedling Magazine (all years), available at: www.grain.org/seedling/ (accessed 13 November 2009).

Self Employed Women's Association (SEWA) (2011) at: www.sewa.org (accessed 16 December 2011).

Sen, A. (1981) *Poverty and Famine; An Essay on Entitlement and Deprivation*, Oxford, Clarendon Press.

Sen, A. (1990) 'More than 100 Million Women are Missing', *The New York Review of Books*, December 20.

Sen, A. (2004) *Disability and Justice*, Presentation to World Bank Conference, Washington USA.

Shiva, V. (1991) 'The Green Revolution in the Punjab', *Ecologist* 21(2), available at: http://livingheritage.org/green-revolution.htm (accessed 30 October 2009).

Shiva, V. (2000) *Stolen Harvest*, London, Zed Books.

Shiva, V. (2001) *Protect or Plunder*, London, Zed Books.

Sirkin, H. L. (2010) 'The New World Powers', *Global Finance Magazine* (www.gfmag.com/archives/110–january-2010/9928–cover-story-global-leaders-in-waiting.html#axzz0z8Ofulxz (accessed 9 September 2010).

Sklair, L. (2005) 'Generic Globalisation, Capitalist Globalisation, and Beyond: A Framework for Critical Globalization Studies', in Richard P. Appelbaum and William I. Robinson (eds) *Critical Globalization Studies*, Routledge, Oxford, United Kingdom, pp. 55–64.

Slocum, R. (2006) 'Whiteness, Space and Alternative Food Practice', *Geoforum* 38(3):520–533.

Smallman-Raynor, M. and Phillips, D. (1999) 'Late Stages of Epidemiological Transition: Health Status in the Developed World', *Health and Place* 5: 209–22.

Smyke, P. (1991) *Women and Health*, London, Zed Books.

Sodjinou, R. *et al.* (2008) 'Obesity and Cardio-metabolic Risk Factors in Urban Adults of Benin: Relationship with Socio-economic Status, Urbanisation, and Lifestyle Patterns', *BMC Public Health* 8(84): 1471–2458 www.biomedcentral.com/1471-2458/8/84 (accessed 18 November 2010).

Soil Association, UK 'Food Futures' at: www.soilassociation.org/Whyorganic/Climatefriendlyfoodandfarming/Foodfutures/tabid/565/Default.aspx (accessed 13 November 2009).

Song, Y. (1999) 'Feminization of Maize Agricultural Production in Southwest China', *Biotechnology and Development Monitor* 3: 6–9.

Sreenivasan, G. and Grinspun, R. (2002) 'Global Trade/Global Poverty: NGO Perspectives on Key Challenges for Canada', Canadian Council for International Co-operation, online at: www.ccic.ca/_files/en/what_we_do/002_global_trade_paper_2.pdf (accessed 18 November 2010).

Stacey, C. (2009) Ethics of Eating Meat, at: www.bbc.co.uk/food/food_matters/meatethics.shtml.

Steel, C. (2008) *Hungry City: How Food Shapes Our Lives*, London, Chatto and Windus.

Steger, M. B. (2009) *Globalization: A Very Short Introduction*, Oxford, Oxford University Press.

Stiglitz, J. (2002) *Globalization and Its Discontents*, London, Penguin.

Strange, S. (1997) *Casino Capitalism*, New York, Blackwell.

Subasinghe, R. (2005) Contribution of Aquaculture to Food Security FAO, 2005–2010. World Inventory of Fisheries. Issues: Fact Sheets. In: *FAO* Fisheries and Aquaculture Department [online]. Rome, available at: www.fao.org/fishery/topic/14886/en (accessed 21 November 2010).

Sudarshan, R. M. and Sinha, S. (2011) 'Making Home-Based Work Visible: A Review of Evidence from South Asia', Urban Policies Research Report, No 10 WIEGO, available at: www.inclusivecities.org/research/RR10_Sudashan%20revert2.pdf (accessed 16 December 2011).

Sustain: The Alliance for Better Food and Farming, at: www.sustainweb.org/ (accessed 13 November 2009).

Szuster, B. W. (2003) Shrimp Farming in Thailand's Chao Phraya River Delta, at: www.iwmi.cgiar.org/assessment/files/word/projectdocuments/chaophraya/szuster.pdf (accessed 21 November 2010).

Talbot, J. M. (2004) *Grounds for Agreement: The Political Economy of the Coffee Commodity Chain*, Rowman & Littlefield Publishers, Inc.

Taylor, J. (2010) Mid-year Global Petfood Sales Trends: Industry Players are Revamping their Offerings to Cater to the Changing Global Pet Trade at www.petfoodindustry.com/5553.html (accessed 17 November 2010).

Terragni, L., Torjusen, H. and Vittersø, G. (2009) 'The Dnamics of Alternative Food Consumption: Contexts, Opportunities and Transformations' in Anthropology of Food, at: http://aof.revues.org/index6400.html (accessed 10 August 2011).

Thekkudan, J. and Tandon, R. (2009) 'Women's Livelihoods, Global Markets and Citizenship', Working Paper No. 336, Institute of Development Studies, Sussex, at www.ntd.co.uk/idsbookshop/details.asp?id=1138 (accessed 20 January 2011).

Thomas, S. C. (2006) China's Economic Development from 1860 to the Present: The Roles of Sovereignty and The Global Economy, at: www.forumonpublicpolicy.com/archive07/thomas.pdf (accessed 19 November 2010).

Tsikata, D. (2010) 'Introduction', in D. Tsikata and P. Golah (eds) '*Land Tenure, Gender and Globalisation: Research and Analysis from Africa, Asia and Latin America*, Ottawa, International Development Research, pp. 1–34, at: www.idrc.ca/openebooks/463-5 (accessed 12 January 2011).

Tsikata, D. and Golah, P. (eds) (2010) 'Land Tenure, Gender and Globalisation: *Research and Analysis from Africa, Asia and Latin America', Ottawa*, International Development Research, at: www.idrc.ca/openebooks/463-5/ (accessed 12 January 2011).

Turner, S. and Caouette, D. (2009) 'Agrarian Angst: Rural Resistance in Southeast Asia', *Geography Compass* 3(3): 950–75, online at: http://onlinelibrary.wiley.com/doi/10.1111/j.1749-8198.2009.00227.x/pdf (accessed 25 October 2010).

UN (2006) Niger Delta Human Development Report, at: http://hdr.undp.org/en/reports/nationalreports/africa/nigeria/nigeria_hdr_report.pdf (accessed 16 December 2011).

UN (2010) WomenWatch Rural Women and Development, at: www.un.org/womenwatch/feature/idrw (accessed 15 December 2011).

UN Habitat (2010) The State of the World's Cities, 2010/2011: Bridging the Urban Divide, at: www.unhabitat.org/pmss/listItemDetails.aspx?publicationID=2917 (accessed 1 February 2011).

UNCTAD (2010) Report on Global and Regional Foreign Direct Investment Trends, at: www.unctad.org/en/docs//webdiaeia20111_en.pdf (accessed 8 March 2011).

UN-Habitat (2008) State of the World's Cities: Bridging the Urban Divide, at: http://books.google.co.uk/books?id=broGtXylVF8C&pg=PA43&lpg=PA43&dq=Slum+life+in+Kibera,+Nairobi:+the+female+experience&source=bl&ots=BMyktt30Nl&sig=BoU1a0E9sZeK7d-r8bkbpmpG7pc&hl=en#v=onepage&q=Slum%20life%20in%20Kibera%2C%20Nairobi%3A%20the%20female%20experience&f=false (accessed 16 December 2011).

UNICEF (2006) The State of the World's Children: Excluded and Invisible, at: www.childinfo.org/files/The_State_of_the_Worlds_Children_2006.pdf (accessed 12 December 2011).

UNICEF (2009) The State of the World's Children, at: www.unicef.org/sowc09 (accessed 12 December 2011).

UNICEF (2011) The State of the World's Children: Adolescence and Age of Opportunity, at: www.unicef.org/publications/index_57468.html (accessed 12 December 2011).

United Nations Development Report (2010) at: http://hdr.undp.org/en/media/HDR_2010_EN_Chapter3_reprint.pdf (accessed 21 November 2010).

United Nations Development Report (2010) at: http://hdr.undp.org/en/reports/global/hdr2010/ (accessed 6 January 2011).

United Nations High Commission for Refugees (UNHCR) Annual Reports, at: www.unhcr.org/pages/49c3646c4b8.html (accessed 15 February 2011).

United Nations Human Settlements Programme (Un-Habitat) 'State of the World's Cities 2010–2011: Cities for All', Bridging the Urban Divide.

USDA (2010) Local Food Systems: Concepts, Impacts, and Issues, available at: http://permanent.access.gpo.gov/lps125302/ERR97.pdf (accessed 10 August 2011).

Via Campesina (2007) Declaration of Nyelení, at: www.viacampesina.org/en/index.php?option=com_content&task=view&id=282&Itemid=38 (accessed 16 December 2011).

Via Campesina Organisation, at: http://viacampesina.org/main_en/ (accessed on 13 November 2009).

Vidal, J. (2010) Billionaires and Mega-Corporations Behind Immense Land Grab in Africa at www.informationclearinghouse.info/article24965.htm (accessed 27 September 2010).

Vorley, B. (2003) Food Inc., Corporate Concentration from Farm to Consumer, London, International Institute for Environment and Development at www.ukfg.org.uk/docs/UKFG-Foodinc-Nov03.pdf (accessed 30 October 2009).

Waldick, L. (2009) 'Promoting Traditional Foods for Nutrition in Indigenous Communities', Canadian Coalition for Global Health Research, at: www.ccghr.ca/docs/CINE_E.pdf (accessed 30 January 2011).

Wallerstein, E. (1974) The Modern World-System, at: http://marriottschool.byu.edu/emp/WPW/Class%209%20-%20The%20World%20System%20Perspective.pdf

Walsh, B. (2009) 'Getting Real about Food', *Time Magazine*, Friday, 21 August, 2009, at: www.time.com/time/health/article/0,8599,1917458,00.html (accessed 27 August 2010).

Watson, R. (2008) quoted in Vidal, J. 'Change in farming can feed the world-report', *The Guardian* 16 April 2008, at: www.guardian.co.uk/environment/2008/apr/16/food.biofuels

Watts, M. (2009) 'Crude Politics: Life and Death on the Nigerian Oil Fields', Department of Geography, University of California at Berkeley, available at: http://oldweb.geog.berkeley.edu/ProjectsResources/ND%20Website/NigerDelta/WP/Watts_25.pdf (accessed 16 December 2011).

Watts, M. J. and Bohle, H. G. (1993) 'The Space of Vulnerability: The Causal Structure of Hunger and Famine', *Progress in Human Geography* 17: 43–6.

Weis, T. (2007) '*The Global Food' Economy*, London, Zed Books.

Weis, T. (2010) The Accelerating Biophysical Contradictions of Industrial Capitalist Agriculture', *Journal of Agrarian Change* 10(3): 315–41.

Weiser, S. D. *et al.* (2007) 'Food Insufficiency is Associated with High-risk Sexual Behavior among Women in Botswana and Swaziland', *PLos Medicine* 4(10) www.plosmedicine.org/article/info%3Adoi%2F10.1371%2Fjournal.pmed.0040260 (accessed 16 January 2011).

White, K. *et al.* (2004) At the Crossroads: Will the Aquaculture Fulfill the Promise of the Blue Revolution, at: www.seaweb.org/resources/documents/reports_crossroads.pdf (accessed 25 November 2010).

Whitehead, A. (2010) 'Foreword', in D. Tsikata and P. Golah (eds) *Land Tenure, Gender and Globalisation: Research and Analysis from Africa, Asia and Latin America*, Ottawa, International Development Research, ibid.

WHO (2002) 'Diet, Nutrition and the Prevention of Chronic Diseases Report of the Joint WHO/FAO Expert Consultation', WHO Technical Report Series, No. 916 (TRS 916), at: www.who.int/dietphysicalactivity/publications/trs916/intro/en (accessed 8 December 2011).

Wilkinson, J. (2009) 'The Globalization of Agribusiness and Developing World Food Systems', *Monthly Review*, September 2009 at: http://monthlyreview.org/090907wilkinson.php (accessed 11 October 2010).

Willem te Velde, D. (2009) 'The Global Financial Crisis and Developing Countries: Taking Stock, Taking Action', ODI Briefing Paper No 54, at: www.odi.org.uk/resources/docs/3705.pdf (accessed 12 December 2011).

Willis, K. (2005) *Theories and Practices of Development*, London, Routledge.

Willis, K. (2011) *Theories and Practices of Development*, 2nd edition, London, Routledge.

Wittman, H. (2009) 'Reworking the Metabolic Rift: La Vía Campesina, Agrarian Citizenship, and Food Sovereignty', *Journal of Peasant Studies* 36(4): 805–26 at: http://pdfserve.informaworld.com/706439__917020256.pdf (accessed 1 March 2011).

Wolf, E. R. (1982) *Europe and the People without History*, Berkeley, University of California Press.

Women in Employment Globally Organising (WIEGO) Informal Workers in Focus, at: (www.wiego.org/publications/Informal-Economy-Fact-Sheets.pdf (accessed 6 January 2011).

World Bank (2006) Aquaculture: Changing the Face of the Waters, at: www.beijer.kva.se/ftp/WIOAQUA/WORLDBANK.pdf (accessed 25 November 2010).

World Bank (2008) World Development Report: Agriculture for Development, at: http://siteresources.worldbank.org/INTWDR2008/Resources/WDR_00_book.pdf (accessed 16 December 2011).

World Bank (2010) Rising Global Interest in Farmland and the Importance of Responsible Agricultural Investment, at: www-wds.worldbank.org/external/default/WDSContentServer/WDSP/IB/2010/09/15/000333038_20100915010822/Rendered/PDF/565830BRI0Box31nt1Issues1Note1541v6.pdf (accessed 28 September 2010).

World Bank Development Report (2006) Development and Equity, at: www-wds.worldbank.org/external/default/WDSContentServer/IW3P/IB/2005/09/20/000112742_20050920110826/additional/841401968_200508263001721.pdf; (accessed 28 January 2011).

World Bank. See a Video about a Partnership Project between the World Bank and Chinese Government to Reduce Poverty in Guanxi, One of China's Poor Remote Regions, at: www.worldbank.org/en/news/2010/03/19/results-profile-china-poverty-reduction (accessed 19 July 2011).

World Food Summit (1996) Rome Declaration on World Food Security, at: www.fao.org/docrep/003/w3613e/w3613e00.htm (accessed 6 December 2011).

World Health Organization (2004) World Health Report: Changing History, available online at: www.who.int/whr/2004/en/index.html

World Health Organization (2005) China: Health Poverty and Economic Development, at: www.who.int/macrohealth/action/CMH_China.pdf

World Health Organization (2005) China: Health, Poverty and Economic Development, at: www.wpro.who.int/NR/rdonlyres/A1F18401-BE93-44EF-9F76-55DDA2C6E12D/0/hped_en.pdf (accessed 20 January 2011).

World Health Organization (2009) Obesity, at: www.who.int/topics/obesity/en/ (accessed on 13 November 2009).

World Health Organization (2010) Hidden Cities: Unmasking and Overcoming Health Inequities in Urban Settings, at: www.hiddencities.org/downloads/WHO_UN-HABITAT_Hidden_Cities_Web.pdf (accessed 2 February 2011).

Worldwatch Report (2011) State of the World; Innovations that Nourish the Planet, at: www.worldwatch.org/sow11 (accessed 6 December 2011).

Xiaoyun, L. *et al.* (2006) 'Impacts of China's Agricultural Policies on Payment for Watershed Service', International Institute for Environment and Development and Department for International Development, United Kingdom, at: www.iied.org/pubs/pdfs/G00343.pdf (accessed on 19 November 2010).

Young, E. M. (1995) 'An Angry Voice from Paradise: Jamaica Kincaid's "*A Small Place*" as a Teaching Resource', *Journal of Geography in Higher Education* 19(1): 91–6.

Young, E. M. (1996) 'Spaces for Famine: A Comparative Geographical Analysis of Famine in Ireland and the Highlands in the 1840s' Trans. IBG (1996) NS 21 pp. 666–80.

Young, E. M. (1997) *World Hunger*, London, Routledge.

Young, E. M. (1999) 'Far-Fetched Meals and Indigestible Discourses: Reflections on Globalisation, Hunger and Sustainable Development', *Ethics, Place and Environment* 2(1): 19–40.

Young, E. M. (2004) 'Globalization and Food Security: Novel Questions in a Novel Context?' *Progress in Development Studies* 4(1): 1–21.

Young, E. M. (2008) 'Globalization and Girths' paper presented at the European Open Science Forum, Barcelona.

Young, E. M. (2010) 'Deadly Diets', *Geography* 95(2, Summer): 60–9.

Zaninka, P. (2003) 'Uganda', in J. Nelson and L. Hossack (eds) *Indigenous People and Protected Areas in Africa: From Principle to Practice*, Moreton-in Marsh, UK Forest People Programme.

Index